Design of Artificial Human Joints & Organs

Subrata Pal

Design of Artificial Human Joints & Organs

Springer

Subrata Pal
Biomedical Engineering
Jadavpur University
Kolkata, India

ISBN 978-1-4614-6254-5 ISBN 978-1-4614-6255-2 (eBook)
DOI 10.1007/978-1-4614-6255-2
Springer New York Heidelberg Dordrecht London

Library of Congress Control Number: 2012953608

Printed on acid-free paper

Springer is part of Springer Science+Business Media (www.springer.com)

Preface

This book is dedicated to my parents, who are no longer in this mortal world, but whose love and care inspired me to break newer grounds and introduce the study of, and research in, biomedical engineering at Jadavpur University at the postgraduate level in the School of Bioscience and Engineering.

Biomedical engineering is the study of the human machine in health and disease. The human organ system is vast, and this text is intended to discuss the basics of the normal systems preciously. Due to aging, diseases, or trauma, body parts may need to be replaced appropriately with surgical intervention using manmade materials.

The human body has all the characteristics of a wonderful machine, but it is much more than a machine. The movement of the body generates forces in various work situations and also internally at various joints, muscles, and ligaments. It is essential to figure out the forces' moments, pressure, and so forth, which are usually dealt with in biomechanics. The mechanical characterizations of the hard and soft tissues are presented systematically using the principles of solid mechanics. The viscoelastic properties of the tissue, unlike many engineering materials, are also discussed.

The design science and methodology are essential from the concept to the realization of the blueprint of the component required to be replaced. Then that blueprint begins its transformation into a product. That is the manufacturing aspect of the prosthesis. We have discussed that in Chap. 24. Young readers need to visit some factories that have casting, forging, and sophisticated facilities for machining using computerized numerically controlled machineries. The market is open, and implants and artificial endoprostheses constitute a huge worldwide market. India's contribution is too little and may be improved to a great extent. This is discussed in Chap. 23.

This book will also serve as a text for students of mechanical engineering and biomedical engineering. The pedagogy is simple enough for those who are learning the subject for the first time.

Many materials were collected from the Internet and are open source for use in this text. I acknowledge with grateful thanks those great peers, researchers spread all over the world. Some of their excellent illustrations and great ideas on Wikipedia were used for the development of this subject area, and I acknowledge them with heartfelt thanks.

I also acknowledge with grateful thanks Prof. S. Guha, Chair Professor of Bio-Engineering at IIT, Kharagpur, for providing input for artificial heart design and Prof. C. P. Sharma of Srichitra Institute of Medical Sciences for his encouragement.

I also express my heartfelt thanks to my previous Ph.D. scholars, who now have faculty positions at different IITs, BESU, and colleges in India and the United States for their technical and intellectual input since we started working in this exciting area in 1980 in the U.S. Thanks are due to all my beloved technical staff of the School of Bioscience and Engineering, which was built through years of painstaking hard labor.

I had the opportunities to work in medical school in USA and in India and interact with professionals which helped me to visualise the requirement for this book.

I also thank my wife and daughter for their forbearance and understanding during the long hours of my absence from home, even during holidays, while I was preparing the text. Thanks are due to Saibal Mandal for his careful preparation of the typescript several times.

I also acknowledge the DST Govt. of India, and specifically AICTE New Delhi, with gratefulness for their support for writing this book. Much information was generated using the equipment and software they provided. The School of Education Technology, Jadavpur University, and KPCMC and Hospital are also acknowledged for providing office space and peaceful facilities for my work. Suggestions for improvement will be gratefully acknowledged.

Kolkata, India Subrata Pal

Contents

1 Overview of Human System and Its Artificial Replacement 1
1.1 Human Body's Superficial Anatomy 1
1.2 Human Body Features 1
1.3 Body Size, Type, and Proportion 2
1.4 A Brief Outline of the Organization of the Human System 3
1.4.1 Tissue + Tissue = Organ (Different Tissues Together Form an Organ) 3
1.4.2 Organ + Organ = Organ System 4
1.5 Major Organ Systems 4
1.5.1 Integumentary System 5
1.5.2 Muscular System 5
1.5.3 Skeletal System 5
1.5.4 Cardiovascular System 7
1.5.5 Respiratory System 7
1.5.6 Urinary System 7
1.5.7 Reproductive System 9
1.5.8 Nervous System 11
1.5.9 Digestive System 11
1.5.10 Lymphatic System 12
1.5.11 Immune System 12
1.6 Common Names of Internal Organs (in Alphabetical Order) 14
1.7 Brief Idea of Artificial Organs 14
1.7.1 Different Types of Organs 15
1.7.2 Brain 15
1.7.3 Cardia 16
1.7.4 Corpora Cavernosa 16
1.7.5 Ear 16
1.7.6 Eye 16
1.7.7 Heart 17
1.7.8 Limbs 17
1.7.9 Liver 17

1.7.10 Lungs ... 18
1.7.11 Pancreas ... 18
1.7.12 Bladder ... 18
1.7.13 Ovaries ... 18
1.7.14 Beyond Restoration ... 19
1.7.15 Timeline of Successful Transplants ... 19
References ... 20
Exercises ... 20

2 Mechanical Properties of Biological Materials ... 23
2.1 Introduction ... 23
2.2 Structural Versus Material Properties ... 23
2.2.1 Anisotropy and Nonhomogeneity ... 24
2.2.2 Viscoelastic Properties ... 24
2.2.3 Viscosity ... 25
2.3 Testing Procedures ... 25
2.4 Bones ... 26
2.4.1 Composition ... 26
2.4.2 Structure ... 27
2.5 Material Properties and Related Behavior ... 29
2.6 Cartilage ... 32
2.6.1 Composition ... 32
2.7 Material Properties and Related Behavior ... 33
2.8 Ligaments ... 34
2.8.1 Composition ... 34
2.9 Material Properties and Related Behavior ... 36
2.9.1 Ligaments Have Characteristics of Strain-Rate Sensitivity, Stress-Relaxation, Creep, and Hysteresis ... 36
2.10 Correlation Between Structure and Function ... 37
2.10.1 Ligament–Bone/Tendon–Bone Insertions ... 38
Reading List ... 39
Problems ... 39

3 Basics of Design Process ... 41
3.1 Introduction ... 41
3.2 Adoptive and Adaptive Design ... 41
3.2.1 Introduction ... 41
3.3 Introduction to Machine Design ... 42
3.4 Principle of Science or Mechanism to Be Used ... 43
3.5 Safety of Products ... 47
3.6 Manufacturability ... 47
3.7 Standardization ... 48
3.8 Customization ... 48
For Further Details, the Following Books May Be Consulted ... 49
Exercises ... 49

4 Biomaterials and Its Characterization ... 51
4.1 Introduction ... 51
4.2 Biomineralization ... 52
4.3 Abalone Shell ... 54
4.4 Self-Assembly ... 55
4.5 Structural Hierarchy ... 56
4.6 Applications ... 56
4.7 Compatibility ... 57
For Further Reading, These Papers May Be Consulted ... 58
4.8 Biopolymers ... 59
4.8.1 Applications ... 60
4.8.2 Sources ... 60
4.8.3 Applications ... 60
4.8.4 Sources ... 60
4.8.5 Applications ... 61
4.8.6 Benefits ... 61
4.9 Polymers for Healthcare ... 62
4.9.1 Sterilization ... 62
4.9.2 Compatibility with Sterilization Method of Various Industrial Biopolymers (Solviva) ... 62
4.9.3 Polyether Ether Ketone (PEEK) ... 63
4.9.4 Applications ... 64
4.10 Bioceramics ... 64
4.10.1 History ... 64
4.10.2 Current Status ... 65
4.10.3 Future Trends ... 65
4.11 Bioceramic Materials ... 65
4.11.1 Bioinert ... 65
4.11.2 Bioactive ... 66
4.12 Hydroxyapatite: Medical Uses ... 66
4.13 Bioactive Glass ... 67
4.13.1 Compositions ... 67
4.13.2 Mechanism of Bioactivity ... 67
4.13.3 Applications ... 68
For Further Study Consult the Reading List (Ceramics and Glass) ... 68
4.14 Metals as Biomaterials ... 69
4.15 Signs of Inferior Tissue Response ... 71
4.16 Stress-Shielding Phenomena ... 71
4.17 Shape-Memory Alloy ... 72
Problems ... 73

5 Dental Implants: Their Design and Manufacture ... 75
5.1 Introduction ... 75
5.1.1 Success Criteria ... 75
5.2 Anatomy of Human Jawbones ... 77
5.3 Dental Implants ... 78

5.4 Components of Dental Implant 81
5.5 Types of Implants in Use 81
5.5.1 Subperiosteal Implants 81
5.5.2 Transosteal Implants 82
5.5.3 Endosteal Implants 83
5.5.4 Biocompatibility and Implant Design 84
5.6 Biting Force 86
5.7 Implant Shape 87
5.8 Surface Characteristics 87
5.9 Bone Factors 88
5.10 Loading Conditions 89
5.10.1 Prosthetic Considerations 90
5.10.2 The Type of Prosthetic Reconstruction 90
5.10.3 The Occlusal Scheme 90
5.11 The Number, Distribution, Orientation, and Design of Implants 90
5.12 Placement of Implants into Extraction Sites 91
5.13 Clinical and Biomechanical Considerations for Fixed Tooth Replacement 92
5.13.1 Bone Tissue Factors 92
5.13.2 Soft Tissue Factors 92
5.13.3 Number of Implants 93
5.13.4 Crown-to-Root Ratio 93
5.13.5 Implant Orientation 94
5.14 Advantages of Permanent Tooth Replacement 94
5.15 Manufacture of Implants and Property Enhancement 94
References 95
Problems 96
Glossary of Dental Terminology 96

6 The Design of a Bone Fracture-Fixation Device 101
6.1 Introduction 101
6.2 External Fixation 101
6.2.1 The Ilizarov Apparatus 105
6.3 Internal Fracture Fixation 106
6.3.1 Wires 106
6.3.2 Braided Multifilament Wire 106
6.3.3 Pins 107
6.3.4 Screws 107
6.3.5 Plates 109
6.3.6 Intramedullary Nails 111
6.3.7 Use of Machine Elements in External Fixators 114
6.4 Other Considerations 114
6.5 Materials Involved in Bone–Plate Design 114
6.6 Composite Materials 115

6.7 Bioactive Fixation Using Bioactive Materials 117
6.8 Compositions 119
6.9 Strong Interfacial Bond with Bone 119
Suggested Study for Fracture Fixation 119
Problems 121

7 The Shoulder Joint and Its Artificial Replacement 123
7.1 Introduction to Joint Replacement 123
7.2 The Shoulder Joint and Its Artificial Replacement 124
7.2.1 Introduction 124
7.2.2 The Different Joints 125
7.2.3 Musculature 126
7.2.4 Ligaments 127
7.2.5 Glenohumeral Joint 127
7.2.6 Glenoid 128
7.2.7 Humerus 128
7.2.8 Kinematics of the Glenohumeral Joint 129
7.2.9 Geometrical Center and Center of Rotation 130
7.3 Humeral Translation Relative to the Scapula 130
7.4 Degrees of Freedom (DOF) 131
7.5 Range of Motion 132
7.6 Total Shoulder Replacement (Glenohumeral Arthroplasty) 134
7.6.1 Reasons for Shoulder Arthroplasty 134
7.7 Types of Shoulder Prostheses 135
7.8 Fully Constrained Type 136
7.9 Nonconstrained Type 138
7.10 Fixation Procedures 143
7.11 Causes of Failure 144
7.11.1 Loosening of the Component 145
7.11.2 Anterior Instability 146
7.11.3 Superior Instability 146
7.11.4 Posterior Instability 146
7.11.5 Inferior Instability 146
7.11.6 Rotator Cuff Tears 146
7.11.7 Intraoperative Fractures 147
7.11.8 Infection 147
7.11.9 Neural Injuries 147
References 147
Problems 148

8 The Elbow Joint and Its Artificial Replacement 149
8.1 Introduction 149
8.1.1 Movements 149
8.2 Muscles, Arteries, and Nerves 151
8.3 Parts of the Joint 151
8.4 Ligaments 152

8.5 Synovial Membrane 152
8.6 Carrying Angle 153
8.7 Diseases of the Elbow 153
8.7.1 Tendonitis 153
8.7.2 Fractures 153
8.7.3 Arthritis 153
8.8 Elbow Replacement 154
8.9 Elbow Replacement Surgery 156
8.10 Life Expectancy of Prosthetic Elbow Joint 156
References 157
Problems 157

9 The Wrist Joint and Its Artificial Replacement 159
9.1 Anatomy 159
9.2 Rheumatoid Arthritis of the Hand and Wrist 159
9.3 Treatment 161
9.3.1 Forces Generated in the Wrist Joint 161
9.4 Implants: Components of a Wrist Arthroplasty 163
9.5 Surgery 164
9.6 Postsurgical Care 165
References 165
Problems 166

10 The Finger Joint and Its Artificial Replacement 167
10.1 Introduction 167
10.2 Diseases of the Finger Joint 168
10.2.1 Symptoms 170
10.2.2 Nonsurgical Treatment 171
10.3 Finger Joint Replacement 171
10.3.1 Introduction 171
Problems 174
References 175

11 The Hip Joint and Its Artificial Replacement 177
11.1 Introduction 177
11.2 Anatomy of the Hip Region 177
11.2.1 Articulation 177
11.2.2 Femoral Neck Angle 178
11.2.3 Capsule 180
11.2.4 Ligaments 180
11.2.5 Blood and Nerve Supply 181
11.2.6 Muscles and Movements 181
11.3 Hip Joint Replacement 182
11.3.1 History 184
11.4 Modern Developments 185
11.5 Costs 186

11.6 Aseptic Loosening of Hip Prosthesis 186
11.7 Techniques of Surgery 187
11.8 Alternatives to Hip Replacement 187
11.8.1 Apollo Hip System: Simple, Straight-Stem Design for the Low-Demand Patient 188
11.8.2 Dr. K. H. Sancheti's Hip Prosthesis 190
11.9 Design Modification Using Hollow Section 191
11.9.1 Stress Shielding 191
11.9.2 Hollow Design 192
References & Reading List 193
Problems 194

12 The Knee Joint and Its Artificial Replacement 195
12.1 Anatomy 195
12.1.1 Cruciate Ligaments 195
12.1.2 Anterior Cruciate Ligament 196
12.1.3 Action of Muscles 197
12.1.4 Knee Joint Stabilization 197
12.2 Knee Replacement, or Knee Arthroplasty 199
12.2.1 Introduction 199
12.3 History 200
12.3.1 Technique 201
12.4 Variations in Design 202
12.4.1 Partial Knee Replacement 202
12.5 Risks and Complications 203
12.5.1 Loss of Motion 204
12.5.2 Instability 204
12.5.3 Infection 204
12.6 Developmental Work on a Congruent Tibial Component 205
12.6.1 This Design has Two Advantages 207
12.6.2 Disadvantages 208
12.7 Indian Effort 208
References 209
Problems 210

13 The Ankle Joint and Its Artificial Replacement 211
13.1 Introduction 211
13.1.1 Structure 211
13.2 Bones and Joints 211
13.3 Ligaments and Tendons 212
13.4 Muscles 213
13.5 Nerves 214
13.6 Blood Vessels 214
13.7 Artificial Replacement of the Ankle Joint 214
13.8 The Operative Procedure 215

13.9 Complications ... 216
13.9.1 Infection ... 217
13.9.2 Loosening ... 217
13.9.3 Nerve Injury ... 217
Study List ... 217
Problems ... 217

14 The Eye and Its Artificial Replacement ... 219
14.1 Anatomy of the Eye ... 219
14.1.1 Extraocular Muscles ... 221
14.2 Development of the Eye ... 222
14.2.1 Disorders of the Eye ... 224
14.2.2 Disorders of the Eye Muscles ... 224
14.2.3 Disorders of the Cornea, Iris, and Lens ... 224
14.2.4 Disorders of the Retina ... 224
14.2.5 Visual Impairments ... 224
14.3 Contact Lens ... 225
14.3.1 Corrective Contact Lenses ... 228
14.3.2 Cosmetic Contact Lenses ... 228
14.3.3 Therapeutic Contact Lenses ... 229
14.4 Classification of Contact Lenses ... 229
14.4.1 By Construction Material ... 229
14.4.2 By Time of Wear ... 230
14.4.3 By Frequency of Replacement ... 230
14.4.4 By Design ... 230
14.4.5 Implantation ... 231
14.5 Manufacturing of Contact Lenses ... 231
14.5.1 Hydrogel Materials ... 232
14.5.2 Contact Lens Prescriptions ... 232
14.6 Complications ... 232
14.6.1 Eyelid ... 233
14.6.2 Conjunctiva ... 233
14.6.3 Cornea ... 233
14.7 Usage ... 233
14.7.1 Care ... 234
14.8 Current Research ... 234
14.9 Intraocular Lens ... 235
14.9.1 History ... 237
14.9.2 Materials Used for Intraocular Lenses ... 237
14.9.3 Intraocular Lenses for Correcting Refractive Errors ... 238
14.9.4 Types of PIOLs ... 239
14.9.5 Accommodating IOLs ... 239
14.9.6 Criticisms ... 240
14.9.7 Candidates ... 240

14.10 LASIK Eye Surgery 241
14.10.1 What Is LASIK? 241
14.10.2 The LASIK Procedure 241
14.10.3 Phacoemulsification 242
14.10.4 Demographics 242
14.10.5 Diagnosis/Preparation 244
14.10.6 Aftercare 245
14.10.7 Risks 245
14.10.8 Alternatives 246
14.11 Bionic Eye 246
Related Literature for Study 246
Study List 247

15 The Lung and Its Transplantation and Artificial Replacement 251
15.1 Introduction 251
15.2 Respiratory Tract 251
15.2.1 Structure 252
15.3 Path Traced by Inhaled Air 254
15.4 Gaseous Exchange 254
15.5 Common Lung Diseases 255
15.6 Lung Transplantation 256
15.7 Types of Lung Transplants 257
15.7.1 Lobe 257
15.7.2 Single-Lung Transplant 257
15.7.3 Double-Lung Transplant 257
15.7.4 Heart–Lung Transplant 257
15.8 Design of Artificial Lungs 258
15.9 Design Overview 258
For Detailed Study, Consult the Following Papers 260
Problems 260

16 Design of the Total Artificial Heart 261
16.1 Introduction 261
16.2 Artificial Heart 261
16.2.1 CardioWest Temporary Total Artificial Heart 261
16.2.2 AbioCor Replacement Heart 262
16.3 Origins 262
16.3.1 Early Development 264
16.4 Early Designs of Total Artificial Hearts 264
16.5 First Clinical Implantation of a Total Artificial Heart 265
16.6 First Clinical Applications of a Permanent Pneumatic Total Artificial Heart 265
16.7 The Development of Permanent, Implantable, Electrically Powered Artificial Hearts 266
16.8 First Clinical Application of an Intrathoracic Pump 267

16.9 First Clinical Application of a Paracorporeal Pump 267
16.9.1 Recent Developments 267
16.10 Total Artificial Heart 268
16.11 Heart Assist Devices 268
16.11.1 Ventricular Assist Device (VAD) 268
16.12 Total Artificial Heart: Aortic Pumping System, Indian Initiative 269
16.12.1 Novel Features 269
16.13 Existing Prior Technologies 270
16.13.1 Diaphragm Pump 270
16.13.2 Pulsatile Blood-Pumping System 271
16.13.3 Orthotopic Total Artificial Heart 272
16.13.4 Heart Assist Devices, Systems, and Methods 273
16.14 Disadvantages of Prior Developed Systems 274
16.15 Artificial Heart Pump 274
16.16 Multistaged Multi-Actuated Pump System (Fig. 16.4) 275
16.16.1 Heart Pump Prototype Model 277
16.16.2 Advantages of the Artificial Heart 277
Reading List 279
Problems 280

17 The Design of Heart Valves 281
17.1 Introduction 281
17.2 The Heart 281
17.3 Heart Valves 284
17.4 Cardiac Cycle 286
17.5 Cardiac Output 287
17.6 Artificial Heart Valves 291
17.6.1 Potential Risks of Artificial Heart Valves 291
17.6.2 Lifestyle Considerations 292
17.6.3 Longevity and Replacements 293
17.7 Design of Valves 293
17.7.1 Valve Descriptions 294
17.7.2 Mechanical Valves 294
17.7.3 Caged Ball Valve 295
17.7.4 Tilting Disk Valve 295
17.7.5 Bileaflet Valves 297
17.8 Biological Valves 298
17.8.1 Homograft, Autograft 299
17.8.2 Autologous Pericardial Valve 300
17.8.3 Porcine Heterograft Valves (Stented) 300
17.8.4 Porcine Heterograft Valves (Unstented) 301
17.8.5 Bovine Pericardial Heterograft Valves 301
17.8.6 Complications 302
17.8.7 The Comparative Clinical Performance 302

17.9 Structural Deterioration of Biological Valves 303
17.9.1 Porcine Valves 303
17.9.2 Pericardial Valves 303
17.9.3 Homografts 303
17.9.4 Pulmonary Autograft 304
17.9.5 Materials of Construction and Manufacture 304
17.9.6 Assembly 305
17.9.7 Sterilization and Packaging 305
17.9.8 Quality Control 305
17.9.9 The Future 306
17.10 Failure Modes in Mechanical Valves 306
17.11 Choice of Valves 307
17.12 Future Developments 307
List for Further Study 308
Problems 308
Example from the Literature for Valve Area Calculation 308
Aortic Valve Area Calculation 308

18 The Kidney and Its Artificial Replacement 311
18.1 Introduction 311
18.2 How Do Kidneys Help Maintain Health? 313
18.3 Diseases 313
18.3.1 Chronic Kidney Disease (CKD) and Related Facts 313
18.3.2 What Is Chronic Kidney Disease? 314
18.3.3 What Causes CKD? 314
18.3.4 The Symptoms of CKD 314
18.4 Dialysis 315
18.4.1 Kidney Failure 315
18.4.2 Hemodialysis 316
18.4.3 Peritoneal Dialysis 316
18.5 The Artificial Kidney 317
18.5.1 Parallel Flow Dialyzers 317
18.6 Coil Hemodialyzer 317
18.6.1 Hollow Fiber Hemodialyzer 319
18.7 Performance Analysis of Dialyzers 319
18.7.1 Ultrafiltration Rate 321
18.7.2 Residual Blood Volume 322
18.7.3 Priming Volume 322
18.7.4 Pyrogenicity 322
18.7.5 Leakage Rate 322
18.8 The Hemodialysis Machine and Its Parts 323
18.9 Conclusions 323
18.9.1 Dialysis Does Not Help Cure Kidney Disease 323

18.10 Recent Developments in Artificial Kidneys ... 324
18.10.1 Nanotechnology ... 325
18.10.2 Portable Dialysis Machine ... 326
List for Further Reading ... 328
Exercises ... 328

19 Skin and the Design of Artificial Skin ... 329
19.1 Anatomy of the Skin ... 329
19.2 Modeling of Soft Biological Tissues ... 330
19.1.1 The Stress–Strain Relationship ... 331
19.3 Mechanical Properties of Soft Tissue ... 332
19.4 Fitting Hyperelastic Material Models to the Experimental Data ... 332
19.5 Artificial Skin ... 333
19.6 Tissue Expanders ... 335
References ... 337
Exercises ... 337

20 The Artificial Pancreas ... 339
20.1 Endocrine Physiology ... 339
20.2 Artificial Pancreas ... 340
20.3 Intensive Insulin Therapy and Insulin Pump ... 341
20.4 Bioengineering Approach to an Artificial Pancreas ... 343
20.5 Gene Therapy Approach ... 344
20.6 Medical Equipment Approach ... 345
20.6.1 Development of Continuous Blood Glucose Monitoring ... 345
20.6.2 Feedback of Real-Time Blood Glucose Data to an Insulin Pump for Basal Control ... 346
20.7 First Clinical Tests: Implantable Insulin Pumps and Continuous Glucose Sensors ... 347
20.8 Insulin and Amylin Combination ... 348
20.9 Feedback of Real-Time Blood Glucose Data to an Insulin Pump for Bolus Control ... 348
20.10 Glucagon Combination ... 349
20.11 Research Around the World ... 350
Reading List ... 350
Problems ... 351

21 The Liver and Its Artificial Replacement ... 353
21.1 Introduction ... 353
21.2 Blood Flow ... 354
21.3 Biliary Flow and the Biliary Tree ... 354
21.4 Lobes ... 355
21.5 Synthesis ... 355
21.6 Breakdown ... 356

21.7 Other Functions ... 356
21.8 Bioartificial Liver Device ... 356
21.8.1 Function ... 357
21.8.2 Hollow Fiber System ... 357
21.8.3 Comparison to Liver Dialysis ... 357
21.9 Progress Toward an Artificial Liver Transplant ... 358
References and for Further Study ... 359
Problems ... 359

22 Female Organs and Their Artificial Replacements ... 361
22.1 Introduction ... 361
22.2 Breast Implants ... 362
22.2.1 History ... 363
22.2.2 Indications ... 363
22.3 Patient Characteristics ... 364
22.4 Types of Implants ... 365
22.5 Silicone Gel Implants ... 366
22.5.1 First Generation ... 366
22.5.2 Second Generation ... 366
22.5.3 Third and Fourth Generations ... 367
22.5.4 Fifth Generation ... 367
22.6 Implant Pocket Placement ... 367
22.7 Failure and Rupture ... 369
22.8 Implants and Mammography ... 369
22.8.1 Breast Implant Manufacturers ... 370
22.8.2 Artificial Breast Prostheses ... 370
22.9 Artificial Vagina ... 371
22.9.1 Veterinary Use ... 371
22.10 Usage as Sex Toy ... 371
22.10.1 Types ... 372
22.11 Materials ... 373
References ... 373
Problems ... 375

23 The Medical Device Market and Ethical Issues of Implants ... 377
23.1 Introduction to the Medical Device Market ... 377
23.2 Ethical Issues of Implants ... 379
23.2.1 The Need for Implants ... 380
23.3 Treatment Modality ... 380
23.4 Transplants and Implants ... 381
23.5 Implant Failure ... 382
23.6 Clinical Trials ... 384
23.7 Ethical Issues of Dental Implants ... 384
23.8 Cost ... 385
23.9 Moral Uncertainties ... 385
23.10 Religion and Morals ... 386

23.11 Ethical Issues of Transplants 387
23.11.1 Availability 389
23.12 Xenografts as a Solution 390
23.13 Life Expectancy 391
Suggested References for Further Details 393
Exercises 393

24 The Manufacturing, Testing, and Sterilization of Implants 395
24.1 Introduction 395
24.2 Casting 396
24.3 Forging 397
24.4 Metal-Shaping Machine Tools 397
24.5 Manufacturing Implants at a High Speed 399
24.6 Rapid Prototyping 400
24.7 Nonconventional Machining 402
24.8 Nondestructive Testing 403
24.9 Sterilization 404
24.9.1 Physical Agent 405
24.10 Conclusions 408
For Further Reading, the Following References Are Given 408
For Computer-Aided Design and Manufacture 408
For Sterilization of Implants 409
Problems 409

Index 413

Chapter 1
Overview of Human System and Its Artificial Replacement

1.1 Human Body's Superficial Anatomy

The objective of this chapter is to briefly review the human system to refresh the reader's knowledge about 11 systems of the wonderful human machine from an engineer's point of view. We will discuss the very basic anatomy of the system.

Superficial, or surface, anatomy is important in human anatomy as the study of anatomical landmarks that can be readily identified from the contours or other reference points on the surface of the body [1]. With knowledge of superficial anatomy, physicians gauge the position and anatomy of the associated deeper structures.

Common names of well-known parts of the human body, from top to bottom, are listed below:

- Head—forehead—jaw—cheek—chin
- Neck—shoulders
- Arm—elbow—wrist—hand—fingers—thumb
- Spine—chest—thorax
- Abdomen—groin
- Hip—buttocks—leg—thigh—knee—calf—heel—ankle—foot—toes

The eye, ear, nose, mouth, teeth, tongue, throat, Adam's apple, breast, penis, scrotum, clitoris, vulva, and navel are also superficial structures. This section is about the human body as a whole. Components of the superficial body parts are indicated in Fig. 1.1.

1.2 Human Body Features

The entire structure of the human body consists of a head, neck, torso, two arms, and two legs. By the time the human reaches adulthood, the body consists of close to 10 trillion cells, the basic unit of life. These cells are arranged systematically to eventually form the whole body.

S. Pal, *Design of Artificial Human Joints & Organs*,
DOI 10.1007/978-1-4614-6255-2_1,

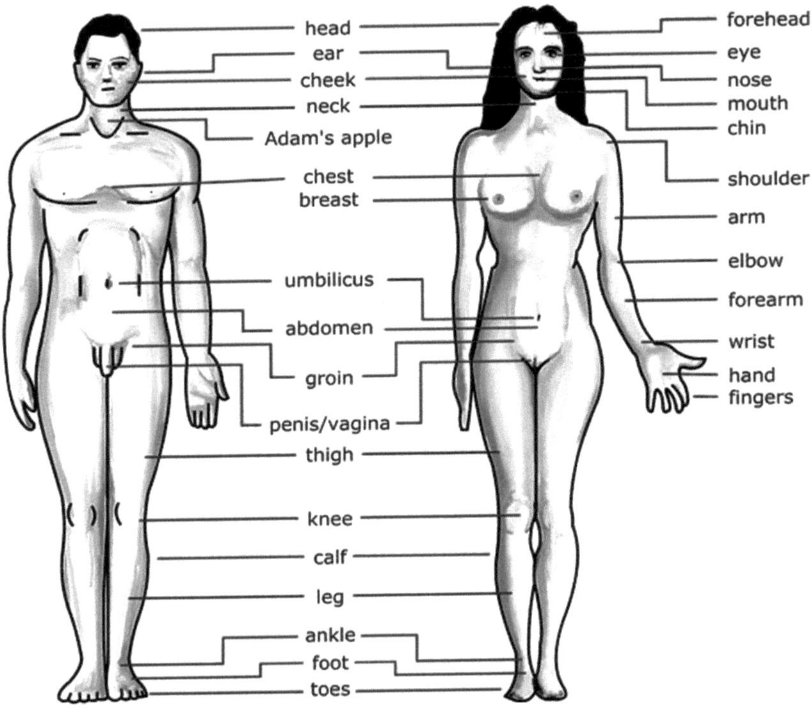

Fig. 1.1 Common names of human body parts

1.3 Body Size, Type, and Proportion

In developed countries, the average height of an adult male human is about 1.7–1.8 m (5′7″ to 5′11″) and that of an adult female is about 1.6–1.7 m (5′3″–5′7″). Generally, these heights are about 5–10% lower in Eastern countries, but height depends on race as well. This size is determined first by genes and second by diet. Body type and body composition are influenced by postnatal factors such as diet and exercise. In India, due to multiple ethnicities, there is a wide range of variation from Kashmir to Kanyakumari and from Kolkata to Kutch. Table 1.1 shows the constituent elements of the human body for an average male who weighs 60 kg.

Anatomy is subdivided into gross anatomy and microscopic anatomy [1]. Gross anatomy (also called topographical anatomy, regional anatomy, or anthroponomy) is the study of anatomical structures that can be seen by unaided vision. Microscopic anatomy is the study of minute anatomical structures assisted with microscopes, which includes **histology** (the study of the organization of tissues) [1] and **cytology** (the study of cells). Anatomy, physiology (the study of function), and **biochemistry** (the study of the chemistry of living structures) are complementary basic medical sciences that are usually taught together (or in tandem).

Table 1.1 Constituents of the human body in a normal man weighing 60 kg

Constituent	Weight (kg) [1]	Percent of atoms (%) [1]
Oxygen	38.8	25.5
Carbon	10.9	9.5
Hydrogen	6.0	63
Nitrogen	1.9	1.4
Calcium	1.2	0.3
Phosphorus	0.6	0.2
Potassium	0.2	0.06

In some of its facets, human anatomy is closely related to embryology, comparative anatomy, and comparative embryology through common roots in evolution. For example, much of the human body maintains the ancient segmental pattern that is present in all vertebrates, with basic units being repeated, which is particularly obvious in the vertebral column and in the ribcage, and can be traced from very early embryos. The history of anatomy has been characterized over a long period of time by a continually developing understanding of the functions of organs and structures in the body. Methods have also advanced dramatically, proceeding from an examination of animals, through dissection of preserved cadavers (dead human bodies), to technologically complex techniques developed in the twentieth century.

1.4 A Brief Outline of the Organization of the Human System

The human body consists of the organ system, organs, tissue, and cells. Cells are the unit of living entity. Cells have their own activities in the body. Groups of cells working together form a **tissue**. There are four types of tissues:

- **Epithelial tissue** covers and protects underlying tissue. When we look at the surface of our skin, we see epithelial tissue.
- **Nerve tissue** sends electrical signals through the body. Nervous tissue is found in the brain, nerves, and sense organs. There is a total network of nerve tissue in the body, connecting and coordinating the activities of various organs and systems.
- **Muscle tissue** is made up of cells that can contract and relax to produce movement; muscle tissue covers the entire body.
- **Connective tissue** joins, supports, protects, insulates, nourishes, and cushions organs. It also keeps organs together in place.

1.4.1 Tissue + Tissue = Organ (Different Tissues Together Form an Organ)

Two or more type of tissues working together form an organ. One type of tissue alone cannot do all of the things that several types working together can do. Figure 1.2 shows the stomach organ, which is made up of all four types of tissue just described.

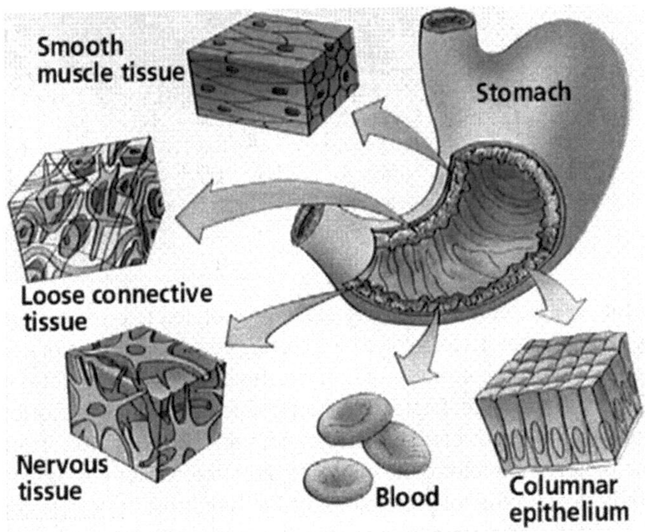

Fig. 1.2 The stomach is an organ having all types of tissues

1.4.2 Organ + Organ = Organ System

Organs working together make up an organ system. A failure of one organ system affects others and may lead to the failure of others.

1.5 Major Organ Systems

The human body consists of the following major organ systems:

1. Integumentary system
2. Muscular system
3. Skeletal system
4. Cardiovascular system
5. Respiratory system
6. Urinary system
7. Reproductive system
8. Nervous system
9. Digestive system
10. Lymphatic system
11. Immune system

Fig. 1.3 Normal human skin

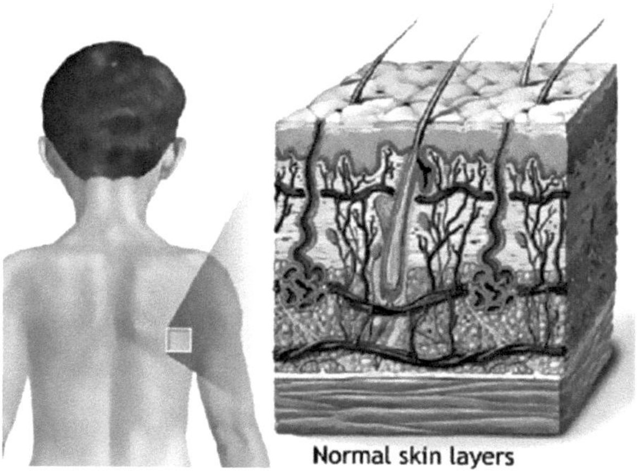

1.5.1 Integumentary System

The integumentary system is the largest organ system in the human body and is responsible for protecting the body from most physical and environmental factors. The largest organ in the body is the skin. The integument also includes appendages, primarily the sweat and sebaceous glands, hair, nails, and arrectores pili (tiny muscles at the root of each hair that cause goosebumps) (Fig. 1.3).

1.5.2 Muscular System

This system moves our bones and helps us to walk, run, and perform all the movement-related activities of daily living. Muscles are connected to the bone through **tendons**, and bones are connected to another bone at the joints through **ligaments** (Fig. 1.4).

1.5.3 Skeletal System

The skeletal system consists of bones interconnected with ligaments and the muscle tendons. This system frames, houses, and protects body parts and organs. The human musculoskeletal system consists of the human skeleton, made by bones attached to other bones with joints, and skeletal muscle attached to the skeleton by tendons.

Bones: An adult human has approximately 206 distinct bones. A list of the number of bones in different body parts is given below; the nomenclature appears in Fig. 1.5.

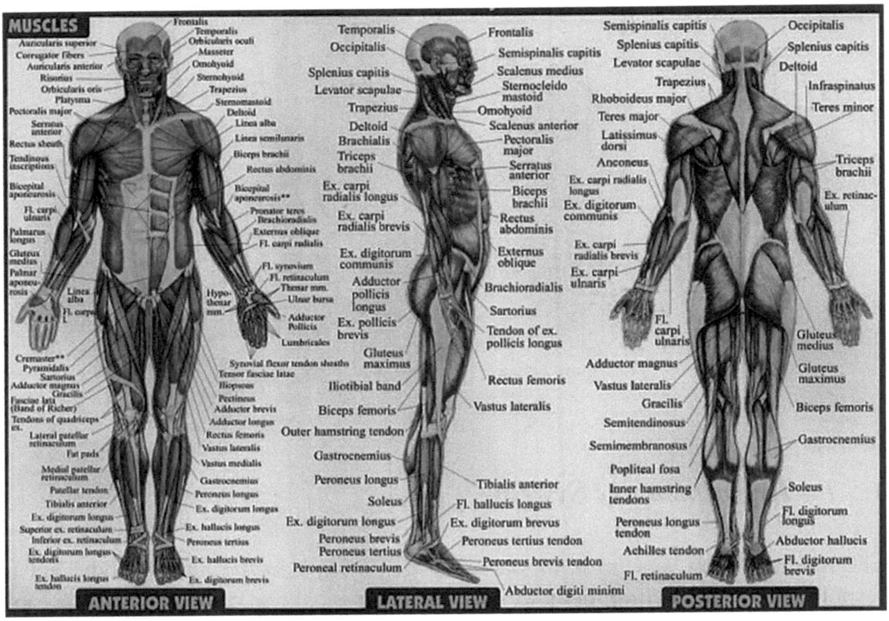

Fig. 1.4 Details of human muscular system

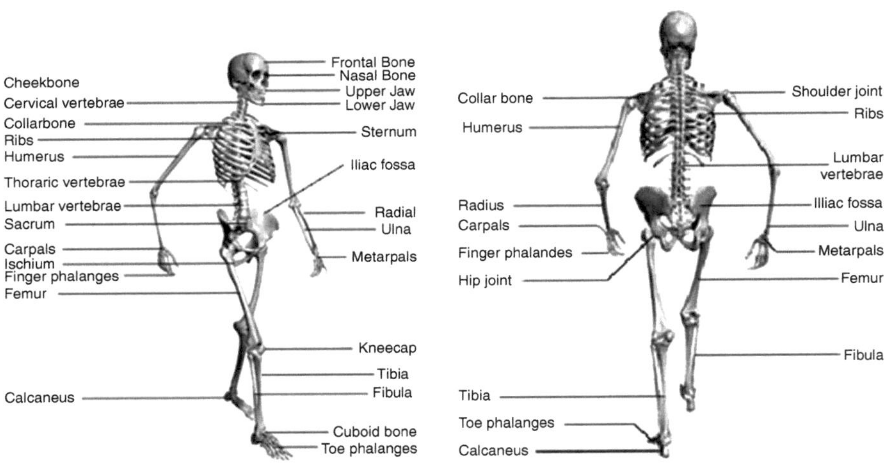

Fig. 1.5 Human skeletal system, anterior (*front*) and posterior (*back side*) views

Spine and vertebral column (26)
Cranium (8)
Face (14)
Hyoid bone, sternum, and ribs (26)
Upper extremities (64), lower extremities (62)

Long bones are bigger at the joints, with spongy bone inside filled with marrow, blood, and body fluid to absorb shock during walking and other activities. The middle part is narrower and cortical (dense bone), a tubular type filled with marrow, blood, and fluid. Blood is produced in this cavity.

1.5.4 Cardiovascular System

The cardiovascular system comprises the heart, veins, arteries, and capillaries. The primary function of the heart is to circulate the blood and, through the blood, oxygen and vital minerals to the tissues and organs that comprise the body. The left side of the main organ (left ventricle and left atrium) is responsible for pumping blood to all parts of the body, while the right side (right ventricle and right atrium) pumps only to the lungs [2, 3]. The heart itself is divided into three layers, called the endocardium, myocardium, and epicardium, which vary in thickness and function (Fig. 1.6) [3].

1.5.5 Respiratory System

A regular supply of oxygen from air is vital to life. Air enters through the nose and mouth, where it is filtered. It then passes down the trachea to the lungs. The trachea has two main branches, the left and right bronchi, which divide into a network of smaller bronchioles. These lead into small air sacks called alveoli. The two lungs are shaped like inverted cones, with a wide base and a narrow top. The right lung has three lobes, while the left has only two to accommodate the heart.

The lungs contain about 2,400 km of airways, and each lung has total surface area of 180 m^2 (Fig. 1.7).

1.5.6 Urinary System

The urinary system removes excess fluid and soluble substances from blood circulation via the kidneys. Some fluids are reabsorbed back into the blood, while excess water and waste products are expelled from the body as urine. Each kidney is about 12 cm long and contains two layers of tissue: an outer cortex and an inner medulla. Urine continuously trickles down two tubes called ureters into the bladder, which is an elastic sac that stores urine until it can be expelled out. The bladder can hold up to 0.5 l of urine.

The urinary tract consists of the organs, tubes, and muscles that work together to make, move, store, and release urine, the liquid waste of the human body. The upper urinary tract includes the kidneys, which filter wastes and extra fluid from the blood, and the ureters, which carry urine from the kidneys to the bladder. The lower urinary

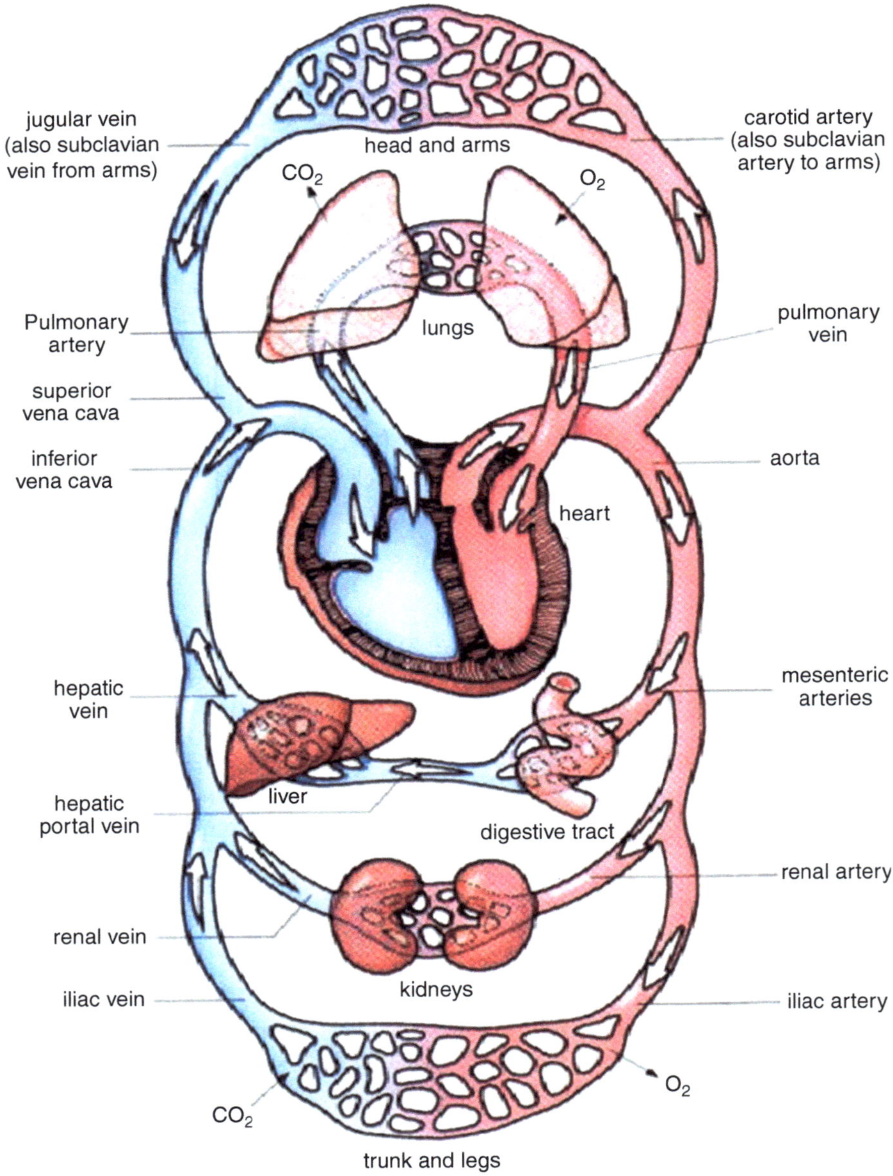

Fig. 1.6 A schematic diagram of the circulatory system

tract includes the bladder, a balloon-shaped muscle that stores urine, and the urethra, a tube that carries urine from the bladder to the outside of the body during urination. Doctors who specialize in kidney problems are called nephrologists. Doctors who specialize in problems of the organs and tubes that transport urine from the kidneys to outside the body are called urologists. These problems may involve cancers or

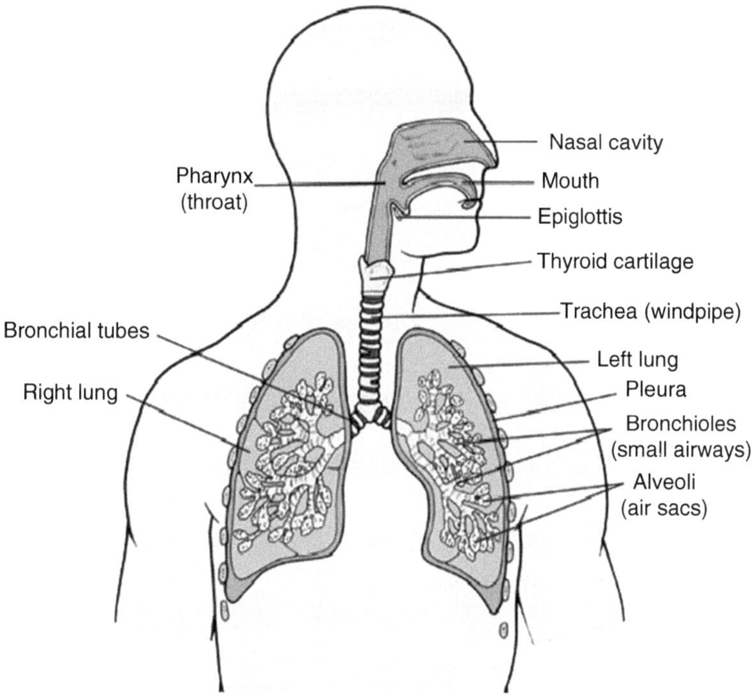

Fig. 1.7 Simplified respiratory system

growths of these organs, including the kidneys, ureters, bladder, and testes, or may involve abnormalities in storing or releasing urine (Fig. 1.8).

1.5.7 Reproductive System

Human reproduction takes place as internal fertilization by sexual intercourse. During this process, the penis ejaculates seminal fluid, containing millions of sperm, into the female's vagina. The sperm then travels through the vagina and cervix into the uterus or fallopian tubes for fertilization of one ovum, the largest cell in the body, visible to the naked eye.

1.5.7.1 Male Reproductive System

The human male reproductive system is a series of organs located outside the body and around the pelvic region of a male that contribute toward the reproductive process. The primary direct function of the male reproductive system is to provide the male gamete or spermatozoa for fertilization of the ovum.

The major reproductive organs of the male can be grouped into three categories. The first category is sperm production and storage. Production takes place in the

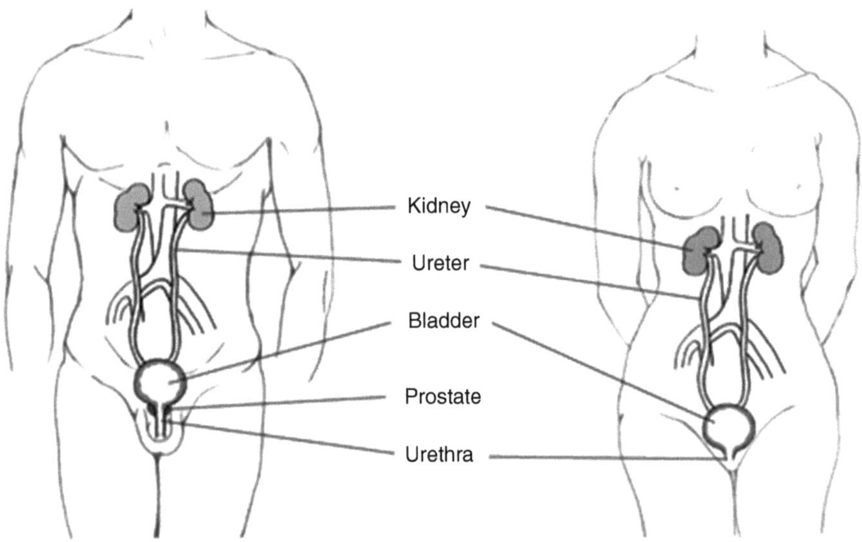

Fig. 1.8 Male and female urinary systems

testes, which are housed in the temperature-regulating scrotum; immature sperm then travel to the epididymis for development and storage. The second category is the ejaculatory fluid-producing glands, which include the seminal vesicles, prostate, and the vas deferens. The final category includes those used for copulation and discharge of the spermatozoa (sperm) within the female organ; these include the penis, urethra, vas deferens, and Cowper's gland.

1.5.7.2 Female Reproductive System

The human female reproductive system is a series of organs primarily located inside the body and around the pelvic region of a female that contribute toward the reproductive process. The human female reproductive system contains three main parts: the vagina, which acts as the receptacle for the male's sperm; the uterus, which holds the developing fetus; and the ovaries, which produce the female's ova. The breasts are also an important reproductive organ during the parenting stage of reproduction.

The vagina meets the outside at the vulva, which also includes the labia, clitoris, and urethra; during intercourse, this area is lubricated by mucus secreted by Bartholin's glands. The vagina is attached to the uterus through the cervix, while the uterus is attached to the ovaries via the fallopian tubes. At certain intervals, typically approximately every 28 days, the ovaries release an ovum, which passes through the fallopian tube into the uterus. The lining of the uterus, called the endometrium, and unfertilized ova are shed each cycle through a process known as menstruation (Figs. 1.9 and 1.10).

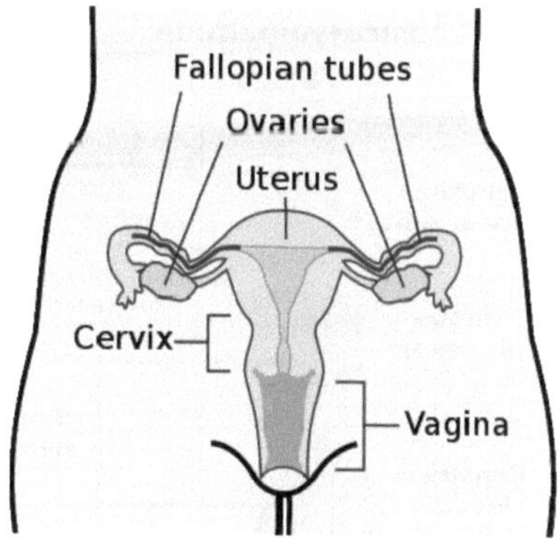

Fig. 1.9 Female reproductive organs

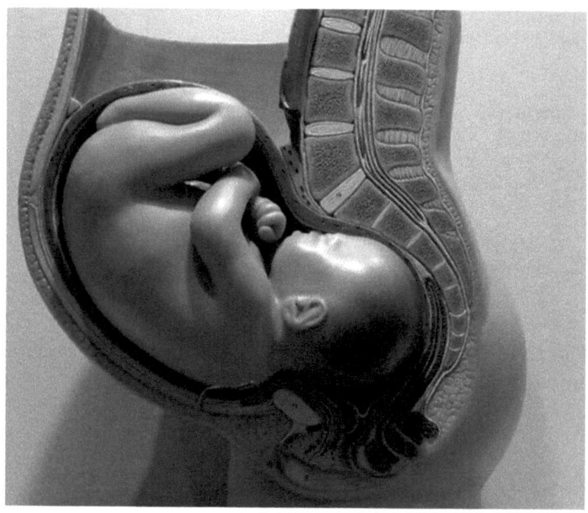

Fig. 1.10 A matured baby inside the uterus

1.5.8 Nervous System

The nervous system and the human brain comprise a network of specialized cells that communicate information about the organism's surroundings and itself. They receive and send electrical messages through a network of nerves through the muscular system (Fig. 1.11).

1.5.9 Digestive System

The digestive system and the human gastrointestinal tract break down food into nutrients. The digestive system provides the body's means of processing food and transforming nutrients into energy. It consists of the organs marked in Fig. 1.12.

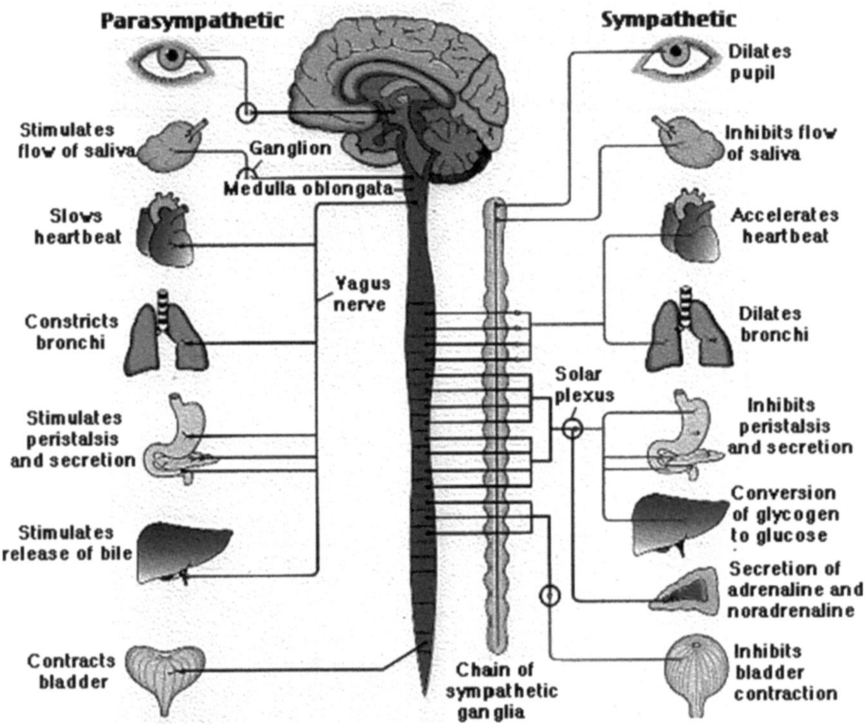

Fig. 1.11 The schema of the nervous system and its innervations with various organs. The organs are easily identifiable

1.5.10 Lymphatic System

The main function of the lymphatic system is to extract, transport, and metabolize lymph, the fluid found in between cells. The lymphatic system is very similar to the circulatory system in terms of both its structure and its most basic function is to carry body fluid (Fig. 1.13).

1.5.11 Immune System

The immune system is the body's defense against infectious organisms and other invaders. Through a series of steps called the immune response, the immune system attacks organisms and substances that invade body systems and cause disease. The immune system is made up of a network of cells, tissues, and organs that work together to protect the body. The cells involved are white blood cells, or leukocytes,

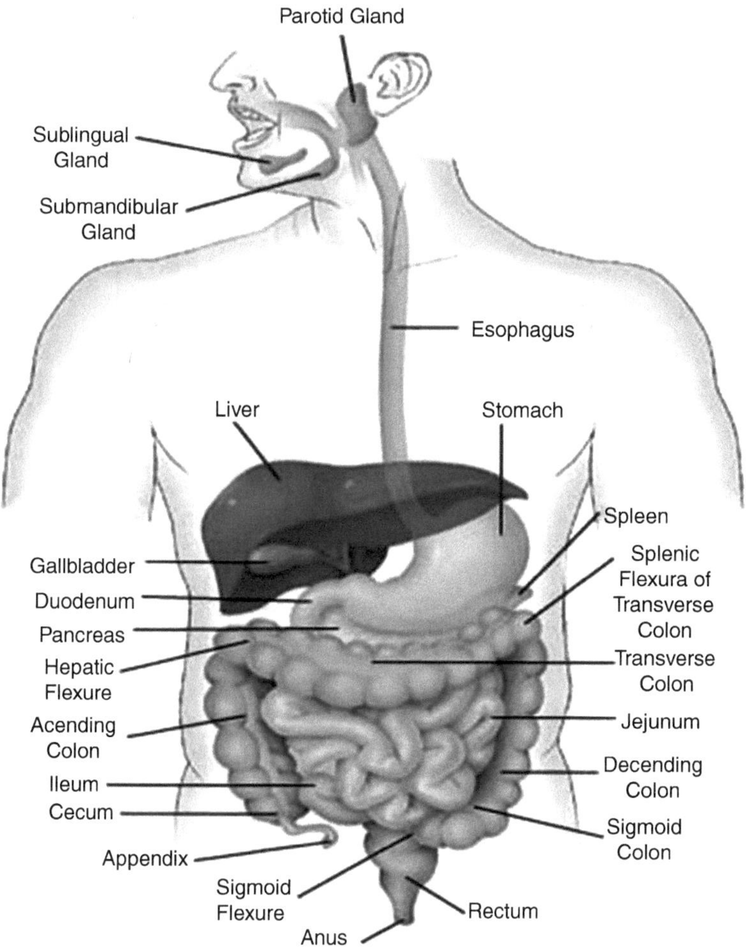

Fig. 1.12 Digestive system

which come in two basic types that combine to seek out and destroy disease-causing organisms or substances.

Leukocytes are produced or stored in many locations in the body, including the thymus, spleen, and bone marrow. For this reason, they are called the lymphoid organs. There are also clumps of lymphoid tissue throughout the body, primarily as lymph nodes, that house the leukocytes.

The leukocytes circulate through the body between the organs and nodes via lymphatic vessels and blood vessels. In this way, the immune system works in a coordinated manner to monitor the body for bacteria or viruses or substances that might cause problems, such as infection, fever, and pain.

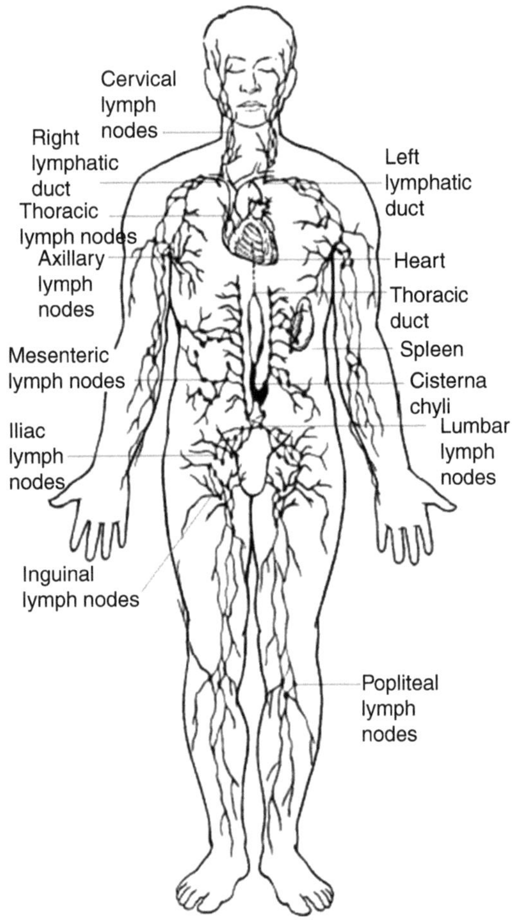

Fig. 1.13 Lymphatic system and the nodes

1.6 Common Names of Internal Organs (in Alphabetical Order)

Adrenals—appendix—bladder—brain—esophagus—eyes—gallbladder—heart—intestines—kidney—liver—lungs—ovaries—pancreas—parathyroids—pituitary—prostate—spleen—stomach—testicles—thymus—thyroid—uterus—veins

A further description of areas of the human brain include amygdala—brain stem—cerebellum—cerebral cortex—limbic system—medulla—midbrain—pons.

1.7 Brief Idea of Artificial Organs

The principal aim of this book is to discuss the design of some major artificial human joints and organs, their fixation modalities, and also the reasons for using such devices.

An artificial organ is a manmade device that is implanted, or integrated into, a human body to replace a natural organ with the purpose of restoring a specific function or a group of related functions so the patient may return to as normal a life as possible. The replaced function need not necessarily have to be related to life support, but often it is.

Implied by this definition is the fact that the device must not be continuously connected to a stationary power supply or other stationary resources, such as filters or chemical processing units. (Periodic rapid recharging of batteries, refilling of chemicals, and/or cleaning/replacing of filters would exclude a device from being called an artificial organ.) Thus, a dialysis machine, while a very successful and critically important life support device that completely replaces the functions of a kidney, is not an artificial organ. At this time, an efficient, self-contained artificial kidney has not become available.

Reasons to construct and install an artificial organ, an extremely expensive process initially, which may entail many years of ongoing maintenance services not needed by a natural organ, might include

- Life support to prevent imminent death while awaiting a transplant (e.g., artificial heart, kidney)
- Considerable improvement of the patient's ability for self-care and ambulation (e.g., artificial joints, limb)
- Improvement of the patient's ability to interact socially (e.g., cochlear implant, IOL)
- Cosmetic restoration after cancer surgery or accident

The use of any artificial organ by humans is almost always preceded by extensive experiments with animals. Initial testing in humans is frequently limited to those either already facing death or who have exhausted every other treatment possibility. (Rarely, testing may be done on healthy volunteers who are scheduled for judicial execution.)

Although they are not typically thought of as organs, one might also consider the replacement of bones and joints of the human body or other animals.

1.7.1 Different Types of Organs

There are now many artificial organs that have been implanted in humans, with varying degrees of success. In this section, we give a brief overview of the various types of artificial organs that have been developed over the years due to the tremendous efforts of various scientists worldwide. Some are discussed in this book, but not all are included in this text.

1.7.2 Brain

Brain pacemakers, including deep brain stimulators, send electrical impulses to the brain in order to relieve depression, epilepsy, tremors from Parkinson's disease, and other conditions such as increased bladder secretions. Rather than replacing existing neural networks to restore function, these devices often serve by disrupting the output of existing malfunctioning nerve centers to eliminate symptoms.

1.7.3 Cardia

Artificial cardia can be used to fight, among other diseases, esophageal cancer, achalasia, and gastroesophageal reflux disease. This pertains to gastric repairs, specifically of the valves at either end of the stomach.

1.7.4 Corpora Cavernosa

To treat erectile dysfunction of penis, both corpora cavernosa can be irreversibly surgically replaced with manually inflatable penile implants. This is a drastic therapeutic surgery meant only for men suffering from complete impotence that has resisted all other treatment modalities.

An implanted pump in the groin or scrotum can be manipulated by hand to fill these artificial cylinders, normally sized to be direct replacements for the natural corpus cavernosa, from an implanted reservoir in order to achieve an erection. Though technically novel, it may not be socially acceptable in everyday society.

1.7.5 Ear

Cochlear Implant
While natural hearing, to the level of musical quality, is not typically achieved with the use of cochlear implants, most recipients are pleased, with some finding it useful enough to return to their surgeon with a request to do implantation in the other ear.

1.7.6 Eye

Visual Prosthetic
The most successful function-replacing artificial eye so far is actually an external miniature digital camera with a remote unidirectional electronic interface implanted on the retina, optic nerve, or other related location inside the brain. The present state of the art yields only very partial functionality, such as recognizing levels of brightness, swatches of color, and/or basic geometric shapes, proving the concept's potential. While the living eye is indeed a camera, it is also much more than that.

Various researchers have demonstrated that the retina performs strategic image preprocessing for the brain. The problem of creating a 100% functional artificial electronic eye is even more complex than what is already obvious. The steadily increasing complexity of the advances in the artificial connection to the retina, optic nerve, or related brain areas, combined with ongoing advances in computer science, is expected to dramatically improve the performance of this technology.

For the person whose damaged or diseased living eye retains some function, other options superior to the electronic eye may be available.

1.7.7 Heart

The use of artificial hearts has been considered a success but still is limited to patients awaiting transplants whose death is imminent. The current state-of-the-art devices are unable to reliably sustain life beyond about 18 months.

Artificial pacemakers are electronic devices that can intermittently augment (defibrillator mode), continuously augment, or completely bypass the natural living cardiac pacemaker as needed. They are so successful that they have become very popular and easily implanted.

Ventricular assist devices are mechanical circulatory devices that partially or completely replace the function of a failing heart, without the removal of the heart itself.

1.7.8 Limbs

Artificial Limb

Artificial limbs include arms with semifunctional hands, some even fitted with working opposable "thumbs" plus two "fingers," and legs with shock-absorbing feet capable of allowing a trained patient to run. While the meaning of "full mobility" is debated, steady progress is being made. Energy consumption is an important criterion for determining the success of an implant.

1.7.9 Liver

Liver Dialysis and Devices

HepaLife, a U.S.-based company, is developing a bioartificial liver device intended for the treatment of liver failure using stem cells. The artificial liver, currently under development, is designed to serve as a supportive device, either to allow the liver to regenerate upon acute liver failure, or to bridge the patient's liver functions until a transplant is available. It is only made possible by the fact that it uses real liver cells, and even then, it is not a permanent substitute for a liver.

On the other hand, the researchers Dr. Colin McGucklin, professor of regenerative medicine at Newcastle University, and Dr. Nico Forraz, senior research associate and clinical sciences business manager at Newcastle University, say that pieces of artificial liver could be used to repair livers injured in the next five years. These artificial livers could also be used outside the body in a manner analogous to the dialysis process used to keep patients alive whose kidneys have failed.

1.7.10 Lungs

Some almost fully functional, artificial lungs promise to be a great success in the near future.

1.7.11 Pancreas

For the treatment of diabetes, numerous promising techniques that may be called an artificial pancreas are currently being tested, including some that incorporate donated living tissue housed in special materials to prevent the patient's immune system from killing the foreign live components.

1.7.12 Bladder

Artificial bladders represent a unique success in that these are autologous laboratory-grown living replacements, as opposed to most other artificial organs, which depend upon electromechanical contrivances and may or may not incorporate any living tissue.

1.7.13 Ovaries

Reproductive-age patients who develop cancer often receive chemotherapy or radiation therapy, which damages oocytes and leads to early menopause. An artificial human ovary has been developed at Brown University with self-assembled micro tissues created using novel 3D Petri dish technology. The artificial ovary will be used for the purpose of the *in vitro* maturation of immature oocytes and the development of a system to study the effect of environmental toxins on folliculogenesis.

1.7.14 Beyond Restoration

It is also possible to construct and install an artificial organ to give its possessor abilities that are not naturally occurring. Research is proceeding, particularly in areas of vision, memory, and information processing; however, this idea is still in its infancy.

Some current research focuses on restoring inoperative short-term memory in accident victims and lost access to long-term memory in dementia patients. Success here would lead to widespread interest in applications for persons whose memory is considered healthy to dramatically enhance their memory far beyond what can be achieved with mnemonic techniques. Given that our understanding of how living memory actually works is incomplete, it is unlikely this scenario will become reality in the near future.

One area of success was achieved in 2002 when a British scientist, Kevin Warwick, had an array of 100 electrodes fired into his nervous system in order to link his nervous system into the Internet. With this in place, he carried out a series of experiments, including extending his nervous system over the Internet to control a robotic hand, a form of extended sensory input and the first direct electronic communication between the nervous systems of two humans.

Another idea with significant consequences is that of implanting a language translator for diplomatic and military applications. While machine translation does exist, it is presently neither good enough nor small enough to fulfill its promise.

This might also include the existing (and controversial when applied to humans) practice of implanting subcutaneous "chips" (integrated circuits) for identification and location purposes. An example of this is the RFID tags made by VeriChip Corporation.

1.7.15 Timeline of Successful Transplants

- 1857: First successful cornea transplant, by Eduard Zirm (Olomouc Eye Clinic, now Czech Republic)
- 1954: First successful kidney transplant, by Joseph Murray (Boston, MA, USA)
- 1966: First successful pancreas transplant, by Richard Lillehei and William Kelly (Minneapolis, MN, USA)
- 1967: First successful liver transplant, by Thomas Starzl (Denver, CO, USA)
- 1967: First successful heart transplant, by Christiaan Barnard (Cape Town, South Africa)
- 1981: First successful heart/lung transplant, by Bruce Reitz (Stanford, CA, USA)
- 1983: First successful lung lobe transplant, by Joel Cooper (Toronto, ON, Canada)
- 1986: First successful double-lung transplant, by Joel Cooper (Toronto, ON, Canada)

- 1995: First successful laparoscopic live-donor nephrectomy by Lloyd Ratner and Louis Kavoussi (Baltimore, MD, USA)
- 1997: First successful allogeneic vascularized transplantation of a fresh and perfused human knee joint by Gunther O. Hofmann a German Surgeon
- 1998: First successful live-donor partial pancreas transplant by David Sutherland (Minneapolis, MN, USA)
- 1998: First successful hand transplant by Jean-Michel Dubernard (Lyon, France)
- 1999: First successful tissue-engineered bladder transplanted by Anthony Atala (Boston, MA, USA)
- 2005: First successful partial face transplant (France) by B. Devauchelle in Amiens
- 2006: First jaw transplant to combine donor jaw with bone marrow from the patient, by Eric M. Genden (Mount Sinai Hospital, New York)
- 2008: First successful complete full double-arm transplant by Edgar Biemer, Christoph Höhnke, and Manfred Stangl (Technical University of Munich, Germany)
- 2008: First baby born from transplanted ovary, by James Randerson, in London, UK
- 2008: First transplant of a human windpipe using a patient's own stem cells, by Paolo Macchiarini (Barcelona, Spain)
- 2008: First successful transplantation of near total area (80%) of face (including palate, nose, cheeks, and eyelid), by Maria Siemionow (Cleveland, OH, USA)
- 2010: First full facial transplant, by Dr. Joan Pere Barret and team (Hospital Universitari Vall d'Hebron, Barcelona, Spain)
- 2011: First double-leg transplant, by Dr. Cavadas and team (Valencia's Hospital La Fe, Spain)

References

1. Burton G (2000) Chemical storylines, p. 3, edn 2, illustrated. Heinemann. Pp 312. ISBN 0435631195, 9780435631192
2. "Cardiovascular System." U.S. National Cancer Institutes
3. Gray H (1918) Introduction page, Anatomy of the human body, 20th edn. Lea & Febiger, Philadelphia, http://www.bartleby.com/107/1.html. Retrieved 27 March 2007 [In public domain]
4. Gray's Anatomy, the anatomical basis of clinical practice, Expert Consult - Online, Susan Standring, PhD, DSc 40th Edition

Exercises

1. List the various organs in the human body and indicate what percent of total blood circulates through it.
2. How can you compare the organs of the body with some engineering machines and systems? Write down your arguments. Hint: The brain is like a computer.

3. Compare the nervous system with an electronic feedback control system.
4. Which system makes the human machine different from other machines?
5. Which major organs or joints in the human body can be successfully replaced? Indicate the approximate service life of such organs and the approximate cost of replacing them.

Chapter 2
Mechanical Properties of Biological Materials

2.1 Introduction

When we want to design, that is, prepare a drawing and fabricate something to replace a desired body part or malfunctioning organ(s) due to disease processes, trauma, or surgical removal, it is necessary to understand the real nature and biomechanical characteristics of those anatomical parts, e.g., tissues and organs. Anthropometry is the science and practice of measuring the size and shape of the human body and its parts. To measure the properties of biological materials and tissues are also anthropometry's tasks. Biomechanics, in turn, studies the structure and function of biological systems using the methods of mechanics. The composition and behavior of bones, cartilages, and ligaments have been studied for many years. However, although we know much about these tissues, newer and better measurement techniques continuously improve the available data. It should be remembered that there are biological variations and environmental factors that significantly affect the mechanical properties of biological tissues.

This chapter introduces the terms and procedures involving biomaterials and tries to identify the last consensus data regarding the hard, but at the same time deformable, tissues relevant in the study of the human joints' behavior. In this group, we include skeletal bones, articular cartilage, and ligaments.

2.2 Structural Versus Material Properties

A biological tissue is often described in terms of its structural and material properties. **Structural properties** characterize the tissue in its intact form. Important structural properties are represented by a relationship between force and deformation, or stress and strain, and must be understood in order to predict how a tissue will behave in vivo.

Material properties characterize the behavior of the material comprising the tissue and to a first approximation are independent of the size of the tissue. The material

S. Pal, *Design of Artificial Human Joints & Organs*,
DOI 10.1007/978-1-4614-6255-2_2, © Springer Science+Business Media New York 2014

properties are usually expressed in terms of the stress–strain relationship of the material. The strength of a material, which is the breaking or ultimate strength under different modes of loading, such as tension, compression, torsion, or bending, will be different, as will the corresponding modulus of elasticity or stiffness, except bending.

The **stiffness** of a material represents the material's ability to resist deformation. Stiffness is commonly characterized by the slope of the linear region of a stress–strain curve, also referred to as **Young's modulus** when tested **under tension**. To describe the slope of other regions of the stress–strain curve, a tangent modulus is often defined. A **tangent modulus** should have associated with it a strain value or a range of strains. There can be different kinds of moduli depending on the loading types (e.g., shear modulus, compression modulus). The larger the stiffness, the greater the force required to cause a given deformation. If the stress in a material is directly proportional to the strain for strains up to the elastic limit, the material is called a **Hookean** material.

2.2.1 Anisotropy and Nonhomogeneity

Ideal materials are isotropic and homogeneous. A material is called isotropic when its properties are the same in each of three coordinate axes' (x, y, z) direction. Tensile and compressive properties may be different, but each respective property must be the same in three directions. A material is said to be homogeneous if it is made of the same material throughout. Biological tissues are **anisotropic** and **nonhomogeneous**.

2.2.2 Viscoelastic Properties

Biological tissues are viscoelastic materials; their behavior is both viscous, meaning time- and history-dependent, as well as elastic. A viscoelastic material possesses characteristics of **stress-relaxation, creep, strain-rate sensitivity**, and **hysteresis**. Force-relaxation (or **stress-relaxation**) is a phenomenon that occurs in a tissue stretched and held at a fixed length. Over time the stress developed within the tissue continually declines. Stress-relaxation is force- or strain-rate–sensitive. In general, the higher the strain or loading rate, the larger the peak force/stress and subsequently the greater the magnitude of the force-relaxation. In contrast to stress-relaxation, which occurs when a tissue's length is held fixed, is creep. **Creep** occurs with time when a constant force/stress is applied across the tissue. If subjected to a constant tensile force, then a tissue elongates with time. The general shape of the displacement-time curve depends on the past loading history (e.g., peak force, loading rate).

Another time-dependent property is **strain-rate sensitivity**. Different tissues show different sensitivities to strain rate. For example, there may be little difference in the stress–strain behavior of ligaments subjected to tensile tests varying in strain rate over 3 decades, while bone properties may change considerably

Additionally, the loading and unloading curves obtained from a force-deformation test of biological tissues do not follow the same path. The difference in the calculated area under the loading and unloading curves is termed the area of **hysteresis** and represents the energy lost due to internal friction in the material. The amount of energy liberated or absorbed during a tensile test is defined as the integral of the force and the displacement. Hence, the maximum energy absorbed at failure equals the area under the force–displacement curve.

Y. C. Fung [1], in his text *QLV* (*Quasilinear Viscoelastic*) *Theory*, suggested that if a step increase in elongation is imposed on the specimen, the stress developed will be a function of time (t) as well as of the material's stretch ratio (λ). The history of the stress response, called the relaxation function [$K(\lambda, t)$], is assumed to be of the form [$K(\lambda, t) = G(t)*T(\lambda)$] in which $G(t)$ is a normalized function of time, called the reduced relaxation function, and $T(\lambda)$ is a function of the stretch ratio alone, called the elastic response. Fung also proposed a function for defining the elastic response of the material under tension conditions.

2.2.3 Viscosity

The viscosity of a fluid is a measure of the fluid's resistance to flow. The viscosity of water is used as a reference to calculate other fluids' viscosity and is considered to be 1. The capsule of diarthrodial joints is normally filled with a fluid of viscosity 10 called synovial fluid. This fluid helps to reduce friction and wear of articulating surfaces. Just for comparison, the viscosity of olive oil, for example, is 84 [2].

2.3 Testing Procedures

Structural properties of biological tissues are usually determined through some form of mechanical testing (e.g., tensile tests, compressive tests, bending and torsion tests). Customized workstations utilizing force transducers, clamps, and an actuator to control the distance between clamps are commonplace. Commercial systems are also available and vary in design depending on the type of tissue being studied (e.g., macroscopic vs. microscopic, hard tissue vs. soft tissue, etc.) and the type of loading rates required. Instron and MTS are the two most common suppliers of mechanical testing systems. Currently, one UK-based company, McMesin, is also supplying such a machine. Most systems allow either force control or length control. See the pictures in Fig. 2.1.

Mechanical testing of tissue in vivo is very difficult and hence not commonly performed. Some of the techniques that have been utilized include (1) buckle transducers to monitor tendon and ligament forces, (2) telemetried pressure sensors to measure joint contact pressure, and (3) strain gauges to quantify bone and ligament strain. Some noninvasive approaches have also been employed. Ultrasound techniques

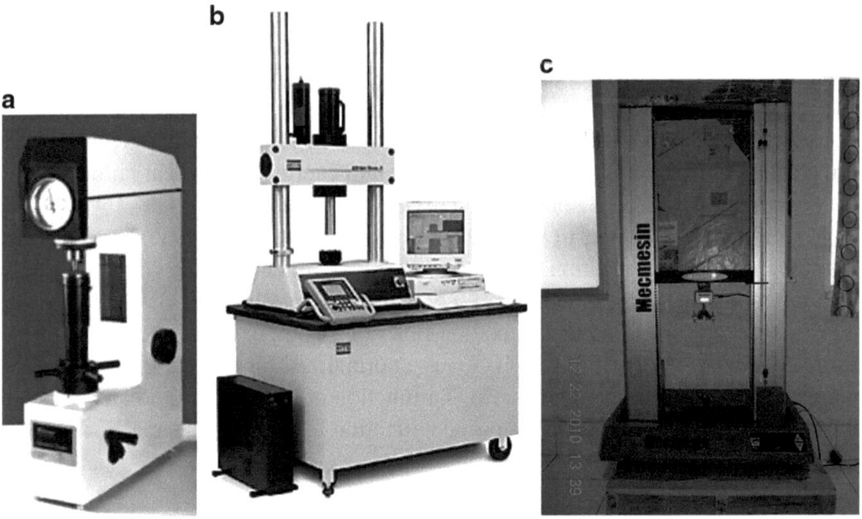

Fig. 2.1 Mechanical testing system: (**a**) hardness tester; (**b**) and (**c**) axial and bending load tester

have been used to detect changes in the speed of sound in different tissues, and these changes have been correlated with the tissue's elastic properties.

Various imaging techniques have also been used to quantify tissue geometry and deformation [2].

2.4 Bones

2.4.1 Composition

Bone is a composite material consisting of both fluid and solid phases. Two main solid phases, one organic and another inorganic, give bones their hard structure. An organic extracellular collagenous matrix is impregnated with inorganic materials, especially hydroxyapatite $Ca_{10}(PO_4)_6(OH)_2$ (consisting of the minerals calcium and phosphate). Unlike collagen, apatite crystals are very stiff and strong. However, a bone's strength is higher than that of either apatite or collagen because, similar to what happens with concrete, the softer component prevents the stiff one from brittle cracking, while the stiff component prevents the soft one from yielding. The organic material gives bone its flexibility, while the inorganic material gives bone its resilience.

Calcium and phosphate account for roughly 65–70% of a bone's dry weight. Collagen fibers compose approximately 95% of the extracellular matrix and account for 25–30% of the dry weight of bone. Surrounding the mineralized collagen fibers is a ground substance consisting of protein, polysaccharides, or glycosaminoglycans (GAGs), primarily in the form of complex macromolecules called proteoglycans. The GAGs serve to cement together the various layers of mineralized collagen fibers. Water accounts for up to 25% of the total weight of bone, with about 85% of the water being located in the organic matrix around the collagen fibers and ground substance. The other 15% is located in canals and cavities that house the bone cells.

2.4.2 Structure

Bone is identified as either **cancellous** (also referred to as trabecular or spongy) or **cortical** (also referred to as compact); see Figs. 2.2a and b. Cortical bone is roughly four times the mass of cancellous bone, in any long bone. The basic material comprising cancellous and compact bone appears identical; thus, the distinction between the two is the degree of porosity and the organization. The porosity of cortical bone ranges from 5 to 30%, while cancellous bone's porosity ranges from 30 to 90%. Bone porosity is not fixed and can change in response to altered loading, disease, and the aging process.

The fibrous layer covering all bones is the **periosteum**. This membrane covers the entire bone except the joint surfaces, which are covered with articular cartilage.

There are numerous terms used to describe the complex architecture of bone at a finer resolution. Both cortical and cancellous bone may contain two types of basic architecture, **woven** and **lamellar**. Bone can also be described as **primary** or **secondary** bone; regions within cortical bone are often described as either **haversian** or **laminar**. Details about this may be found in any textbook on the biomechanics of bones.

In the human **femur**, there is a remarkable adaptation of the inner structure of the bone to the mechanical requirements due to the load on the femur head. The various parts of the femur taken together form a single mechanical structure wonderfully well adapted for the efficient, economical transmission of the loads from the acetabulum to the tibia. The bony material is arranged in the paths of the maximum internal stresses, which are thereby resisted and transmitted with the greatest efficiency, and hence with a maximum economy of material. The inner structure and external form of human bone are closely adapted to the mechanical conditions existing at every point in the bone. The inner architecture of normal bone is determined by definite and exact requirements of mathematical and mechanical laws to produce a maximum of strength with a minimum of material.

The cancellous bone of the upper femur to the lower limit of the lesser trochanter is composed of two distinct groups of trabeculae arranged in a nonlinear path: One has its origin in the medial (inner) side of the shaft and curving upward in a fan-like radii to the opposite side of the bone; the other originates in the outer portion of the shaft and runs upward and medially to end in the upper surface of the greater trochanter, neck, and head. These two systems cross each other at 90° angles.

In the shaft, the inner architecture is configured in order to economize for resisting shearing stresses, bending moment, and axial stress. Its structure serves to secure great strength with a relatively small amount of material (Fig. 2.3).

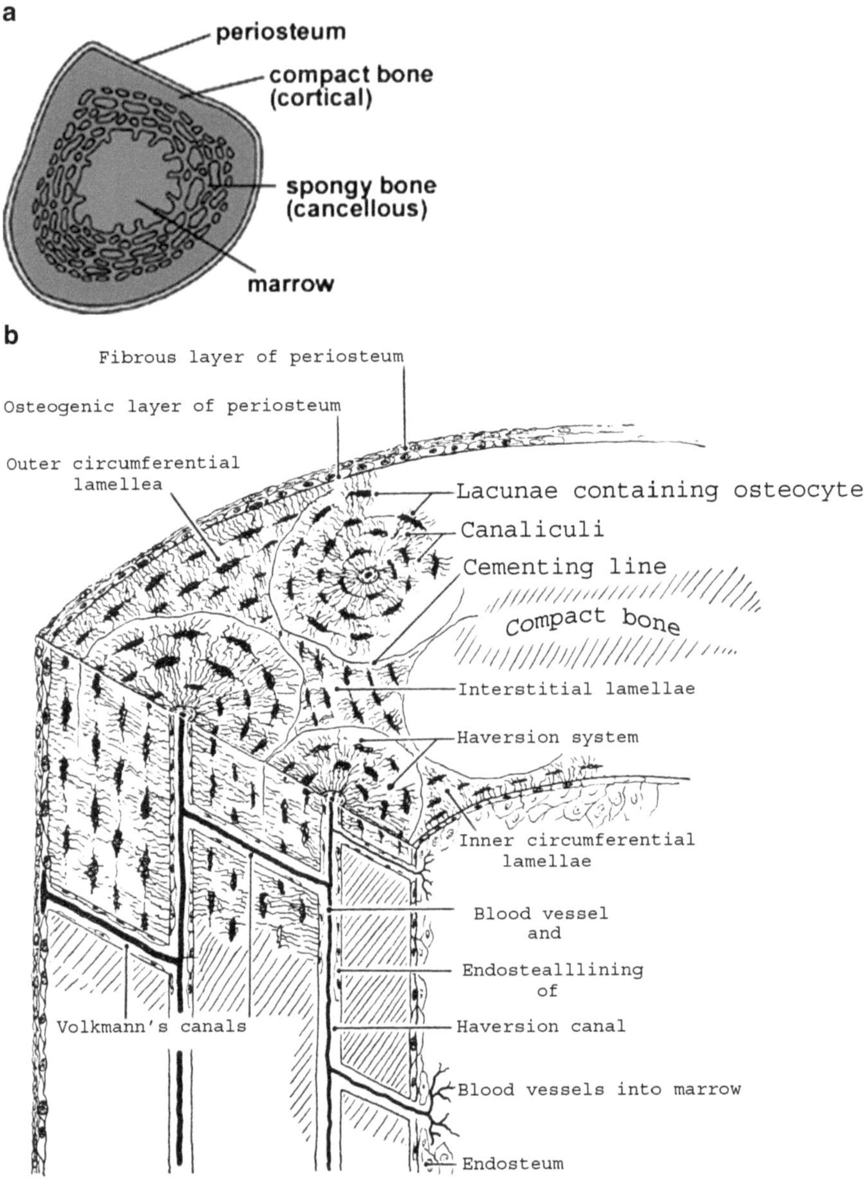

Fig. 2.2 (**a**) Cancellous and compact bone; (**b**) the conceptual structure of compact bone adopted from Ham with modification (1969)

In the **pelvis**, the thicker parts of the bone consist of cancellous tissue, enclosed between two layers of compact tissue; the thinner parts, as at the bottom of the acetabulum and the center of the iliac fosse, are usually semitransparent and composed entirely of compact tissue. It may be noted that the properties of bone vary from species to species, race to race, region to region, male to female, young to old,

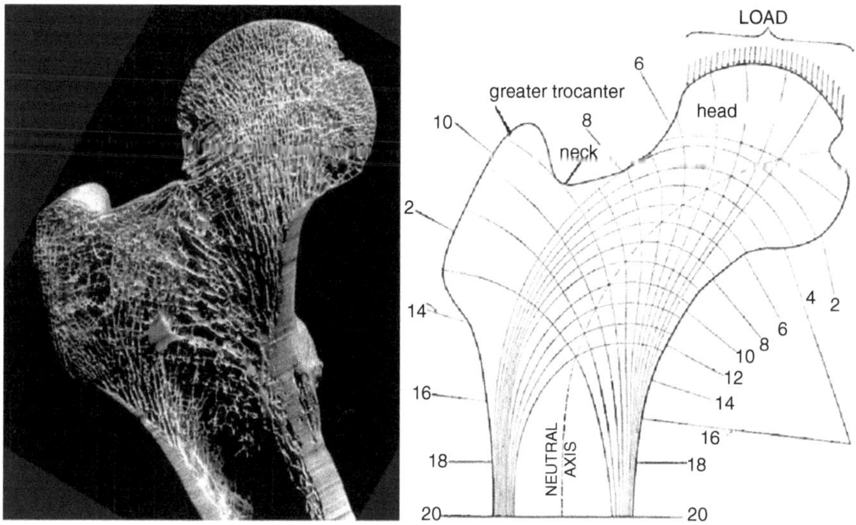

Fig. 2.3 Photograph of the upper femur in coronal section (*left*), and lines of stress (*right*), based upon the mathematical analysis of the right femur [7]

fresh to dry or embalmed, and direction to direction. In the same body there is a regional variation, and bone remodels itself according to the stress generated within it during activities. To illustrate these facts, Table 2.1 shows some illustrative values of the compact bone properties of various species in wet conditions and when loaded parallel to the axis.

2.5 Material Properties and Related Behavior

The mineral content of bone affects its mechanical property. Higher mineralization makes the bone stronger and stiffer (higher modulus of elasticity), but it lowers the toughness; that is, it is less capable of absorbing shock and strain energy. The organic phase makes it more pliable and shock-absorbing, which are desirable for athletes.

Cancellous bone is actually extremely anisotropic and nonhomogeneous. Cortical bone, on the other hand, is approximately linear elastic, transversely isotropic, and relatively homogenous. The material properties of bone are generally determined using mechanical testing procedures; however, ultrasonic techniques have also been employed. Force-deformation (structural properties) or stress–strain (material properties) curves can be determined using tests. However, the properties of bone and most biological tissues depend on the freshness of the tissue. These properties can change within a matter of minutes if allowed to dry out in the open. Cortical bone, for example, has an ultimate strain of around 1.2% when wet and about 0.4% if the water content is not maintained. Thus, it is very important to keep bone specimens wet in lactated Ringer's solution or normal saline water during testing.

Table 2.1 Mechanical properties of wet compact bone in various mode of loading parallel to axis

Bone	Horses	Cattle	Pigs	Human (20–39 years)
Ultimate tensile strength (MPa)				
Femur	121 ± 1.8	113 ± 2.1	88 ± 1.5	124 ± 1.1
Tibia	113	132 ± 2.8	108 ± 3.9	174 ± 1.2
Humerus	102 ± 1.3	101 ± 0.7	88 ± 7.3	125 ± 0.8
Radius	120	135 ± 1.6	100 ± 3.4	152 ± 1.4
Ultimate percentage elongation				
Femur	0.75 ± 0.008	0.88 ± 0.020	0.68 ± 0.010	1.41
Tibia	0.70	0.78 ± 0.008	0.76 ± 0.028	1.50
Humerus	0.65 ± 0.005	0.76 ± 0.006	0.70 ± 0.033	1.43
Radius	0.71	0.79 ± 0.009	0.73 ± 0.032	1.50
Modulus of elasticity in tension (GPa)				
Femur	25.5	25.0	14.9	17.6
Tibia	23.8	24.5	17.2	18.4
Humerus	17.8	18.3	14.6	17.5
Radius	22.8	25.9	15.8	18.9
Ultimate compressive strength (MPa)				
Femur	145 ± 1.6	147 ± 1.1	100 ± 0.7	107 ± 4.3
Tibia	163	159 ± 1.4	106 ± 1.1	
Humerus	154	144 ± 1.3	102 ± 1.6	
Radius	156	152 ± 1.5	107 ± 1.6	
Ultimate percentage contraction				
Femur	2.4	1.7 ± 0.02	1.9 ± 0.02	1.85 ± 0.04
Tibia	2.2	1.8 ± 0.02	1.9 ± 0.02	
Humerus	2.0 ± 0.03	1.8 ± 0.02	1.9 ± 0.02	
Radius	2.3	1.8 ± 0.02	1.9 ± 0.02	
Modulus of elasticity in compression (GPa)				
Femur	9.4 ± 0.47	8.7	4.9	
Tibia	8.5		5.1	
Humerus	9.0		5.0	
Radius	8.4		5.3	
Ultimate shear strength (MPa)				
Femur	99 ± 1.5	91 ± 1.6	65 ± 1.9	54 ± 0.6
Tibia	89 ± 2.7	95 ± 2.0	71 ± 2.8	
Humerus	90 ± 1.7	86 ± 1.1	59 ± 2.0	
Radius	94 ± 3.3	93 ± 1.8	64 ± 3.2	
Torsional modulus of elasticity (GPa)				
Femur	16.3	16.8	13.5	3.2
Tibia	19.1	17.1	15.7	
Humerus	23.5	14.9	15.0	
Radius	15.8	14.3	8.4	

Bone shows a linear range in which the stress increases in proportion to the strain. The slope of this region is defined as Young's modulus, or the elastic modulus. An illustration of the material properties of bone relative to other materials is shown in Fig. 2.4.

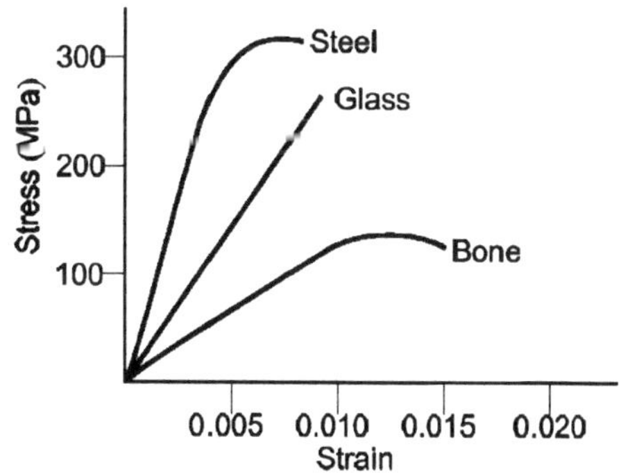

Fig. 2.4 Comparative properties of different materials

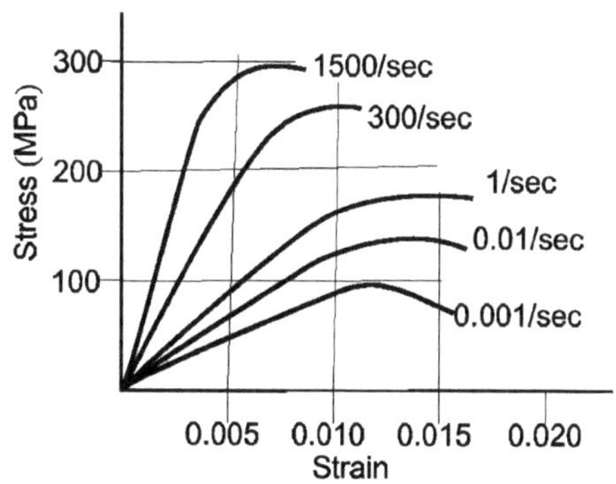

Fig. 2.5 Strain-rate sensitivity of cortical bone. As the strain rate increases, the ultimate strength increases and the ultimate strain decreases

Bone is strain rate–sensitive (Fig. 2.5) and tends to be more strain rate–sensitive than other biological tissues. This has implications for bone–ligament and bone–tendon injuries. The optimal strain rate for energy absorption is around 0.1–1 per second.

Material properties of the two types of bone differ. Cortical bone is more than 2 decades stiffer than cancellous bone. It can sustain greater stress but less strain before failure. Cancellous bone can sustain strains of 75% before failing in vivo, but cortical bone will fracture if the strain exceeds 2%. Cancellous bone has a greater capacity to store energy compared to compact bone since it is porous and filled with fluid, including blood, marrow, and body fluid.

Three important parameters that characterize some of the mechanical properties of bone–ultimate force, maximum deformation to failure, and the energy that it can store before failing–can be obtained from a force-deformation curve. The ultimate force represents the maximum load that the bone can sustain before it breaks. The ultimate force varies depending on the type of load applied (e.g., tensile, compressive, shear) and the loading rate. The deformation at failure is self-explanatory and also depends on the loading rate and direction. The energy absorbed before failing can be calculated from the area under the force-deformation curve and therefore depends on both the ultimate force and the ultimate strain. Children's bones tend to absorb more energy before failure compared to adults (as much as 45% more).

Children's bones are weaker but more compliant (children's bones can be 68% as stiff as adult bone).

2.6 Cartilage

2.6.1 Composition

Articular cartilage, also called hyaline cartilage, is made of a multiphasic material with two major phases: a fluid phase composed of water (68–85%) and electrolytes, and a solid phase composed of collagen fibrils (primarily type II collagen) (10–20%), proteoglycans and other glycoproteins (5–10%), and the chondrocytes (cartilaginous cells). Thirty percent of all cartilage water resides in this interstitial fluid, and this amount does not vary with age. However, there is a significant increase in the total amount of water in degenerating cartilages [3].

This multiphasic system allows fluid flowing from the tissue to the solution surrounding the tissue, and vice versa, through the pores of the collagen–proteolgycan solid matrix. As the fluid passes to the pores, the force exerted on the walls of the pores causes more compaction. Thus, it becomes more and more difficult to squeeze fluid from the tissue with prolonged compression. This nonlinear flow-induced compression effect is very important in the physiology of cartilage not just because it determines cartilage compressive viscoelastic behaviors, but also because it provides the mechanism for energy dissipation (Fig. 2.6).

The thickness of articular cartilage varies with the particular joint and the location within the joint. Generally, it ranges from 0.5 mm in rabbit knee joints to 10.0 mm in the patellofemoral groove of bovine knee joints, and in humans it is thickest over the ends of femur and tibia, ranging from 2–4 mm [4].

The distribution and arrangement of cartilage components are not uniform. Instead, each layer has different biochemical, structural, and cellular characteristics. Some authors consider articular cartilage to have three distinct layers (superficial, 10–20%; middle, 40–60%; and deep, 25–35%) along its depth. Others prefer to divide articular cartilage into four zones: superficial, middle, deep, and calcified (Fig. 2.7). The *superficial zone* is characterized by flattened chondrocytes, relatively low quantities of proteoglycan, and high quantities of collagen fibrils arranged parallel to the articular surface. The *middle zone*, in contrast, has round chondrocytes, the highest level of proteoglycan among the four zones, and a random arrangement of collagen. The *deep zone* is characterized by collagen fibrils that are perpendicular to the underlying bone, and columns of chondrocytes arrayed along the axis of fibril orientation. The *calcified zone* is partly mineralized and acts as the transition between cartilage and the underlying subchondral bone. Considering either three or four layers, based on the depth-related differences in the structural, biochemical, and cellular compositions, it is reasonable to assume that the intrinsic mechanical properties of articular cartilage vary with depth.

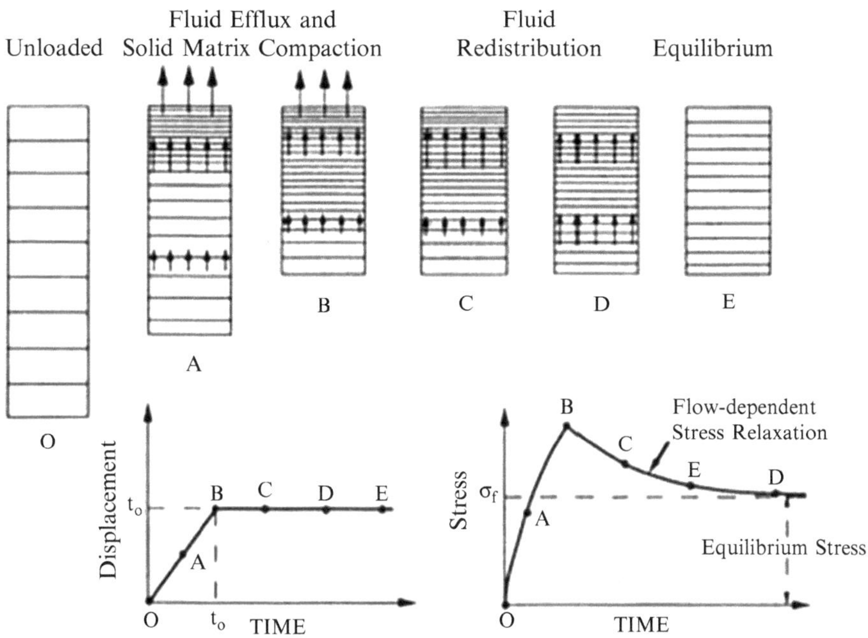

Fig. 2.6 Viscoelastic behavior of cartilage

2.7 Material Properties and Related Behavior

Interactions take place among the fluid, proteoglycan molecules, and various electrostatic charges, providing superior quality of lubrication and shock absorption. The cartilaginous tissue is extremely well adapted to glide. Its coefficient of friction is several times smaller than that between ice and an ice skate. There are electrostatic attractions between the positive charges along the collagen molecules and the negative charges that exist along the proteoglycan molecules. Hydrostatic forces also exist as forces are applied to cartilage and the fluid tries to move throughout the tissue. It is the combined effect of all these interactions that gives rise to the mechanical properties of the material.

Like bone, cartilage is an anisotropic material. The anisotropy results in part from the structural variations noted above. Because of its structure, cartilage is rather porous, allowing fluid to move in and out of the tissue. When the tissue is subjected to a compressive stress, fluid flows out of the tissue. Fluid returns when the stress is removed.

The mechanical properties of cartilage change with its fluid content, thus making it important to know the stress–strain history of the tissue to predict its load-carrying capacity. The material properties also change with pathology. The compressive aggregate modulus for human articular cartilage correlates in an inverse manner with the water content and in a direct manner with proteoglycan content per wet weight. There is no correlation with the collagen content, thus suggesting that proteoglycans are responsible for the tissue's compressive stiffness.

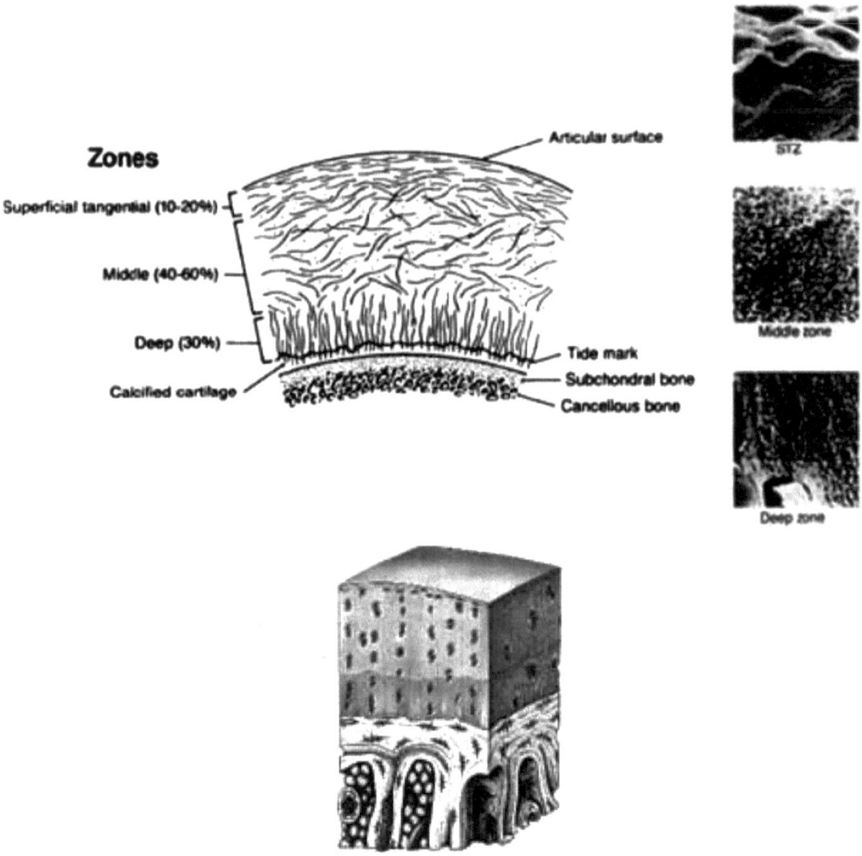

Fig. 2.7 Cartilage zones or layers

2.8 Ligaments

2.8.1 Composition

The major constituents of ligaments are collagen, elastin, glycoproteins, protein polysaccharides, glycolipids, water, and cells (mostly fibrocytes). The greatest quantities of constituents found in ligaments are collagen and ground substance. For practical purposes, the physical behavior of ligaments is usually predicted based on the content and organization of these substances alone [5].

Collagen constitutes 70–80% of the dry weight of ligament, the majority being type I collagen, which is also found in tendon, skin, and bone. Collagen has a relatively long turnover rate, with its average half-life being 300 and 500 days, which is slightly longer than that of bone. Therefore, several months may be required for a ligament to alter its structure to meet changes in physical loading conditions or to repair itself after injury. Water makes up about 60–80% of the wet weight of ligaments.

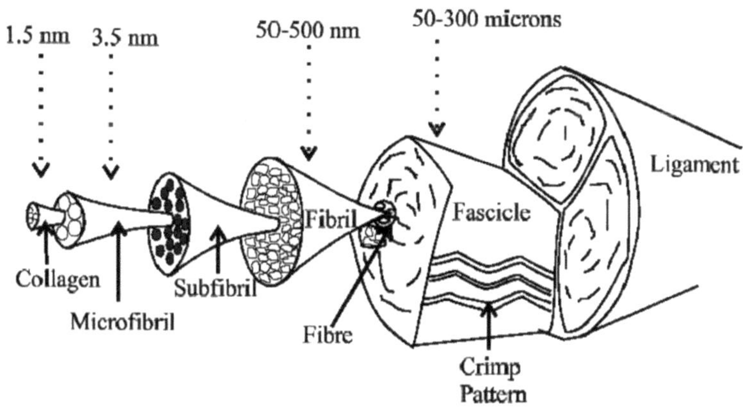

Fig. 2.8 Hierarchical structure of ligaments. The basic structural element is the tropocollagen molecule

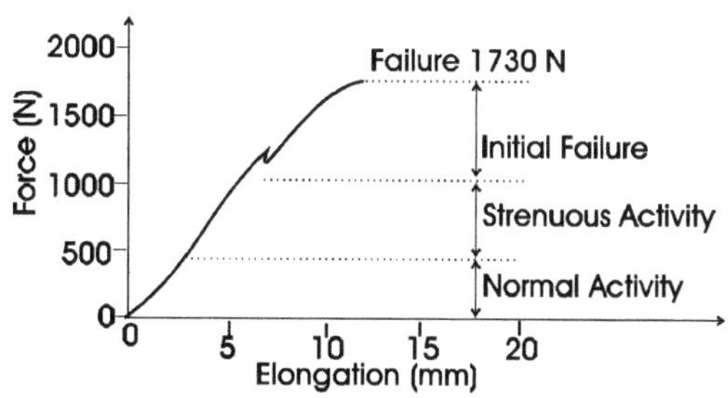

Fig. 2.9 Force-elongation diagram obtained during tensile test of ligaments

A significant amount of this water is associated with the ground substance. On a dry weight basis, the ground substance comprises only about 1% of the total tissue mass. The ground substance likely provides lubrication and spacing, which aid in the sliding of fibers. In addition, the presence of ground substance is a source of ligament viscoelastic behavior.

Closely packed, parallel collagen fiber bundles are oriented to provide motion and stability for the musculoskeletal system (Fig. 2.8). Properties can change according to strain rate, temperature, hydration, maturation, aging, immobilization, exercise, and healing.

The structural properties of isolated ligaments and bone–ligament–bone preparations are normally determined via tensile tests. In such a test, a ligament, tendon, or bone–ligament–bone complex is subjected to a tensile load applied at constant rate. A typical force-elongation curve can be obtained from a tensile test, as shown in Fig. 2.9. The force-elongation curve is initially upwardly concave, but the slope becomes nearly linear in the prefailure phase of tensile loading. The force-elongation curve represents structural properties of the ligament. That is, the shape of the curve depends on the geometry of the specimen tested (e.g., tissue length and cross-sectional area).

2.9 Material Properties and Related Behavior

Although significant advances have been made in the biology, biochemistry, and mechanics of soft tissue, there is still much work left to be done. There is limited information available on in vivo tissue mechanical characteristics and behavior. Without accurate values of such in vivo information, extrapolations from animal and human in situ bone–ligament–bone testing to the function of intact human ligaments cannot be made confidently. Currently, we know that ligaments are composite, anisotropic structures exhibiting nonlinear time- and history-dependent viscoelastic properties. Described in this section are the mechanical behavior of ligamentous tissue, the physiological origin of this behavior, and the implications of such properties to ligament function during normal joint motion.

As seen above, the force-elongation curve represents structural properties of the ligament. Material properties, in turn, are more generally expressed in terms of a stress–strain relationship (Fig. 2.10).

2.9.1 Ligaments Have Characteristics of Strain-Rate Sensitivity, Stress-Relaxation, Creep, and Hysteresis

Ligaments exhibit significant time- and history-dependent viscoelastic properties. Time-dependent behavior means that during daily activities, ligaments are subjected to a variety of load conditions that affect their mechanical properties. For example, they become softer and less resistant after some minutes of running, returning to normal hardness when the exercise is interrupted. History dependency, in turn, means that frequent intense activities will change the tissue properties on a medium-term basis. For example, the ligaments of an athlete, after 6 months of daily training, will become softer and thus more adapted to the intense exercise, even when he or she is not training. In the same way, if the activities are interrupted for some months, the ligament properties will go back to normal levels. Figure 2.11 illustrates ligament softening, a decrease in peak loads occurring during cyclic testing of ligaments to a constant strain and at a constant strain rate.

Ligaments are also temperature-sensitive, with peak stresses increasing with decreased temperatures. Bone–ligament–bone preparations tested cyclically at

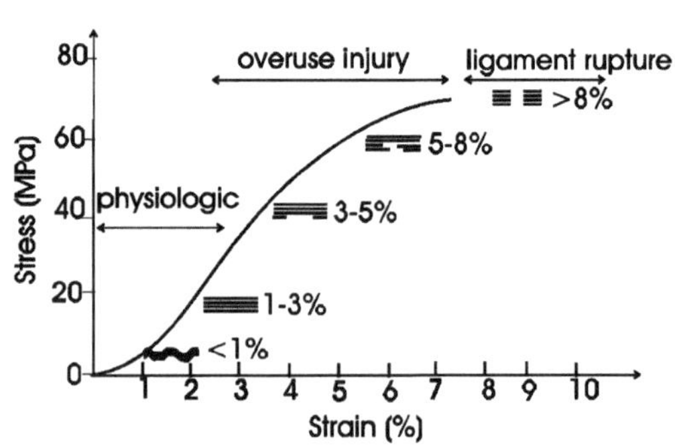

Fig. 2.10 Stress–strain relationship for human ligament

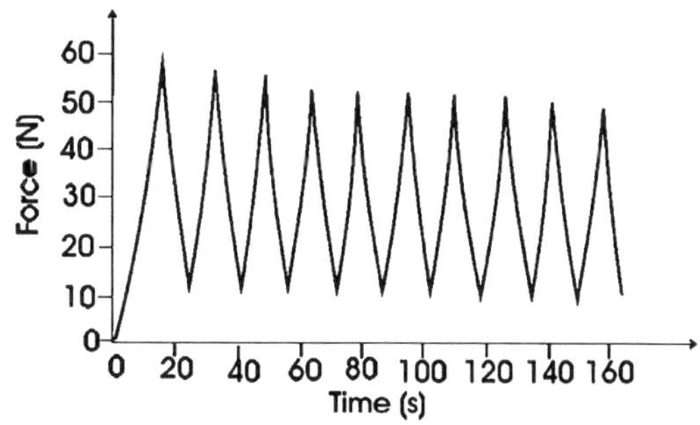

Fig. 2.11 Ligament response to cyclic loading and unloading. Peak loads decreased with each cycle, indicating ligament softening (Woo et.al. 1982)

21 °C show 30% greater peak loads than the same preparation tested at 37 °C. It has been suggested that the temperature of superficial tissues in vivo may be within 2 °C of the skin temperature, which can be 10 °C lower than the body temperature.

There are two age-related processes, maturation and aging, that also affect bone–ligament properties. During maturation, the structure and mechanical properties of collagenous tissues change. The stabilization of collagen with maturity enhances tissue strength, while the loss of water and elastin reduces tissue plasticity [6]. Aging connective tissue undergoes a generalized decrease in water content, which results in a reduction in tissue compliance. The elastic elements become coarser and more easily fractured. However, it is very difficult to distinguish aging effects from effects created by other factors such as disease or changes in activity levels. It has been estimated that regular exercise may retard the physiologic decline associated with aging by as much as 50% [6].

2.10 Correlation Between Structure and Function

The "crimp pattern" and the interaction and cross-linking of elastic, reticular, and collagen fibers of ligaments are critical for normal joint mobility. These features allow ligaments to have a limited range of strains over which they produce minimal resistance to movement. As a result, joints may easily be moved in certain directions and over certain ranges. Additionally, if a joint is displaced toward the outer limit of some normal range of motion, the strain in specific ligaments of that joint increases, causing recruitment of collagen fibers from their "crimp" state to a straightened condition. Fiber recruitment causes the ligament to quickly increase its resistance to further elongation, hence stabilizing the joint.

Another feature of ligaments that may be important for maintaining joint integrity is their neural network. Ligaments contain a variety of sensory receptors that may detect joint position, velocity, and acceleration. This feature may indirectly contribute to maintaining joint integrity by initiating the recruitment (or decruitment) of dynamic stabilizers such as muscles. More work is needed in this area to determine the role of these neural components.

Table 2.2 Comparative properties of biological materials

Density	Bone 1,810 kg/m^3	Cartilage 1,100 kg/m^3	Ligament
Strength (variation factor)	1.35 from the weakest to the strongest place	3 MPa (tensile-stress–strain curve becomes nonlinear)	
Strength (correlation coefficient)	0.4–0.42		
Young's modulus	18 GPa (intermediary between apatite = 165 GPa and collagen = 1.24 GPa)	1–10 MPa (tension) 1 MPa (compression)	1.2–1.8 GPa
Shear modulus	3.5 GPa		
Viscoelasticity	Yes	Yes, time- and history-dependent	Yes, time- and history-dependent
Anisotropy	Cancellous = extremely anisotropic Cortical = transversely isotropic	Anisotropic	Yes, for load deformation or stress–strain behaviur (primarily oriented to resistance of the tensile loads)
Nonlinear behavior	Cancellous = nonlinear Cortical = approximately linear		Nonlinear
Ultimate stress (at failure)	135 MPa (tension) 200 MPa (compression)		50–150 MPa (tension)
Ultimate strain (at failure)			13–18% (varies among different materials)

Collected from many sources from Reading List

2.10.1 *Ligament–Bone/Tendon–Bone Insertions*

Tendons have large parallel fibers that insert uniformly into the bone. Ligament fibers are of smaller diameter than the tendon fibers, which can be either parallel or branching and interwoven. Ligament insertion sites are well suited for dissipating force. As the ligament passes through the insertion site, it is transformed from ligament to fibrocartilage and then to bone.

The bone–ligament junction of younger animals is consistently weaker than that of the ligament substance. The reverse is true for mature animals. This suggests an asynchronous rate of maturation between the bone–ligament junction and that of the ligament substance.

Two different types of insertions exist: *direct*, which is more common, where the tendon or ligament crosses the mineralization front and progresses from fibril to fibrocartilage (usually less than 0.6 mm), to mineralized fibrocartilage (less than 0.4 mm), and finally to bone; *indirect*, which is less common, where it inserts into bone through the periosteum, with short fibers that are anchored to the bone. For quick reference, we have included a table for comparative properties of biological materials (Table 2.2).

Reading List

1. Fung YC (1993) Biomechanics: mechanical properties of living tissues, 2nd edn. Springer, New York
2. Hawkins D. (2001) Tissue mechanics. Human performance laboratory, University of California, Davis. Lecture available at: http://dahweb.engr.ucdavis.edu/dahweb/126site/126site.htm
3. Mow VC et al (1991) Structure and function of articular cartilage and meniscus. In: Mow VC, Hayes WC (eds) Basic orthopedic biomechanics. Raven, New York
4. Fish R. (2000) Sinovial joints, PhD Thesis. Department of Biological Sciences, University of Manchester, Kingston, Canada. Available at: http://www.teachingbiomed.man.ac.uk/student_projects/2000/mmmr7rjf/articula.htm
5. Akeson WH, Woo SL-Y, Amiel D, Frank CB (1984) The chemical basis of tissue repair. In: Funk FJ, Hunter LY (eds) Rehabilitation of the injured knee. CV Mosby, St. Louis, pp 93–104
6. Menard D, Stanish WD (1989) The aging athlete. Am J Sports Med 17(2):187–196
7. Gray H (2000) Anatomy of the human body. Lea & Febiger, Philadelphia, Bartleby.com, 2000. http://www.bartleby.com/107/. (Mar 18, 2002)
8. Fung YC et al (eds) (1972) Biomechanics: its foundations and objectives. Prentice-Hall, Englewood Cliffs, NJ
9. Gallagher RH et al (eds) (1982) Finite elements in biomechanics. Wiley, New York
10. Ratner BD, Hoffman AS, Schoen FJ, Lemons JE (eds) (1996) Biomaterials science: an introduction to materials in medicine. Academic, New York

Problems

1. List a total of five metallic, polymeric, ceramic, and composite biocompatible materials.

   List their compositions and mechanical properties, such as ultimate strength, yield strength, ultimate strain, and hardness. In a single plot, draw their stress–strain diagram under tensile load with approximate scale indicating the yield point, ultimate strength, and modulus of elasticity.
2. List names of some hard and soft tissues in the human body. How do we characterize them? Show their stress–strain diagram.
3. Compare the hardness of different components of teeth. What materials will be suitable for the replacement of teeth?
4. The following figure shows the load deformation diagram of a scaffold of chitosan–hydroxyapatite composite mixed in different proportions. Discuss the effect of

HA on the ultimate strength and the primary and secondary moduli of elasticity. The cross-sectional area may be taken as $15 \times 10\ mm^2$ (based on a 2007 biomedical engineering master's thesis at Jadavpur University). The starting point of each plot is (0, 0).

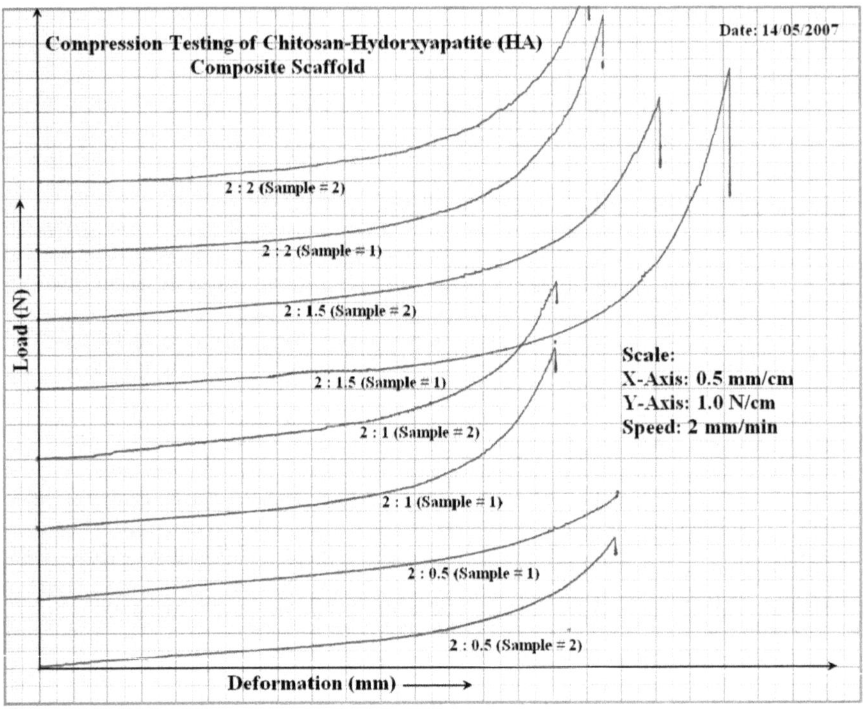

Chapter 3
Basics of Design Process

3.1 Introduction

Traditionally, engineers are trained to deliver products, projects, or services to fit some specified requirements. They work to a required performance and quality, within a given time and budget. This is despite the fact that these decisions may require not only an understanding of the engineering possibilities, but also knowledge of and experience with potential environmental and social consequences.

Engineering decisions have an enormous impact upon the quality of life in the global community—for example, the social and economic effects of building dams in developing countries or new airports on the outskirts of cities. It is essential, therefore, that engineers play a full and significant role in ordering the affairs of society, not merely fulfill the role of technicians carrying out the instructions of others.

In designing nonliving components for the human body for their possible replacement, we need to understand the structure and function of that particular part as a singular body as well as its anatomical location, for instance, a joint or heart valve, intraocular lens, kidney, lungs, or similar organs. Some may be similar or comparable to an existing mechanical part or a chemical process capsule or an electrical system. Macro-, micro-, or nano-sized particulates, proteins, ionic fluids, electrical charge flow, ion transfer, blood, and other fluid flows are the constituents of the environment in which the living parts function, replicate, repair themselves, and survive. The canvas is enormous and full of complex problems.

3.2 Adoptive and Adaptive Design

3.2.1 Introduction

Design is essentially a decision-making process. If we have a problem, we need to design a solution. In other words, to design is to formulate a plan to satisfy a particular need and to create something with a physical reality.

S. Pal, *Design of Artificial Human Joints & Organs*,
DOI 10.1007/978-1-4614-6255-2_3,

Consider, for example, the design of an artificial limb. A number of factors need to be considered first:

(a) The purpose for which the limb is to be designed, such as whether it is to be used for cosmetic purposes or as an active elbow prosthesis.
(b) Whether this is to be designed for an adult or a child; this involves considerations for shape and size.
(c) Material for the limb, its strength, and cost factors.
(d) Finally, the aesthetics of the designed part. Almost everyone is involved in design, in one way or another, in our daily lives because problems are posed and they need to be solved. The basic concept of a part design is a decision-making process in every stage of the design.

Once a critical decision is made, the rest of the design features follow. A bad concept leads to a bad design and a malfunctioning product. Design may be for different products; with the present specialization and knowledge bank, we have a long list of design disciplines, including internal prosthesis design of an artificial joint, artificial bone, teeth, heart, valve, kidney, lungs, and breast, for either replacement of or support to the existing part.

A machine is a combination of bodies with successfully constrained relative motions that are used to transform other forms of energy into mechanical energy or transmit and modify available energy to do some useful work. In this respect, all animals are living machines: They convert chemical energy from food, water, and oxygen (energy available in nature) into mechanical energy and heart energy; we may call it a living prime mover.

This modification or transformation of energy requires a number of machine elements, some small and some large. Machine design involves primarily designing these elements so that they may transmit the forces safely and perform their task successfully.

Consider the following simple mechanisms: a person hand-carrying water, pumping a tube well, moving a trolley, or grinding wheat; a cow pulling a cart; or an elephant or camel carrying a load on its back.

In each of these mechanisms, some useful work is being obtained with certain combinations of a number of machine parts.

Designing these mechanisms would involve first designing these elements when the part or organ is diseased or malfunction.

3.3 Introduction to Machine Design

There are several types of designs, such as adaptive design, which is based on the existing design, for example, standard products or systems adopted for a new application that is for a specific person or living being. Hip joint replacement for humans was adapted for other animals, including the dog and horse.

In developmental design, we start with an existing design, but finally a modified design is obtained to address the problems associated with the given type of design. Austin Moore's hip prosthesis was modified to Peterson's hip and then to Charnley's cemented hip. This type of design is an entirely new one but based on existing scientific principles. Some scientific invention is involved, such as using bone cement, and requires creative thinking to provide a better solution. There are many factors to be considered while attacking a design problem. In many cases, these are common-sense approaches to solving a problem. Some of these factors are as follows:

(a) What device or mechanism is to be used? This would decide the relative arrangement of the constituent elements.
(b) Materials of construction.
(c) Forces, pressure on the elements from the existing body part.
(d) Size, shape, and space requirements for putting a prosthesis in place with a minimum of invasion.
(e) The method of manufacturing the components and their assembly.
(f) How will it operate?
(g) Reliability and safety aspects from strength, wear, and degeneration in an ionic and 37 °C temperature environment.
(h) Inspectability using any medical imaging technique.
(i) Maintenance, cost, and life expectancy of the designed product.
(j) The final weight of the product is also a major concern; preferably it should be close to the body part it will replace. For metallic implants, it is really away from the tissue density.

3.4 Principle of Science or Mechanism to Be Used

Which principle of science or mechanism to use is best decided by understanding the problem thoroughly. Designers need to talk with physicians, surgeons, anatomists, and manufacturers. Sometimes a particular function can be achieved by a number of means or by using different mechanisms, and the designer has to decide which one is the most effective under the circumstances. A conceptual design or layout diagram may be made to crystallize the thoughts regarding the relative arrangement of the elements.

Material. This is a very important aspect of any design, especially for living systems. It should be nontoxic and have good strength and stiffness or flexibility depending on the problem at hand. A wrong choice of material may lead to failure, endanger life, lead to over- or undersized products, or create unaffordable items. The choice of materials is thus dependent on suitable properties of the material for each component, their suitability for fabrication or manufacture, quality assurance, and cost. The materials should be free of contaminants and toxicants. They should follow the state standards (Bureau of Indian Standard, BIS) or ASTM or International (ISO) standards, depending on the country.

Load. External loads cause internal stresses in the elements; these stresses must be determined or estimated accurately since they will be used in determining the minimum component size. Loading may be due to (1) energy transmission by a body part; (2) dead weight; (3) inertial forces; (4) thermal effects; (5) frictional forces. In other ways, loads may be classified as follows:

1. **Static load**, which does not change in magnitude and direction and normally increases gradually to a steady value;
2. **Dynamic load, which** (a) changes in magnitude with time; for example, during locomotion, running varying forces were generated at the joints; (b) changes in direction; for example, the load on joints while going up or down stairs.

Static loads are a constant dead weight. **Dynamic loading** varies with time. Vibration and shock loading are types of dynamic loading. In the human body system, we encounter dynamic loading regularly while walking briskly, jogging, engaging in sporting activity, entering a slow-moving train, or detraining. In these situations, the loading may increase to several times (3–5 times) the normal loading.

Then we need to determine the size, shape, and space requirements and the weight. A preliminary analysis would give an approximate size, but if a standard element is to be chosen, the next-larger size must be selected. Shapes of standard elements are known, but for nonstandard elements, shapes and space requirements must depend on the available space in a particular inner space in the human body. An X-ray or other imaging layout drawing may be useful to arrive at an initial shape and size. Weight is important depending on the application. This means that the material chosen must have the required strength, yet it usually must be light. Then the designer needs to consider the biocompatibility, or the low level of inclusion of nonmetallic materials within limits stated by recognized state or world standards. The material must be hemocompatible and tissue-compatible. Care must always be taken to ensure that the designed elements may be manufactured with ease, within the available facilities, and at low cost.

In terms of how it will operate in the final stage of the design, a designer must ensure that the component may be put in place surgically with ease. The material must be sterilizable with existing methods without losing its properties. However, in many other cases, a sequence of operations is to be specified. This sequence must not be complicated, and the operations should not require excessive blood loss and injury to the tissue.

With time-tested design considerations, the sequences have been made user-friendly. As in any other product, these products also go through continuous innovation and development. Reliability and safety are important factors in any design. A designed part should work effectively and reliably. The probability that an element or a machine will not fail in use is called the reliability. Reliability lies between $0 \leq R < 1$. To ensure this, every detail should be examined. Possible overloading, adverse environment (e.g., a pacemaker should not be placed near an MRI or high-tension line), wear of elements, excessive heat generation as in bone cement polymerization, and other detrimental factors must be avoided. There is no single answer for this, but an overall safe design approach and care at every stage of the design, manufacture, packaging, and sterilization processes would result in a reliable implant. Safety is a matter of paramount importance in healthcare-related products.

Implants must be designed to serve mankind, to heal rather than to harm or kill. Drug Controller of India and FDI regulations ensure that the manufacturer is liable for any damage or harm arising out of a defective product. Using a safety factor only in design does not ensure the product's overall reliability. Maintenance, cost, and aesthetics in external products and safety are often interlinked. Good maintenance ensures the prosthesis will work well. Often a regular maintenance schedule is maintained and a thorough checkup of moving and loaded parts is carried out to avoid catastrophic failures.

Low friction and wear are to be maintained by the proper choice of materials. Lubrication in joints is limited by body fluid. This is a major aspect of design since wherever there are moving parts, friction and wear are inevitable. High friction leads to an increased loss of energy and pain. Wear debris of parts leads to toxic effects and may enter the circulatory system and cause loosening of the prostheses and premature failure Cost is essentially related to the choice of materials, which in turn depends on the stresses developed in a given condition. Although in many cases aesthetic considerations are not essential aspects, especially in endoprosthesis design, ergonomic aspects also must be taken into considerations for ease of use and placement of the replaced part. The patient should not consume sizable extra energy.

These items are to be handled carefully at all times to avoid scratches and should be properly packaged for sterilization. Sterilization has a shelf-life, and the date of expiry is important. It may have to be resterilized.

For the specific item design, for example, for any joint replacement, we will proceed as follows:

1. Select the joint, study its anatomy, and analyze it biomechanically or refer to a book, such as *Textbook of Biomechanics* [4], and other literature to determine the maximum forces and the mechanism of the joint. For easy understanding, we may think of a hemi-hip joint, i.e., only the femoral component.
2. Study the degrees of freedom and the range of movement from any textbook on anatomy.
3. Get X-rays from two reference planes and measure the marrow cavity in true scale.
4. Decide what type of prosthesis you will design.
5. For a hemi-hip, we will have a ball and a stem going into the marrow cavity.
6. The ball size will be closer to the existing hip head, and it should have a shoulder to avoid subsidence, as the diseased head part will be removed surgically.
7. The stem will go into the marrow cavity and it will be uncemented, that is, push-fit. Then we will have to ensure that it remains stable. Thus, we should

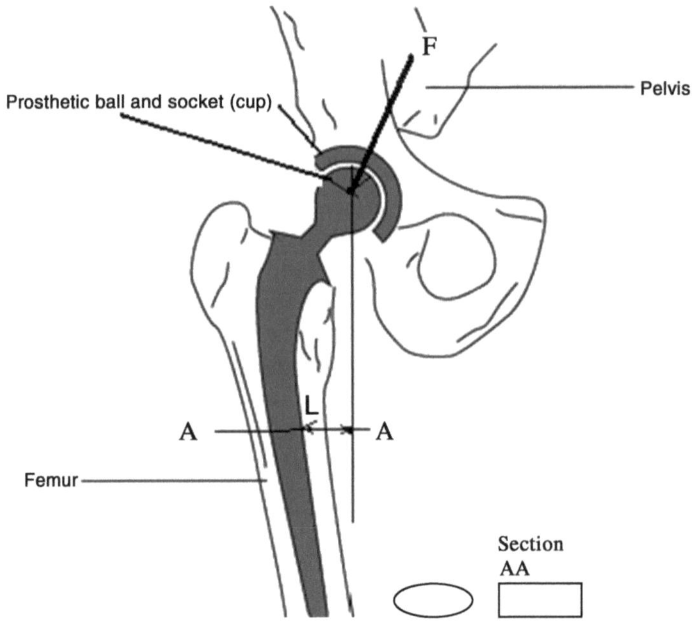

Fig. 3.1 A hip prosthesis placed inside a femur. The prosthesis may be of rectangular or elliptical section, with the major axis as shown in section A–A. A rounded-corner rectangular section is a preferred design. F passes through the center of the ball

choose a rectangular cross section and tapered on both sides. To avoid tilting, we should have a sufficient length, say one third of the femur length.

8. Now the ball and stem should be connected through a shoulder, which should be sturdy.
9. Now the material may be AISI-316L stainless steel or Ti6Al4V or Co-Cr-Mo alloy, as these are biocompatible and are typically used. We will see their properties from handbooks or the Internet. In recent years, the use of stainless steel has been greatly reduced.
10. The force on the joint may be taken as five to six times the body weight.
11. Then sketch the system, and apply a distributed load or pressure on the surface equal to two thirds the area of the sphere, i.e., $8/3\pi R^2$, where R is the radius of the sphere. If you use the finite element method of analysis, this load may be distributed on the nodes. Otherwise, a concentrated load at a 30° angle to the vertical may be taken (can be found from the X-ray of the joint) and a force going radially through the center of the ball. Then consider the offset. The hip prosthesis is eccentrically loaded and subjected to **direct compression** and **bending moment**. The internal part will be under compression and the external part under tension. Figure 3.1 depicts the model.

The stress could be calculated based on the following equations:

$$\sigma = M / Z + - F / A, \tag{3.1}$$

where σ = bending stress, M = bending moment = FL, Z = section modulus, F = joint force, at the center, A = area of cross section of ellipse or rectangle.

The area A is to be expressed in one parameter, e.g., $A = 2bxb$, where b the is smaller side; Lx, Ly = moment arm in the X- and Y-direction, respectively; Z and A are geometrical parameters, which may be converted into one unknown variable.

*If σ is the allowable stress, then Eq. (3.1) could be solved and the dimensions determined. This will be the minimum area and the chosen size should be close to the marrow cavity, which can be found from an X-ray of the femur.

12. Compare the stress–strain with the maximum allowable stress of the chosen material. A high level of safety factors (steps 5–8) is suggested based on the yield strength. There are many vulnerabilities, such as dynamic loading and body fluid at a 37 °C environment. Too much contact stress with bone may lead to bone necrosis and aseptic loosening of the prosthesis. This a very simple approach.
13. Next, we have to plan for the item's manufacture. It may be produced by forging and machining. Some local manufacturers use welding to connect the stem and the shoulder part, which is not advisable. The articulating surface of the ball, which will be in direct contact with cartilage of the acetabular cavity, should be polished to a mirror finish.
14. Today metal-to-metal and ceramic-to-ceramic articulations are also being used in total hip joint replacement surgery to limit wear debris production. This debris causes loosening of the prosthesis when it enters the joint space.

3.5 Safety of Products

Safety aspects of a product are of paramount importance, especially for biomedical products that are directly implanted in the human system in vivo. There are several stringent criteria set up by the various national bodies to protect patients from any adverse effects. These are biocompatibility, strength and stiffness, wear resistance or minimal wear debris production, and long-term survivability of the product in vivo, just to mention a few.

3.6 Manufacturability

Manufacturability is a very important criterion while a product is being designed. A designer should have an idea about the way his product design should be given shape. He may consult shop engineers or technicians while designing his product.

Common manufacturing methods are casting, forging, and machining, which are separate from other fabricating processes such as welding, screwing, and riveting. Today there are many sophisticated manufacturing techniques like CAD-CAM, using computer and associated hardware. There are many fine machining and automated techniques for manufacturing. Plastics were also manufactured by molding and hot pressing and casting. Ceramic parts are manufactured by slip casting, high-temperature heating, and machining as well. These are specialty areas.

3.7 Standardization

Whenever possible, standard product regulations should be followed, e.g., thread profile, thread pitch and diameter, nut shape, tooling, etc. This is essential for the interchangeability of products and ease of choosing surgical tools and equipment during surgery. Standardization helps in international marketing and exchangeability. **Interchangeability** between parts is also an important concept. For this purpose, concepts of fits and tolerance are important for "mating" parts, meaning where one part goes inside another part.

3.8 Customization

Customization means producing something for a specific subject based on her anatomical features. For example, a hip prosthesis will fit a specific person. It will be a bit more expensive, but the patient will feel better and have near-normal activities. Usually, CT-scan data of the bony part are obtained; then, using software, the bony cavity is replicated and the prosthesis is produced using computer-aided manufacturing techniques (Fig. 3.2a). We illustrate some examples in Fig. 3.2. We designed a customized hip prosthesis for a female patient. Two surgeons from Kolkata RamKrishna Mission Hospital helped us to get the relevant CT data for the patient. We designed it using MIMICS and other software, and the prosthesis was manufactured in a Kolkata-based precision manufacturing company using a CNC milling machine and various pieces of software to give it an appropriate shape as generated by MIMICS. The orthopedic surgeons studied the pros and cons of using the model in the patient but have not implanted it yet, as we want to be sure about its efficacy before its in vivo trial. For that reason, we are examining a goat femur model and have manufactured one such in vitro study for a fresh goat femur, as shown in Fig. 3.2b. This model was tested under load, and when X-rayed, it showed an excellent fit.

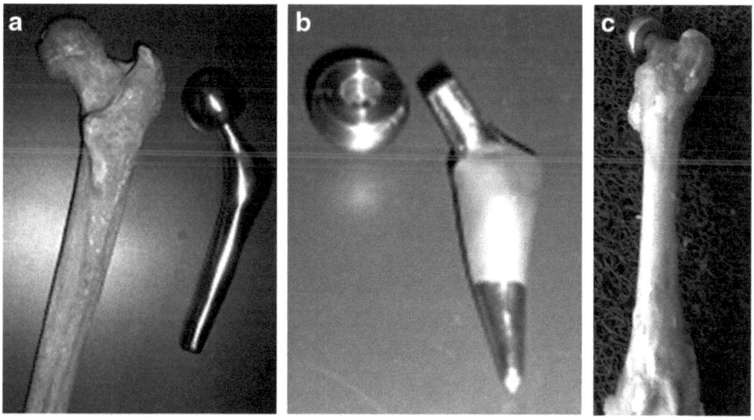

Fig. 3.2 (**a**) Human femur and customized hip prosthesis; (**b**) goat hip prosthesis for animal trial; (**c**) the prosthesis is installed in the goat femur. These studies are based on a DST-Govt. of India funded project of the author's in 2010

For Further Details, the Following Books May Be Consulted

1. Shigley J, Mischke C, Brown T (2004) Standard handbook of machine design. McGraw Hill, New York
2. Malhotra MM, Subramaninian R (1994) Textbook in applied mechanics. New Age International, New Delhi
3. For static stress analysis, any text on the strength of materials or applied mechanics may be consulted
4. Pal S (2009) Text book of biomechanics. Viva Books, New Delhi

Exercises

1. How do you propose to analyze the stress and strain of a modular-type hip prosthesis? Assume a suitable dimension from any modular-type hip prosthesis. (This is described in Sect. 3.4 of this chapter.)
2. Animal hip joint prostheses are in demand in these days. How one can approach the design of a (1) dog hip prosthesis and (2) a horse hip prosthesis? Consider the dynamic effect of the movement of these animals. In fact, one student, Sandip Bag, wrote a thesis on the canine hip under the guidance of this author, in 2006. (Hint: UHMWPE & Ceramic composite).

Chapter 4
Biomaterials and Its Characterization

4.1 Introduction

At the time of this publication, the development of biomaterials as a science is nearly 50 years old. The study of this subject is known as biomaterials science and involves engineering as well as development of products. It has shown a steady and strong growth over the years, with business groups investing large amounts of money into the development of new products. Biomaterials science is highly interdisciplinary in nature, involving medical sciences, biosciences, chemistry, tissue engineering, and materials science principles (Fig. 4.1).

The definition for the term "biomaterial" has been widely accepted as the following:

- A biomaterial is any material, natural or manmade, that comprises a whole or part of a living structure or biomedical device that performs, augments, or replaces a natural function.
- A biomaterial is a nonviable material used in medical device, so it is intended to interact with a biological system without harming it.

A biomaterial is essentially a material that is used and adapted for a medical application. Biomaterials may have a benign function, such as being used for a heart valve, or may be bioactive with a more interactive functionality, such as hydroxyapatite-coated hip implants. Biomaterials are also used every day in dental applications, surgery, and drug delivery (a construct with impregnated pharmaceutical products can be placed into the body, which permits the prolonged release of a drug over an extended period of time).

The definition of a biomaterial does not just include manmade materials that are fabricated out of metals, polymers, or ceramics. A biomaterial may also be an autograft, allograft, or xenograft used as a transplant material.

S. Pal, *Design of Artificial Human Joints & Organs*,
DOI 10.1007/978-1-4614-6255-2_4, © Springer Science+Business Media New York 2014

Fig. 4.1 The iridescent nacre inside a nautilus shell (adopted from Wikipedia)

4.2 Biomineralization

In the biological world, biomineralization (e.g., silicification) is quite common and occurs in bacteria, single-celled organisms, plants (e.g., petrified wood), and animals (invertebrates and vertebrates). Crystalline minerals formed in this type of environment often show exceptional mechanical properties (e.g., strength, hardness, fracture toughness) and tend to form hierarchical structures that exhibit microstructural order over a range of length or spatial scales. The minerals are typically crystallized from an environment that is undersaturated with respect to certain elements like silicon, calcium, and phosphorus, which are readily oxidized under conditions of neutral pH and low temperature (0–40°C). Formation of the mineral may occur either within or outside the cell wall of an organism, and specific biochemical reactions for mineral deposition exist that include lipids, proteins, and carbohydrates. The significance of the cellular machinery cannot be overemphasized, and it is with advances in experimental techniques in cellular biology and the capacity to mimic the biological environment that significant progress is currently being reported (Fig. 4.2) [1–4].

Examples include silicates in algae and carbonates in invertebrates, and calcium phosphates and carbonates in vertebrates. These minerals often form structural features such as seashells and the bone mineral in mammals and avian species. Organisms have been converted into mineralized skeletons for nearly 600 million years. The most common biominerals are the phosphate and carbonate salts of calcium that are used in conjunction with organic polymers such as **collagen and chitin** to give shape and mechanical strength to bones and shells. Other examples include copper, iron, and gold deposits involving bacteria [7].

Thus, most natural (or biological) materials are complex composites whose mechanical properties are often outstanding, considering the weak constituents from which they are made. These complex structures, which have evolved from hundreds of millions of years of evolution, are inspiring materials scientists

Fig. 4.2 Sand from Pismo Beach, California, including quartz, shell, and rock fragments (Wikipedia)

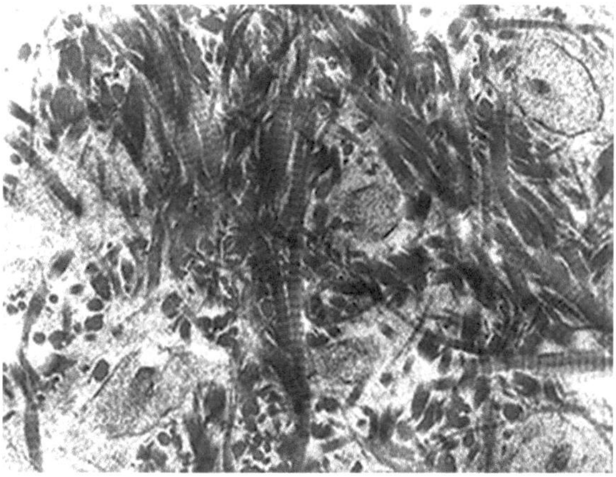

Fig. 4.3 Collagen fibers of woven bone

interested primarily in the design of novel materials with exceptional physical properties for high performance in adverse conditions. Their defining characteristics, such as hierarchy, multifunctionality, and the capacity for self-healing, are currently being investigated by many authors (Figs. 4.3 and 4.4) [8–11].

The basic building blocks begin with the 20 amino acids and proceed to polypeptides and polysaccharides. These, in turn, compose the basic proteins, which are the primary constituents of the "soft tissues" common to most biominerals. With well over 1,000 proteins possible, current research emphasizes the use of collagen, chitin, keratin, and elastin. The "hard" phases are often strengthened by crystalline minerals, which nucleate and grow in a biomediated environment that determines the size, shape, and distribution of individual crystals. The most important silicate phases have been identified as hydroxyapatite, silica, and aragonite. Using the classification of Wegst and Ashby, the principal mechanical characteristics and structures of a number of biological ceramics, polymer composites, elastomers, and cellular materials have been recently characterized. Selected systems in each class

Fig. 4.4 SEM ×10,000 magnification of crystalline bone mineral

are being investigated, with an emphasis on the relationship between their microstructure over a range of length scales and their mechanical response (especially fracture toughness) [12–15].

Thus, the crystallization of inorganic materials in nature generally occurs at ambient temperature and pressure. Yet the vital organisms through which these minerals form are capable of consistently producing extremely precise and complex structures. Understanding the processes in which living organisms control the growth of crystalline minerals such as silica could lead to significant advances in the field of materials science and open the door to novel synthesis techniques for nanoscale composite materials, or nanocomposites. Many scientists all over the world are working in this area.

4.3 Abalone Shell

The nacre from the nautilus shell of the abalone recently has become one of the most intensively studied biological structures among materials scientists. The highly ordered microscale aragonite tiles separated by thin nanoscale organic sheets along with a macrostructure of larger periodic growth bands form a hierarchical composite. Early work showed that the overall composite consists of only 5 wt. % organic materials, yet the work to fracture was increased by up to 3,000 times over inorganic $CaCO_3$ crystals as a result of the intricate hierarchy of structural organization. High-resolution SEM observations have since been performed on the microstructure of nacre. Clearly visible in these images are the neatly stacked (or ordered) mineral tiles separated by thin organic sheets along with a macrostructure of larger periodic growth bands, which collectively form a hierarchical composite structure [11, 15, 23].

The mineral formation process, following periods of growth interruption (growth bands), has been described by several authors. Flat pearl implantation as well as a new trepanning method have been used to observe the transitory phases of calcium carbonate that nucleate and grow during this process. An initial random nucleation of the aragonite polymorph is observed. This nucleation and growth process is followed by a transition toward spherulitic growth. During this transition, the structural characteristics of the organism are determined by a combination of mechanical and chemical actions. About 6 weeks after implantation, a steady-state growth of aragonite tiles begins after shorter and more irregular tiles cover the outer surface of the spherulites. Finally, organic scaffolding is observed during the steady-state growth of tiled aragonite.

In addition, observations of mineral growth following the deposition of these membranes confirm the presence of mineral bridges originating from subsurface tiles and extending through the organic matrix. Field emission scanning electron microscopy as well as TEM work on fractured deproteinated nacre have revealed the presence of mineral bridges existing between individual layers of tiles. These nanoscale bridges have been identified as the origins of the extreme fracture toughness of this biomaterial [20, 21, 23].

A joint collaboration research recently at the University of California Santa Barbara and UC San Diego has produced striking results, including high-resolution SEM images of the microstructure of the mother-of-pearl (or nacre) portion of the abalone shell, which exhibits the **highest mechanical strength and fracture toughness of any nonmetallic substance known** [9, 17, 18, 20].

4.4 Self-Assembly

"Self-assembly," or "auto-assembly," is the most common term in use in the modern scientific community to describe the spontaneous aggregation of particles (atoms, molecules, colloids, micelles, etc.) without the influence of any external forces. Large groups of such particles are known to assemble themselves into thermodynamically stable, structurally well-defined arrays, quite reminiscent of one of the seven crystal systems found in metallurgy and mineralogy (e.g., cubic, face-centered cubic, body-centered cubic, hexagonal close packed, etc.). The fundamental difference in equilibrium structure is in the spatial scale of the unit cell (or lattice parameter) in each particular case.

Molecular self-assembly is found widely in biological systems and provides the basis for a wide variety of complex biological structures. This includes an emerging class of mechanically superior biomaterials based on microstructural features and designs found in nature. Thus, self-assembly is also emerging as a new strategy in chemical synthesis and nanotechnology. Molecular crystals, liquid crystals, colloids, micelles, emulsions, phase-separated polymers, thin films, and self-assembled monolayers all represent examples of the types of highly ordered structures that are obtained using these techniques. The distinguishing feature of these methods is self-organization.

4.5 Structural Hierarchy

Nearly all materials can be seen as hierarchically structured, especially since the changes in spatial scale bring about different mechanisms of deformation and damage. However, in biological materials this hierarchical organization is inherent to the microstructure. One of the first examples of this, in the history of structural biology, is the early X-ray scattering work on the hierarchical structure of hair and wool by Astbury and Woods [32]. In bone, for example, collagen is the building block of the organic matrix—**a triple helix with a diameter of 1.5 nm**. These tropocollagen molecules are interlaced with the mineral phase (hydroxyapatite, a calcium phosphate), forming fibrils that curl into helicoids of alternating directions. **These "osteons" are the basic building blocks of bones, with the volume fraction distribution between organic and mineral phase being about 60/40**. In another level of complexity, the **hydroxyapatite crystals are platelets that have a diameter of approximately 70–100 nm and a thickness of 1 nm**. They originally nucleate at the gaps between collagen fibrils. Table 4.1 shows the mechanical properties of human bone.

Similarly, the hierarchy of abalone shell begins at the nano level, with an organic layer having a thickness of 20–30 nm. This layer proceeds with single crystals of aragonite (a polymorph of $CaCO_3$) consisting of "bricks" with a dimension of 0.5 and finishing with layers of approximately 0.3 mm (mesostructure).

Crabs are arthropods whose carapace is made of a mineralized hard component (which exhibits brittle fracture) and a softer organic component composed primarily of chitin. The brittle component is arranged in a helical pattern. Each of these mineral "rods" (1-μm diameter) contains chitin–protein fibrils with approximately a 60-nm diameter. These fibrils are made of 3-nm-diameter canals that link the interior and exterior of the shell. Shrimp and lobsters also have a shell structure on their outer body consisting of chitin–protein fibrils. This author used these waste shells for conversion to chitin and finally to chitosan for the development of various useful products using a wet chemical method, e.g., PVA–chitosan biomembrane for wound healing and biocomposites admixing hydroxyapatite for medicine.

4.6 Applications

Biomaterials are used in a variety of applications, some of which include joint replacements, bone fracture fixation plates, bone cement, artificial ligaments and tendons, dental implants for tooth fixation, blood vessel prostheses, heart valves,

Table 4.1 Mechanical properties of human bone

Tissue	Tensile strength (MPa)	Elastic modulus (GPa)	Ultimate strain%
Cortical bone (longitudinally)	130	12.0	3
Cortical bone (transverse)	60	13.4	1
Cancellous bone	2	0.39	2.5

skin repair devices (artificial tissue), cochlear replacements, contact and intraocular lenses, and breast implants, to name a few.

Many more applications are appearing regularly. Biomaterials must be compatible with the body, and there are often issues of biocompatibility that must be resolved before a product can be sold on the market and used in a clinic. Because of this, biomaterials are usually subjected to the same requirements as new drug therapies [9]. All manufacturing companies are also required to ensure the traceability of all of their products so that if a defective product is identified, others in the same batch may be traced and recalled.

There are international organizations that provide recommendations and standards for the manufacturing and testing of biomaterials (ISO: International Standards Organization; ASTM: American Society for Testing and Materials). There are also national organizations that supervise biomaterial applications in human use, such as BIS in India. One of the best-known and most demanding control organizations is the U.S. Food and Drug Administration (FDA) (Brown et al. 1996). It must be pointed out that the FDA does not regulate the materials used in medical devices, but rather the devices themselves. The biocompatibility of the material is a central factor in devices intended for use inside the body. There are two main factors that determine the biocompatibility of a material: the host reactions induced by the material, and the degradation of the material in the body's environment.

4.7 Compatibility

Biocompatibility is related to the behavior of biomaterials in various environments under various chemical and physical conditions.

The term may refer to specific properties of a material without specifying where or how the material is to be used. For example, a material may elicit little or no immune response in a given organism, and may or may not able to integrate with a particular cell type or tissue. An immune system is a system of biological structures and processes within an organism that protects against disease by identifying and killing pathogens and tumor cells. It detects a wide variety of agents, from viruses to parasitic worms, and needs to distinguish them from the organism's own healthy cells and tissues in order to function properly. The ambiguity of the term reflects the ongoing development of insights into how biomaterials interact with the human body and eventually how those interactions determine the clinical success of a medical device (such as pacemaker or hip prosthesis). Modern medical devices and prostheses are often made of more than one material—so it might not always be enough to talk about the biocompatibility of a specific material.

Also, a material should not be toxic unless specifically engineered to be so, like "smart" drug delivery systems that target cancer cells and destroy them. Understanding the anatomy and physiology of the action site is essential for a biomaterial to be effective. An additional factor is the dependence on specific anatomical sites of implantation. It is thus important, during design, to ensure that the implement will fit complementarily and have a beneficial effect with the specific anatomical area of action.

For Further Reading, These Papers May Be Consulted

1. Omori M, Watabe N (eds) (1980) Mechanisms of biomineralization in animals and plants. Tokai University Press, Tokyo
2. Perry CC (2003) Silicification: the processes by which organisms capture and mineralize silica. Rev Mineral Geochem 54: 291
3. Heuer AH et al (1992) "Innovative materials processing strategies: a biomimetic approach. Science 255:1098. doi:10.1126/science.1546311
4. Lin A, Meyers MA, et al. (2008) Biological materials: structure and mechanical properties. Prog Mat Sci 53
5. Currey JD (1999) The design of mineralised hard tissues for their mechanical functions. J Exp Biol 202(23):3285–3294, PMID 10562511
6. Heuer A, Fink D, Laraia V, Arias J, Calvert P, Kendall K, Messing G, Blackwell J et al. (1992) Innovative materials processing strategies: a biomimetic approach. Science 255:1098. doi:10.1126/science.1546311
7. Whitesides G, Mathias J, Seto C (1991) Molecular self-assembly and nanochemistry: a chemical strategy for the synthesis of nanostructures. Science 254:1312. doi:10.1126/science.1962191
8. Aksay IA et al (2000) Self-assembled ceramics. Ann Rev Phys Chem 51:601. doi:10.1146/annurev.physchem.51.1.601.PMID 11031294
9. Sarikaya M, Gunnison KE, Yasrebi M, Aksay IA (1990) Mechanical property–microstructural relationships in abalone shell. Mater Res Soc 174:109 (conference presentation)
10. Menig R (2000) Quasi-static and dynamic mechanical response of *Haliotis rufescens* (abalone) shells. Acta Materialia 48:2383
11. Lin A, Meyers M (2005) Growth and structure in abalone shell. Mater Sci Engineer A 390:27. doi:10.1016/j.msea.2004.06.072
12. Lin AY, Chen PY, Meyers MA (2008) The growth of nacre in the abalone shell. Acta Biomater 4 (1):131
13. Fritz M, Morse DE (1998) Formation of highly organized biogenic polymer/ceramic composite materials: the high-performance microaluminate of molluscan nacre. Coll Int Sci 3:55
14. Feng QL et al (2000) Crystal orientation, toughening mechanisms and a mimic of nacre. Mater Sci Eng C11:19
15. Meyers M, Lin A, Chen P, Muyco J (2008) Mechanical strength of abalone nacre: role of the soft organic layer. J Mech Behav Biomed Mater 1(1):76–85
16. Whitesides GM et al (1991) Molecular self-assembly and nanochemistry: a chemical strategy for the synthesis of nanostructures. Science 254:1312
17. Pederson T (2006) Present at the flood: how structural molecular biology came about. FASEB J 20:809–810
18. Kammula RG, Morris JM (2001) Considerations for the biocompatibility evaluation of medical devices. Medical Device and Diagnostic Industry, May 2001
19. Stupp SI, Braun PV (1997) Role of proteins in microstructural control: biomaterials, ceramics and semiconductors. Science 277:1242
20. Klemm D, Heublein B, Fink H, Bohn A (2004) Cellulose: fascinating biopolymer/sustainable raw material. Ang Chemie (Intl Edn) 44:3358
21. Chandra R, Rustgi R (1998) Biodegradable polymers. Prog Polym Sci 23:1273
22. Kumar A et al (2007) Smart polymers: physical forms and bioengineering applications. Prog Polym Sci 32:1205
23. Weiner S, Lowenstam HA (1989) On biomineralization. Oxford University Press. ISBN 0-19-504977-2

4.8 Biopolymers

Biopolymers are polymers produced by living organisms. Cellulose and starch, proteins and peptides, and DNA and RNA are all examples of biopolymers, in which the monomeric units, respectively, are sugars, amino acids, and nucleotides. Cellulose is the most common biopolymer and the most common organic compound on earth. About one third of all plant matter is cellulose.

Some biopolymers are biodegradable. That is, they are broken down into CO_2 and water by microorganisms. In addition, some of these biodegradable biopolymers are compostable. That is, they can be put into an industrial composting process and will break down by 90% within 6 months. Biopolymers that do this can be marked with a "compostable" symbol, under European Standard EN 13432 (2000). Packaging marked with this symbol can be put into industrial composting processes and will break down within 6 months (or less). An example of a compostable polymer is PLA film under 20 μm thick: Films that are thicker than that do not qualify as compostable, even though they are biodegradable. A home composting logo may soon be established: This will enable consumers to dispose of packaging directly onto their own compost heap [37–39].

Biopolymers are polymers that are biodegradable. The input materials for the production of these polymers may be either renewable (based on agricultural plant or animal products) or synthetic. There are four main types of biopolymers, based, respectively, on

1. **Starch**
2. **Sugar**
3. **Cellulose**
4. **Synthetic materials.**

Current and future developments in biodegradable polymers and renewable input materials relate mainly to the scaling up of production and the improvement of product properties. Larger-scale production will increase availability and reduce prices.

At present, either renewable or synthetic starting materials may be used to produce biodegradable polymers. Two main strategies may be followed in synthesizing a polymer. One is to build up the polymer's structure from a monomer by a process of chemical polymerization. The alternative is to take a naturally occurring polymer and chemically modify it to give it the desired properties. A disadvantage of chemical modification is, however, that the biodegradability of the polymer may be adversely affected. Therefore, it is often necessary to seek a compromise between the desired material properties and biodegradability.

4.8.1 Applications

Polymers have properties that make them suitable for use in protecting products from moisture, increasing shelf-life, and making products easier to dispense.

Every biopolymer has its own material-specific properties, e.g., barrier properties such as oxygen permeability. The barrier properties are relevant to the choice of biopolymers for the packaging of particular products. Bioplastics have very promising prospects for use in pesticide, for packaging in-flight catering products, and for packaging dairy products.

Thermoplastic starch is unsuitable for packaging liquids. It can sustain only brief contact with water. It has good oxygen barrier properties.

4.8.2 Sources

Starch is a natural polymer that occurs as granules in plant tissue, from which it can easily be recovered in large quantities. It is obtained from potatoes, maize, wheat, tapioca, and similar sources. Starch can be modified in such a way that it can be melted and deformed thermoplastically. The resulting material is thus suitable for conventional plastic-forming processes such as injection moulding and extruding.

4.8.3 Applications

Polylactides decompose harmlessly in the human body and have therefore long been used for medical applications. Examples include surgical implants that do not require operative removal. Until recently, it was not feasible to use polylactides for packaging because of their high price, around US$500/kg.

4.8.4 Sources

The starting material for polyhydroxibutyrate is made from sucrose or starch by a process of bacterial fermentation. Varying the nutrient composition of the bacteria produces differences in the end product. This makes it possible to tune the properties of the material, e.g., its moisture resistance. The polymer can be formed by injection, extrusion, blowing, and vacuum forming.

Polylactides (lactic acid polymers) are made from lactic acid, which is in turn made from lactose (or milk sugar) obtained from sugar beets, potatoes, wheat, maize, etc. Polylactides are water-resistant and can be formed by injection moulding, blowing, and vacuum forming.

4.8.5 Applications

The relatively high price of biodegradable polymers of synthetic substances, e.g., aliphatic aromatic copolyesters, has prevented them from reaching a bigger market. The best-known application is for making substrate mats.

Synthetic compounds derived from petroleum can also be a starting point for biodegradable polymers, e.g., aliphatic aromatic copolyesters. These polymers have technical properties resembling those of low-density polyethylene (LDPE). Although these polymers are produced from synthetic starting materials, they are fully biodegradable and compostable.

4.8.6 Benefits

Besides being available on a sustainable basis, biopolymers have several economic and environmental advantages. Biopolymers could also prove an asset to waste processing. For example, replacing the polyethylene used in coated papers by a biopolymer could help eliminate plastic scraps from occurring in compost.

Consumers have a lively interest in biopolymers too. Conventional plastics are often seen as environmentally undesirable. Sustainable plastics could therefore provide an image advantage.

The major advantage of biodegradable packaging is that it can be composted. But the biodegradability of raw materials does not necessarily mean that the product or package made from them (e.g., coated paper) is itself compostable.

Biopolymers can also have advantages for waste processing. Coated paper (coated with, e.g., polyethylene) is a major problem product for composting. Although such materials are usually banned from inclusion in organic waste under separate collection schemes, some of them usually end up in the mix nonetheless. The paper decomposes, but small scraps of plastic are left over in the compost. The adoption of biopolymers for this purpose would solve the problem.

Widespread interest for biopolymers among consumers is common these days. The environmental benefits of biodegradable packaging must be reflected in cost advantages if large-scale applications are to become feasible. In the short term, it would be preferable to communicate the functional advantages of biodegradable packaging rather than its compostability.

Whether the use of biopolymers will contribute to a more sustainable society is a question that will have to be studied by a representative life-cycle analysis (LCA) of each application and by comparison with existing applications. Representative LCAs are needed at the material and product levels; they must make allowance for future developments in biopolymers and take account of all relevant environmental impacts associated with the complete product life cycle, including the depletion of raw materials, the agricultural use of fertilizers and pesticides, transportation, utilization, and waste disposal.

We will now discuss some practical aspects of biomaterials used in healthcare from an industrial point of view.

4.9 Polymers for Healthcare

Based on biocompatibility testing as defined by ISO 10993–1, healthcare plastics should not demonstrate evidence of cytotoxicity, sensitization, intracutaneous reactivity, or systemic toxicity. These materials can be sterilized using conventional sterilization methods, including **steam, ethylene oxide, vaporized hydrogen peroxide, and gamma radiation**.

4.9.1 Sterilization

Since each application has unique performance and design criteria, it's important to evaluate a material under conditions that best simulate the function of the component or system in its intended use. This includes cleaning and disinfection followed by sterilization in order to assess their cumulative effect on the material.

4.9.2 Compatibility with Sterilization Method of Various Industrial Biopolymers (Solviva)

The methods of sterilization used are steam at 121°C and one atmosphere absolute (ata), or 14.7 psig, ETO gas, vaporized hydrogen peroxide, high-energy gamma radiation, usually 25 kGy or more. The method has to be decided after due consideration of the properties of the materials and the effect of the sterilization on the material.

Solviva Biomaterials has manufactured in compliance with relevant aspects of ISO 13485 and the current Good Manufacturing Practices. Solviva's biomaterial manufacturing processes are carefully validated, and enhanced controls provide product traceability. In addition, all materials are tested in an accredited lab that is ISO 17025 compliant.

Based on biocompatibility testing, the biomaterials should not demonstrate any evidence of cytotoxicity, sensitization, irritation, or acute systemic toxicity. Some of the normal tests and procedures are listed in the following table.

Quality-assured manufacturing	Biocompatibility tests
• Adhere to the relevant aspects of Good Manufacturing Practices • Complaint with the relevant aspects of ISO 13485 manufacturing • ISO 17025 complaint laboratory testing • Process validation • Product traceability	• Genotoxicity ISO 10993:3 • Hemolysis ISO 10993:4 • Cytotoxicity ISO 10993:5 • Local effects of implantation ISO 10993:6 • Sensitization and irritation ISO 10993:10 • Acute systemic toxicity ISO 10993:11 • Chemical analysis ISO 10993:18

4.9.3 Polyether Ether Ketone (PEEK)

PEEK is a colorless organic polymer thermoplastic used in engineering applications. PEEK is a semicrystalline thermoplastic with excellent mechanical and chemical resistance properties that are retained to high temperatures. ItsYoung's modulus is 3.6 GPa and its tensile strength is 90–100 MPa [4], near that of human bone. PEEK has a glass transition temperature at around 143°C (289 °F) and melts around 343°C (662 °F). The thermal conductivity increases nearly linearly versus temperature between room temperature and solidus temperature [5]. It is highly resistant to thermal degradation as well as attack by both organic and aqueous environments. It dissolves completely in concentrated sulfuric acid at room temperature. Its general properties are given below. It is better than metals in many applications, strong but light.

Density	1.3 g/cc or1,320 kg/m^3
Young's modulus (*E*)	3.6 GPa
Tensile strength (σ_t)	90–100 MPa
Elongation at break	50%
Impact notch test	55 kJ/m^2
Glass transition temperature	143°C
Melting point	~343°C
Thermal conductivity	0.25 W/m-K
Water absorption, 24 h (ASTM D 570)	0.1%

4.9.4 *Applications*

PEEK is one of the few plastics compatible with ultrahigh-vacuum applications. PEEK is considered an advanced biomaterial used in medical implants. It is extensively used in the aerospace, automotive, teletronic, and chemical process industries. PEEK's mechanical properties at elevated temperatures have led to its being used in at least two varieties of Reprap extruders as thermal insulation. It has shape-memory properties, and it is being used in orthopedic surgery, in making plates and screws, and in artificial total joint replacement. It is also used in trauma and spinal research (*Handbook of PEEK Biomaterials*, published by Elsevier 2011).

4.10 Bioceramics

Bioceramics and bioglasses are ceramic materials that are biocompatible [1]. Bioceramics are an important subset of biomaterials [2, 3]. Bioceramics range in biocompatibility from the ceramic oxides, which are inert in the body, to the other extreme of resorbable materials, which are eventually replaced by the materials they were used to repair. Bioceramics are used in many types of medical procedures. A primary medical procedure where they are used is implants. In this section we are primarily concerned with rigid materials commonly used as surgical implants, though some bioceramics are flexible. The ceramic materials used are not the same as porcelain-type ceramic materials. Rather, bioceramics are closely related to the body's own materials or are extremely durable metal oxides.

4.10.1 *History*

Prior to 1925, the materials used in implant surgery were primarily relatively pure metals and alloys. The success of these materials was surprising considering the relatively primitive surgical techniques. The year 1925 marked the beginning of the era of better surgical techniques and also the first use of alloys such as Ti-6Al-4V, meaning approximately 6% Al, 4% Va, and 90%Ti.

In 1969, L. L. Hench and others discovered that various kinds of glasses and ceramics could bond to living bone [4, 5]. Hench was inspired with the idea on his way to a conference on materials. He was seated next to a colonel who had just returned from the Vietnam War. The colonel shared that after an injury, the bodies of soldiers would quite often reject the implant. Hench was intrigued and began to investigate materials that would be biocompatible. The final product was a new material he called "Bioglass®." This work inspired a new field called "bioceramics" [6]. With the discovery of Bioglass, interest in bioceramics grew rapidly.

On April 26, 1988, the first international symposium on bioceramics was held in Kyoto, Japan.

4.10.2 *Current Status*

Ceramics are now commonly used in medical fields as dental and bone implants [7, 8]. Artificial teeth and bones are relatively common. Surgical cements are used regularly. Joint replacement prostheses are often coated with bioceramic materials to enhance biointegration, reduce wear, and lessen inflammatory response. Other examples of medical uses for bioceramics are in pacemakers, kidney dialysis machines, and respirators [6].

4.10.3 *Future Trends*

One proposed use for bioceramics is the treatment of cancer. Two methods of treatment have been proposed; treatment through hyperthermia, and radiotherapy. Hyperthermia treatment involves implanting a bioceramic material that contains a ferrite or other magnetic material. The area is then exposed to alternating magnetic fields, which causes the implant and surrounding area to heat up. Alternatively, the bioceramic materials can be doped with β-emitting materials and implanted into the cancerous area [11].

Other trends include engineering the materials for specific tasks. Ongoing research involves the chemistry, composition, and micro and nanostructures of the materials to improve their biocompatibility [10–12].

4.11 Bioceramic Materials

Bioceramic materials are commonly subdivided by their bioactivity, namely, bioactive or bioinert. Bioinert materials are nontoxic and noninflammatory. These materials must be long-lasting, structurally failure-resistant, and corrosion-resistant. Bioceramics additionally must have a low Young's modulus to help prevent cracking of the material.

4.11.1 *Bioinert*

Bioinert materials include the following:

- Oxide ceramics—alumina
- Silica ceramics
- Carbon and fiber
- Synthetic diamonds.

4.11.2 *Bioactive*

The most talked about material over the past several decades is hydroxyapatite (HA) because of its enormous end use, which helps in hard tissue growth and repair. That is why it is called a bioactive material.

4.12 Hydroxyapatite: Medical Uses

Flexible hydrogel–HA composite has a mineral-to-organic matrix ratio approximating that of human bone. Hydroxyapatite can be found in teeth and bones within the human body. Thus, it is commonly used as a filler to replace surgically removed bone or as a coating to promote bone ingrowth into prosthetic implants. Although many other phases exist with similar or even identical chemical makeup, the body responds much differently to them. Coral skeletons can be transformed into hydroxyapatite by high temperatures; their porous structure allows relatively rapid ingrowths at the expense of initial mechanical strength. The high temperature also burns away any organic molecules such as proteins, preventing an immune response and rejection.

Many modern implants, e.g., hip-stem and dental implants, are coated with hydroxyapatite. It has been suggested that this promotes osseointegration. Properties of hydroxyapatite crystals are noted below.

General:	
Category:	Phosphate mineral
Chemical formula:	$Ca_5(PO_4)_3(OH)$
Crystal symmetry:	Hexagonal 6/m—dipyramidal
Unit cell:	a=9.41 Å, c=6.88 Å, Z=2
Identification	
Molar mass:	502.31 g
Color:	Colorless, white, gray, yellow, yellowish-green
Crystal habit:	As tabular crystals and as stalagmites, nodules, in crystalline to massive crusts
Crystal system:	Hexagonal
Cleavage:	Poor on {0001} and {1010}
Fracture:	Conchoidal
Tenacity:	Brittle
Mohs scale hardness:	5
Luster:	Vitreous to subresinous, earthy
Streak:	White
Diaphaneity:	Transparent to translucent
Specific gravity:	3.14–3.21 measured, 3.16 calculated
Optical properties:	Uniaxial (–)
Refractive index:	$n_\omega = 1.651$ $n_\varepsilon = 1.644$
Birefringence:	$\delta = 0.007$

4.13 Bioactive Glass

Bioactive glasses are a group of surface-reactive glass-ceramic biomaterials and include the original bioactive glass, Bioglass. The biocompatibility of these glasses has led to their being extensively investigated for use as implant materials in the human body to repair and replace diseased or damaged bone. Larry Hench and colleagues first developed these materials in the late 1960s at the University of Florida; the glasses have been further developed by Hench's research team at the Imperial College London and other researchers worldwide.

4.13.1 Compositions

There have been many variations subsequently from the original composition, which was approved by the FDA and termed Bioglass. This composition is known as **45S5**. Other compositions are listed here:

Bioglass 45S5: 46.1 mol% SiO_2, 26.9 mol% CaO, 24.4 mol% Na_2O, and 2.5 mol% P_2O_5
Bioglass 58S: 60 mol% SiO_2, 36 mol% CaO, and 4 mol% P_2O_5 Bioglass S70C30: 70 mol% SiO_2 and 30 mol% CaO.

4.13.2 Mechanism of Bioactivity

The mechanisms that activated bioactive glasses to act as materials for bone regeneration or repair have been investigated since the first work of Hench et al. at the University of Florida. Attention was paid to changes in the bioactive glass surface. Five inorganic reaction stages are commonly thought to occur when a bioactive glass is immersed in a physiological environment:

1. Ion exchange, in which modifier cations (mostly Na^+) in the glass exchange with hydronium ions in the external solution
2. Hydrolysis, in which Si–O–Si bridges are broken, forming Si–OH silanol groups, and the glass network is disrupted
3. Condensation of silanols, in which the disrupted glass network changes its morphology to form a gel-like surface layer, depleted in sodium and calcium ions
4. precipitation, in which an amorphous calcium phosphate layer is deposited on the gel
5. mineralization, in which the calcium phosphate layer gradually transforms into crystalline hydroxyapatite, which mimics the mineral phase naturally contained with vertebrate bones.

Later, it was discovered that the morphology of the gel surface layer was a key component in determining the bioactive response. This was supported by studies on bioactive glasses derived from sol–gel processing. Such glasses could contain significantly higher concentrations of SiO_2 than traditional melt-derived bioactive glasses and still maintain bioactivity (i.e., the ability to form a mineralized hydroxyapatite layer on the surface). The inherent porosity of the sol–gel-derived material was cited as a possible explanation for why bioactivity was retained and often enhanced with respect to the melt-derived glass. Subsequent advances in DNA microarray technology enabled an entirely new perspective on the mechanisms of bioactivity in bioactive glasses. Previously, it was known that a complex interplay existed between bioactive glasses and the molecular biology of the implant host, but the available tools did not provide a sufficient quantity of information to develop a holistic picture. Using DNA microarrays, researchers are now able to identify entire classes of genes that are regulated by the dissolution products of bioactive glasses, resulting in the so-called genetic theory of bioactive glasses. The first microarray studies on bioactive glasses demonstrated that genes were associated with osteoblast growth and differentiation, maintenance of extracellular matrix, and promotion of cell–cell and cell–matrix adhesion.

4.13.3 Applications

Bioactive glasses have many applications, but these are primarily in the areas of bone repair and bone regeneration via tissue engineering. Synthetic bone graft materials are used for general orthopedic, craniofacial (bones of the skull and face), maxillofacial and periodontal (the bone structure that supports teeth) repair, also in treating dentine hypersensitivity and promoting enamel remineralization. These are available to surgeons in a particulate form.

Bioactive glasses have also been used in cochlear implants. Finally, bioactive glasses have been investigated in many forms for bone tissue engineering scaffolds, in particular as porous (contains pores into which cells can grow and fluids can travel) three-dimensional scaffolds.

This section on bioactive glass was prepared based on the materials from Wikipedia dated 1-11-2010.

For Further Study Consult the Reading List (Ceramics and Glass)

1. Ducheyne P, Hastings GW (eds) (1984) Metal and ceramic biomaterials, Vol 1. CRC, Boca Raton, FL. ISBN 0-8493-6261-x
2. Shackelford JF (ed) (1999) MSF bioceramics applications of ceramic and glass materials in medicine. ISBN 0-87849-822-2
3. Oonishi H, Aoki H, Sawai K (eds) (1988) Bioceramics. vol. 1. ISBN 0-912791-82-9

4. Hench LL (1991) Bioceramics: from concept to clinic. J Am Ceramic Soc 74(7):1487–1510
5. Yamamuro T, Hench LL, Wilson J (eds) (1990) Handbook of bioactive ceramics, Vol II. CRC, Boca Raton, FL. ISBN 0-8493-3242-7
6. Ruth K (2003) Ceramics: from magic pots to man-made bones. Twenty-First Century Books, Brookfield, CT. ISBN 978 0761325857
7. Muster D (ed) (1992) Biomaterials hard tissue repair and replacement. Elsevier, North Holland. ISBN 0-444-88350-9
8. Kinnari TJ et al (2007) Bacterial adherence to SiO2-based multifunctional bioceramics. J Biomed Mater Res Part A 89A:215–223. doi:10.1002/jbm.a.31943, www.interscience.wiley.com
9. Shackelford JF. ISBN 0-87849-822-2
10. Chai C, Leong KW (2006) Biomaterials approach to expand and direct differentiation of stem cells. Mol Therap 15(3):467–480. doi:10.1038/sj.mt.6300084
11. Zhu X et al (2004) Cellular reactions of osteoblasts to micron and submicron scale porous structures of titanium surfaces. Cell Tissues Organ 178:13–22. doi:10.1159/000081089
12. Hao L, Lawrence J, Chian KS (2005) Osteoblast cell adhesion on a laser modified zirconia based bioceramic. J Mater Sci: Mater Med 16:719–726

4.14 Metals as Biomaterials

Metals and alloys are distinctly different from ceramics and are being treated separately. Metals and alloys have been used in medicine for a long time. On the basis of their interface reactions, materials may be classified as toxic, biologically inactive (nearly inert), bioactive, or resorptive (Hench 1996). Toxic materials are not used in implants. Metallic biomaterials are classified as nearly inert materials. Because of their mechanical strength and stiffness and biocompatibility, metals are superior in load-bearing implants.

The biocompatibility of the implant material is closely related to the reactions between the surface of the biomaterial and the inflammatory host response (Thomsen et al. 1991). There are several factors that contribute to this. These may depend on individual patient characteristics, such as general health, age, tissue perfusion, immunological factors, or implant characteristics, such as surface roughness and porosity, chemical reactions at the surface, corrosion properties of the material, and the toxicity of the individual metals present in the alloy (Klinger et al. 1997). The surgery itself, the technique applied, and biomechanical considerations (stability) modify the inflammatory response.

The releasing agents from the cell may alter the characteristics of the material surface. The surface is also changed due to the influence of proteins absorbed from plasma (Anderson et al. 1990).

After implantation, a coagulating and vascularizing process takes place. The implant is covered by a blood clot containing leukocytes and erythrocytes, thrombocytes, and coagulating proteins. The implant and the surgical trauma trigger an inflammatory reaction that eliminates the damaged tissue, clot, and bacteria. Inflammatory cells, first polymorphonuclear granulocytes and later monocytes, arrive to expurgate the debris and foreign materials. If there is too much foreign substance for granulocytes, monocytes develop into macrophages. If there is a delay in removing the substance, the enzymes of activated macrophages affect the fibroblasts to make a fibrous capsule around the implant. As long as phagocytic activation is maintained, the capsule becomes thicker. In soft tissues, the inert material forms a thin, fibrous encapsule around the implant (Anderson 1996).

The implantation response in bone differs in some ways from that taking place in soft tissue. There are inflammatory and reparative responses, which occur one on top of the other. The reparative response starts 2–3 days after the implantation. The stem cells of bone develop into osteoblasts, which form a layer near the implant together with fibroblasts. Fibroblasts, osteoblasts, and capillaries penetrate into the blood clot, replacing it, and fill the space between the implant and bone (Tarr et al. 1986). After the formation of a collagen-rich extracellular matrix (ECM), mineralization follows. Normally, there are vesicles in ECM and some of them include calcification focuses. The presence of vesicles with biomaterial in the early period is a sign of good primary acceptance. When the membranes of these vesicles rupture, the erupted apatite crystals unite and form calcifying structures (Davies 1991). Early trabecles grow and continue to mineralize, and some of them reach the implant surface. In an optimal situation, the material is covered by bone tissue and not by fibrous capsule. The healing of bone tissue continues as in fracture healing. Remodeling of bone tissue begins after 2 weeks and continues for the lifetime. Woven bone is replaced by functionally oriented lamellar bone.

Changes in the local environment, such as acidity, oxygen content, electric charge, ion concentration, enzymes, growth factors, etc., have effects on the differentiation and migration of stem cells. The attachment of osteogenic stem cells to substrate and the formation of mineralized ECM are essential for the differentiation of osteoblasts (Davies 1991). When the material is biocompatible, there is an abundance of ECM and osteoblasts. This is confirmed by the close attachment and fast proliferation of these cells (Vrouwenvelder et al. 1993). Brånemark et al. (1969) first suggested that titanium may form direct bone contact. They defined this "osseointegration" as direct metal-to-bone contact at the light-microscopic (0.5-μm) level. This definition appears to be somewhat inaccurate, and some clinically based definitions of osseointegration have been suggested (Albrektsson et al. 1993). In optimal situations, however, bone accepts the implant as part of its ECM, and clinically asymptomatic rigid fixation is achieved in bone during functional loading. Such fixation is possible with titanium implants. Other implant metals usually form some fibrous tissue between the bone and the implant and are often called "nearly inert." Because of their mechanical strength and biocompatibility, metals are superior in load-bearing implants. Even chemical bone bonding (*the establishment, by physicochemial processes, of continuity between implant and bone*) is seen with bioactive glasses, but their mechanical properties are inferior to metal biomaterials (Hench 1996).

4.15 Signs of Inferior Tissue Response

Highly toxic material causes tissue necrosis. The signs of subacute toxicity or low tissue tolerance may be manifested in several ways. A large amount of foreign-body giant cells is usually a sign of a prolonged stimulus. Also, the presence of phagocytes at a later time may signify a rejection of the implant. The propagation of lymphocytes or plasma cells may indicate the activation of the immune system against the material. Profuse accumulation of neutrophils is a sign of infection. Vacuolization and resorption of muscle are signs of an inferior tissue response (Williams 1986).

The response of individual cells to material can be considered to be dependent on how well the material mimics the natural (extracellular) environment of the cell. The physical structure of the surface may have an inferior influence on the biological response of the material, which is normally nontoxic and does not release any biologically active substance. Osteolysis, bone resorption, and the formation of a thick fibrous layer between the implant and bone reflect poor biocompatibility. Also, microparticles of a certain size of normally nontoxic materials may trigger an inflammatory response. These particles cause an irritation of phagocytic cells and activate them to produce and release cytokines, proteinases, growth factors, and other proinflammatory factors, finally leading to chronic inflammation, fibrosis, osteolysis, and porosis in bone (Shanbhag et al. 1994; Tang et al. 1996). In the case of aseptic loosening of the prosthesis, wear particles are expected to lead to the formation of a poorly vascularized, synovial-like interface membrane between the prosthesis and bone (Santavirta et al. 1998). The formation of necrotic focuses, granulomas, and osteolysis may finally result in loosening of the prosthesis (Santavirta et al. 1996). The increase in metallic wear increases the surface of the metallic material and the quantity of metal ions. The porous surface increases the surface area, but also particular wear.

4.16 Stress-Shielding Phenomena

Stress-shielding phenomena cause bone in contact with the metal prosthesis to lose its density and become osteoporotic as the metal carries the whole load and bone becomes disuse-osteoporotic due to the wide mismatch in their E-values or stiffness. Table 4.1 gives the specific modulus of elasticity; it can be seen how the bone differ from the alloys. For the alloys, it is nearly ten times, which indicates we need a material that will be closer to bone.

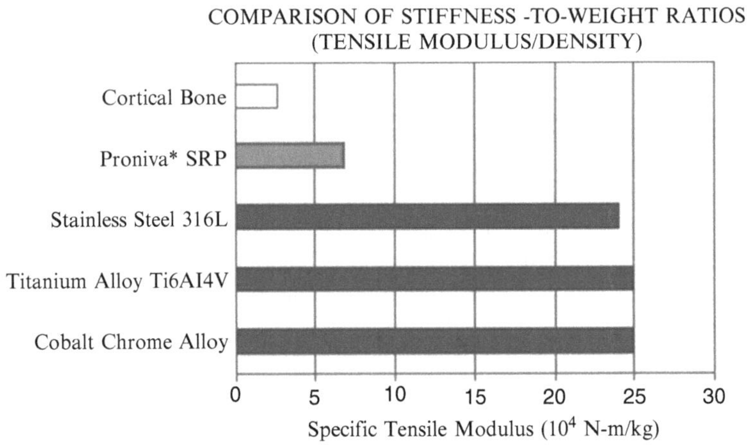

4.17 Shape-Memory Alloy

At the present time, in the developed countries of the world, more than 2.5 million metallic constructions are implanted within bodies in various operations. The use of gold, silver, titanium, stainless steel, and composite materials as implants in medicine has revealed significant shortcomings in all these materials that cannot be eliminated. Their main shortcoming is the following: In traditional methods of fixing implants in the body, the implants are fixed in place with the aid of screws, nuts, and cement, leading to complex systems of attachment to bones and soft tissues. In many cases, such systems are not only inefficient, but cannot even be used because of their cumbersomeness.

The discovery of the phenomenon of shape memory in metals has led to the creation of a new scientific approach in medicine that eliminates the just-described shortcomings. This approach involves the development and study of alloys, constructions, and devices with shape memory and the development of original treatment methods that elevate medicine to a new level.

The use of shape-memory alloys in concrete construction and devices requires special study. The selection or design of alloys with a specific form changes parameters for each problem, with a consideration of construction features and conditions of use, together with general requirements (high corrosion resistance and compatibility with body organs for the shape-memory alloys). Special requirements also must be met: high strength and plasticity, elasticity and rigidity, flexibility wear, and failure resistance. Alloys based on the intermetallic compound TiNi manifest the entire range of properties enumerated here, which makes them among the most promising materials for medical use. Moreover, the shape-memory property provides implants with new qualities: They can change form (up to 15%) upon change in temperature, develop significant stresses (up to 900 MPa) upon form change, manifest superelastic properties in a specific temperature range, and resist failure under repeated mechanical loading (they can withstand cyclical deformation to 5% for more than 20 million cycles; usually, metals yield at less than 0.5% strain).

The first is the use of the ability of such materials to change form with a change in temperature and produce significant stresses over a certain temperature interval. This property is used specifically in surgical traumatology. The material's shape-changing property is used to fixate bone fractures. In treating patients with bone fractures, two basic problems must be solved: restoration of anatomical form of the injured organ and reestablishment of full function. As is well known, it is only when the adjacent segments of the fractured bone are fixed stably relative to each other with proper alignment of the osteogenic sections and maintenance of function of the traumatized bone that we really achieve optimum biophysical conditions for regeneration.

Acknowledgment V. D. Kuznetsov Siberian Pbysicotechnical Iwititute, Tomsk State University. Translated from Izvestl YaV, rsshikh Uchebnykh Zavedenii, Fizika, No.5, pp. 127–132, May 1985, for information on shape-memory alloy.

Problems

1. Hydroxyapatite (HA) and tricalcium phosphate (TCP) are generated simultaneously during a thermal conversion process. How will you identify the HA and TCP and their relative percentages?
2. We have produced HA from the bones and teeth of mammals, sea corals, and eggshells. Suggest how those can be achieved and what equipment and chemicals will be required based upon the chemistry of transformation.
3. HA is a ceramic usually produced by a high-temperature heating process. Hard tissues of animals, like tooth and bone, contain a huge amount of such a mineral. How is that possible in a natural process at a low temperature?
4. What is a stress-shielding phenomenon, and how does it affect the performance of metallic prostheses? Suggest suitable means for solving such problems.
5. What is bioactive glass? State its composition and how those components can be produced. Where can these materials be used gainfully in medical problems?
6. How can PEEK material be used for the development of hip joints?

Chapter 5
Dental Implants: Their Design and Manufacture

5.1 Introduction

Edentulousness due to diseases, trauma, or developmental anomalies is a worldwide problem. Implants have been used to support dental prostheses for many decades, but they are not always received with enthusiastic response from the patients. This situation has changed dramatically with the development of endosseous osseointegrated dental implants. They are the nearest equivalent replacement to the natural tooth and are therefore a useful addition in the management of patients who have missing teeth. There are a number of dental implant systems that offer predictable long-term results backed by good scientific research and clinical trials. First of all, it may be helpful to understand some of the commonly used terms in implantology, given in Table 5.1.

5.1.1 Success Criteria

It is important to establish success criteria for implant systems, and for implants to be tested in well-controlled clinical trials. The minimum success criteria proposed by Albrektsson et al. (I JOMI1986; 1: 11) are depicted in Table 5.2.

The most obvious sign of implant failure is mobility. However, some of the criteria in Table 5.2 apply to the overall requirements of an implant system but are not as useful when judging the success of individual implants. This is well illustrated by considering the radiographic criteria. Bone remodeling occurs in the first year of function in response to occlusal forces and establishment of the normal dimensions of the peri-implant soft tissues. The "ideal" bone level is usually judged against a specific landmark on the implant (such as the implant/abutment junction), and it may differ between implant systems. Subsequently, the bone levels are usually

DOI 10.1007/978-1-4614-6255-2_5,

Table 5.1 Basic terminologies in implant dentistry

Osseointegration	A direct structural and functional connection between ordered, living bone and the surface of a load-carrying implant (Albrektsson et al., Acta Orthopaedica Scand 1981; 52)
Endosseous dental implant	The component that attaches to the dental implant and supports the prosthesis. A transmucosal abutment (TMA) passes through the mucosa overlying the implant. A temporary of healing abutment may be used during the healing of the peri-implant soft tissue before the definitive abutment is chosen
Abutment screw	An intermediate screw to connect the implant to a crown or denture
Single-stage implant surgery	Surgical placement of a dental implant that is left exposed to the oral cavity following insertion. This is the protocol used in nonsubmerged implant systems
Two-stage implant surgery	Initial surgical placement of a dental implant that is buried beneath the mucosa and then subsequently exposed with a second surgical procedure some months later. This is used in submerged implant systems

Table 5.2 Suggested minimum success criteria for dental implants[a]

1.	An individual, unattached implant is immobile when tested clinically
2.	Radiographic examination does not reveal any peri-implant radiolucency
3.	After the first year in function, radiographic vertical bone loss is less than 0.2 mm per annum
4.	The individual implant performance is characterized by an absence of signs and symptoms such as pain, infections, neuropathies, paraesthesia, or violation of the inferior dental canal
5.	As a minimum, the implant should fulfill the above criteria with a success rate of 85% at the end of a 5-year observation period and 80% at the end of a 10-year period

[a]Based on Albrektsson et al. IJOMI 1986,1; 11

more or less stable, and small changes such as 0.2 mm per annum are impossible to measure with conventional radiographs. These specified changes therefore do not apply to individual implants, but to mean (average) changes measured across a large number of implants. For example, a detectable change of 1 mm or more may occur at very few implants in contrast to the majority, which remain unchanged or in a steady state. It is also difficult to stipulate what level of change in an individual implant over a given period of time would constitute failure. A rapid change in bone level may be followed by a long period of stability. On the other hand, progressive or continuous bone loss is a worrying sign of impending failure. An implant with a marked loss of bone may therefore be judged as "surviving" rather than "successful."

The role of modern restorative dentistry is to revive the patient's oral health, comfort, function, and aesthetics by restoring or replacing the dentition and surrounding tissue. Removal and fixed prostheses have more or less achieved this goal, but with a lot of limitations. But people were not happy; they were searching for an alternative, and the implant systems were invented. Oral implants have overcome these limitations and today offer a simple and predictable means of treatment not only for the dental cripple, but even for patients with a single missing tooth, providing excellent aesthetics, comfort, and function.

The ancient history of implants dates back to Egyptian civilization when seashells were literally hammered into the jaw to replace missing teeth; these shells actually worked. Slots were made into the bone and the shells were driven in like little wedges.

In 1952, in a modestly appointed laboratory in the university town of Lund, Sweden, Professor Per-Ingvar Brånemark had a lucky accident—what most scientists call serendipity. Much to his irritation, Dr. Brånemark discovered that it was impossible to recover any of the bone-anchored titanium from the jaw bone while he was using microscopes in his research. The titanium had apparently bonded irreversibly to living bone tissue, an observation that contradicted contemporary theory. His curiosity aroused, Dr. Brånemark subsequently demonstrated that under carefully controlled conditions, titanium could be structurally integrated into living bone with a very high degree of predictability and without long-term soft tissue inflammation or ultimate implant rejection. Dr. Brånemark named the phenomenon "osseointegration." The first practical application of osseointegration was the implantation of new titanium roots in an edentulous patient in 1965. More than 30 years later, the nonremovable teeth attached to these roots were functioning perfectly.

5.2 Anatomy of Human Jawbones

Humans evolved as omnivores, and our teeth reflect this history. We have relatively weak canine teeth compared to cats and even other primates such as baboons. Our premolars and molars are not nearly as efficient as those of ruminants or horses for grinding tough plant material. Nonetheless, our teeth generally serve us well for the kind of food we consume. Figures 5.1 and 5.2 depict the anatomical features of our facial bones and jawbones, indicating the placements of the teeth in their housing. A detailed description is avoided.

These figures are given to learn the basic anatomy in connection with the jawbone and the nomenclature of the teeth, which are important for learning the subject.

Figure 5.1 shows the upper teeth arrangement in the bony cavity of the maxilla and mandible. It also shows the maxillary bone and its relationship with the other parts of bony arcades of the facial bone, such as the naso-ethmoidal complex, zygomatic arch, and floor of the orbit.

Central incisors, canine, premolar, and molar teeth have their individual role in oral function. The mandible has 16 teeth, matching the upper jaw teeth.

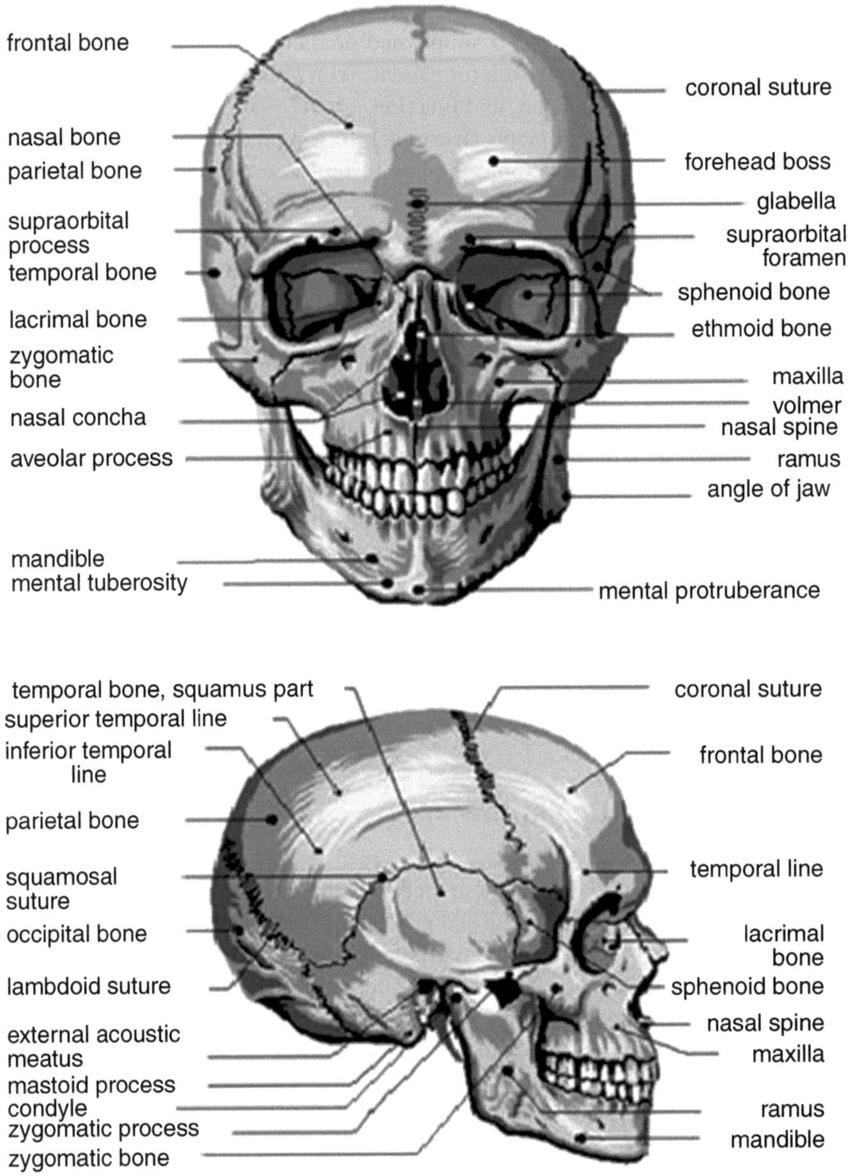

Fig. 5.1 The frontal and lateral views of human head bones, jawbones, and teeth

5.3 Dental Implants

A dental implant is an artificial tooth root that a surgeon places into a jawbone to hold a replacement tooth or bridge. Dental implants are an ideal option for people in good general oral health who have lost a tooth or teeth due to caries, periodontal disease, an injury, or some other reason.

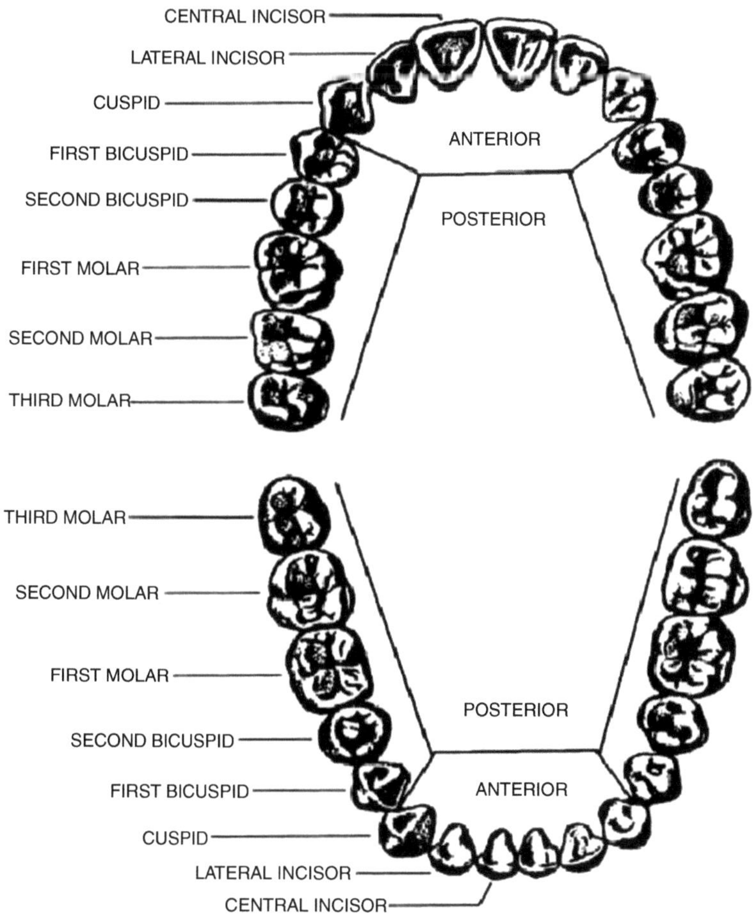

Fig. 5.2 Nomenclature of the teeth in both maxilla (*upper jaw*) and mandible (*lower*), as viewed from inside (*occlusal view*)

While high-tech in nature, dental implants are actually more tooth-saving than traditional bridgework, since implants do not rely on neighboring teeth for support.

Dental implants look and feel so natural that people may forget they ever lost a tooth. If people are missing one or more teeth and would like to smile, speak, and eat again with comfort and confidence, there is good news. Dental implants are teeth that can look and feel just like natural teeth. Under proper conditions, such as placement by a periodontist and diligent patient maintenance, implants can last a lifetime. Long-term studies continue to show improving success rates for implants.

Dental implants are artificial tooth roots. Holes are made into the jawbone and threaded sometimes; the artificial roots are placed or screwed and covered over. The implants then heal and integrate with the jawbone. After healing, an artificial

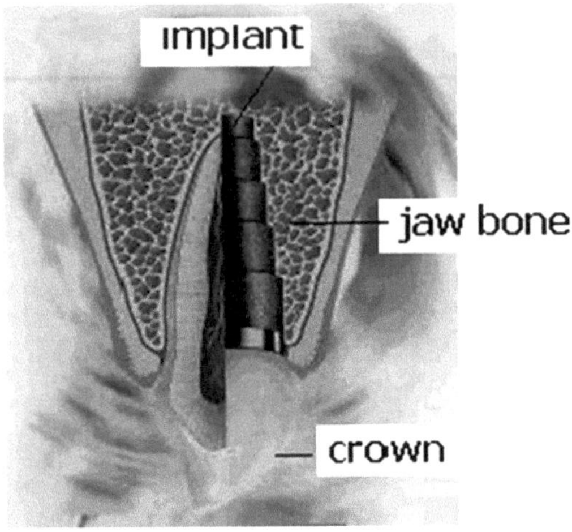

Fig. 5.3 Composite longitudinal sectional view of an ideal screw form dental implant placed in the upper jawbone. Note the spongy nature of the jawbone

top (crown) is placed into the implant, and crowns, bridges, and attachments for dentures can be placed.

Simply stated, a dental implant establishes a foundation for a tooth or a number of teeth to be affixed to a spot where previously there was none.

The term "root form implant" refers to a titanium cylinder that is surgically placed into bone. Root form dental implants have become widely used and accepted for the past 14–15 years in India. There are several types of implant systems, but for our purposes and ease of understanding, this section will deal with the root form implant.

When an individual has lost all of his teeth, usually a denture can be made. Over the years, the bone that supports the denture may be decayed to such a degree that it can no longer support the denture. Most long-term denture wearers will ultimately use adhesives to stabilize their denture. However, they will still lose 40–80% of their chewing ability.

Dental implants provide a stable foundation for the fully edentulous patient to regain nearly 90% of her chewing ability, never having to experience a limited diet or the embarrassment of an ill-fitting denture.

The partially edentulous patient who seeks to be restored through conventional crowns and bridges may not desire to have any of her natural healthy teeth utilized to fill the space of a missing tooth. A dental implant provides a foundation for a single tooth or multiple teeth to be placed.

Experience and competency are critical when working with a dentist or oral surgeon. Implants are more expensive to receive than other traditional forms of conventional dentistry, but the rewards may far outweigh the financial costs. It is essential that implants be given proper care or they can be lost by the same type of bone loss as natural teeth (Fig. 5.3).

5.4 Components of Dental Implant

Many different implant designs exist, and the implants produced by each manufacturer have their own unique characteristics and features. However, the same basic components are shared by different implant types. In addition to the implant body, implants include abutments and various types of screws.

(a) **Abutments**. The abutment is a component that is intermediate between the implant and the restoration. It can usually be separated from the rest of the implant, but in some cases they form an inseparable part of the implant itself. Not all implant types require abutment. In some cases, the crown is screwed directly into the implant. Some abutments are prefabricated by implant manufacturers, while others are custom-made in the laboratory. Abutments may also be angled to counter the inclinations of implants. Special "healing abutments" are sometimes used soon after surgery. They are temporary and will be removed before the restoration is placed.

(b) **Screws**. Screws are used to join two parts together. Crowns are often maintained on abutments or implants with screws. The term "gold screw" is sometimes utilized to refer to the crown-retaining screw. Screws can also be employed to connect abutments to implants. The "cover screw" is a screw that blocks the implant entrance during the healing period after surgery. When first inserted, screw surfaces are somewhat rough to help osteocytes to grow onto themselves.

5.5 Types of Implants in Use

1. The types of implants are classified according to the position of the implant to the tissue:

   (a) Subperiosteal implants
   (b) Supraperiosteal implants
   (c) Transosteal implants
   (d) Endosteal implants: (1) blade implants; (2) cylindrical or root form implants.

5.5.1 Subperiosteal Implants

This type of implant is also known as Ramus Frame implant and can be used if the lower jawbone is too thin for a root form implant. A Ramus Frame implant is embedded in the jawbone in the back corners of the mouth (near the wisdom teeth) and near the chin. Once it is inserted and the tissue heals, a thin metal bar is visible around the top of the gum. Dentures are made that can fit onto this bar. Ramus Frame implants also can stabilize weak jaws and help to prevent them from fracturing.

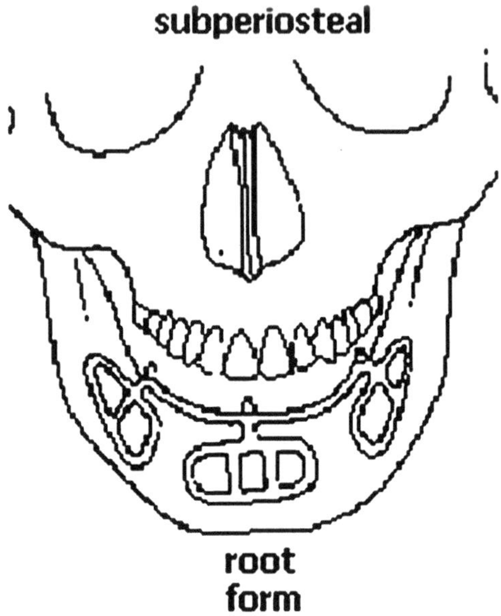

Fig. 5.4 Subperiosteal implant in situ

Subperiosteal implants are placed on the bone. These are placed on top of the jaw with the metal framework's posts protruding through the gum to hold the prosthesis [1]. These types of implants are used for patients who are unable to wear conventional dentures and who have minimal bone height. The implant is placed under the mucoperiosteum over the edentulous ridge to support the prosthesis. But this is not widely accepted because of the cumbersome procedure and very short-term success rate. Failure usually happens because of infection, and bone resorption appears, which causes implant mobility (Fig. 5.4).

5.5.2 *Transosteal Implants*

Transosseous implants originally were designed to be used in people who had very little bone in their lower jaws and who had no teeth in the mandible. However, they are rarely used today because placing them requires extensive surgery, general anesthesia, and hospitalization. Also, their use is limited to the lower jaw. Placing transosseous implants involves inserting two metal rods from below the chin, through the chin bone, until they are exposed inside the mouth.

In 1975, Small introduced the transosteal mandibular staple bone plate, a reconstructive device placed through a submental incision and attached to the mandible with multiple fixation and two transosteal screws to support a full arch prosthesis [2, 3]. Other transosteal implants of historical and practical importance are Cranin

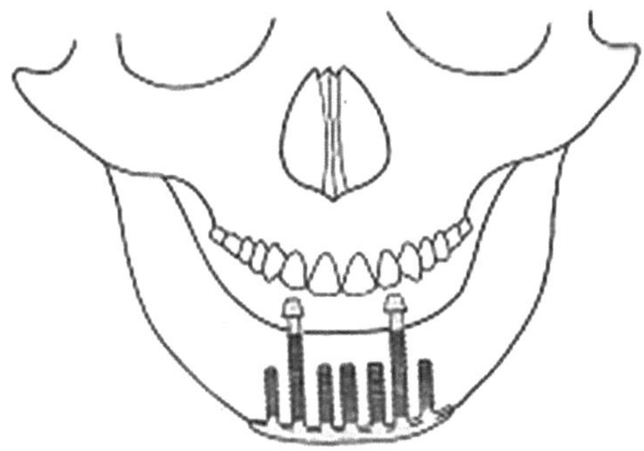

Fig. 5.5 Transosteal implant

and Denninson's single transosteal implant [4] and Bosker and Van Dijk's transmandibular implant [5].The success rate up to 5 years is 94%, and that up to 10–15 years is 87%. Different designs are available. Initial stability is excellent and very useful in atrophic mandible. Disadvantages include that transosteal implants are only possible in anterior mandible, and because of the complex nature of the surgical approach, it is not a very popular procedure.

Potential problems with tooth- and implant-supported fixed partial denture include

1. Breakdown of osseointegration
2. Cement failure on natural abutments
3. Screw or abutment loosening
4. Failure of implant prosthetic component (Fig. 5.5).

5.5.3 Endosteal Implants

Endosteal implants are the most commonly used type of implant. The various types include screws, cylinders, or blades surgically placed into the jawbone. Each implant holds one or more prosthetic teeth. This type of implant is generally used as an alternative for patients with bridges or removable dentures.

Implants are placed within the alveolar and basal bone surgically. These implants can be of root form and blade form implants. Root form implants are cylindrical in shape and may have an external head. **Nominal diameters are 3–6 mm, in intervals of 0.5 mm, and a threaded length of 8–20 mm. Blades are wedge-shaped or rectangular in cross section—2.5 mm wide, 8–15 mm deep, and 15–30 mm long.**

Endosteal implants are subdivided into two categories: single-stage surgery and two-stage surgery. In single-stage surgery, the implant head is projected out of the

Fig. 5.6 Blade form implant

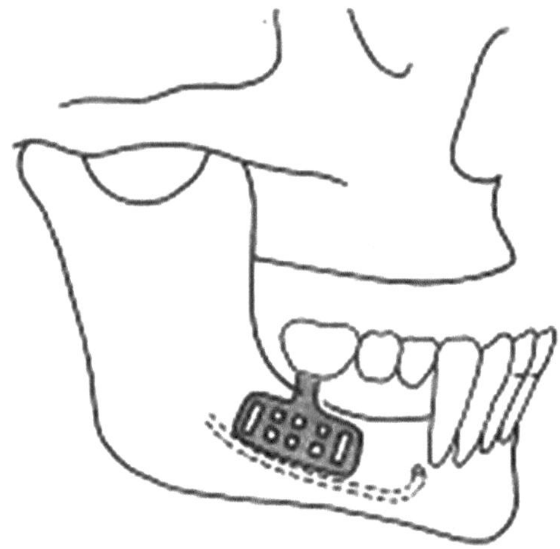

mucosa into the oral cavity, and in two-stage surgery, the implant is placed at the cortical base level and underneath the oral mucosal layer. After six to nine months, the second surgery is done to expose the implant screw, the abutment part is placed, and it is projected through the oral mucosa.

(a) **Blade form implants.** These involve one-stage surgery. Initially, these were very popular and success rates were quite satisfactory; 5-year follow-up might be as high as 80%, but surgical placement is a difficult proposition and failure causes a wide area of defect in the jawbone (Fig. 5.6).
(b) **Cylinders**. A two-stage surgical procedure with cylindrical root form dental implants is the state of the art in implant dentistry. According to the U.S. National Institutes of Health's 1988 consensus conference, 78% of the implant market was cornered by cylindrical implants.

Brånemark was the first to develop a surgical protocol for both threaded and nonthreaded implant systems. Two-stage cylindrical implants enjoy the most widespread popularity (Fig. 5.7).

In single-stage surgery, part of the implant is visible permucosally and a second surgery is not required; usually, a false crown is placed. But in two-stage surgery, the entire implant will be placed subperiosteally, and the second surgery is essential to attach the abutment part for holding the crown in position (Fig. 5.8).

5.5.4 *Biocompatibility and Implant Design*

Implants made of commercially pure titanium have established a benchmark in osseointegration against which few other materials compare. Related materials such as niobium are able to produce a high degree of osseointegration, and, in addition,

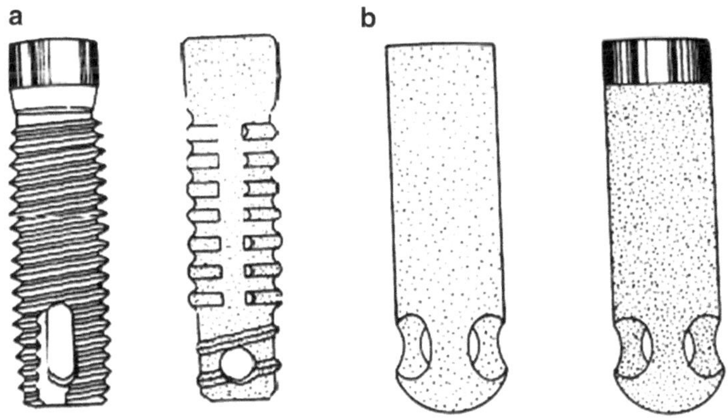

Fig. 5.7 Different types of cylindrical implant: (**a**) single-stage surgery and (**b**) two-stage surgery implants

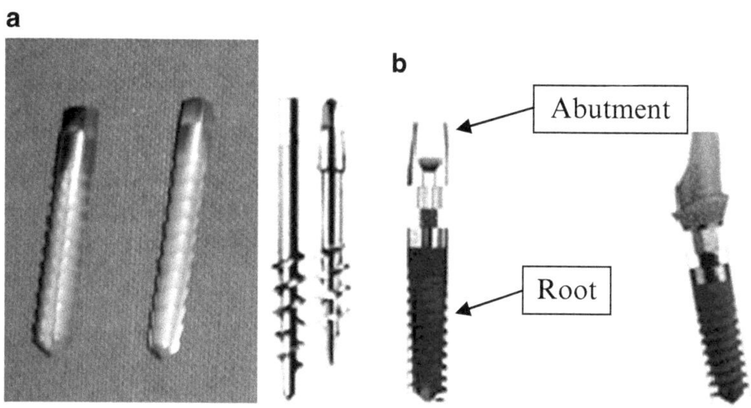

Fig. 5.8 (**a**) Our design with Titan-12; (**b**) implant for single-stage and double-stage surgery

successful clinical results have been reported for some titanium alloys and hydroxyapatite-coated implants. More recently, resorbable coatings have been developed that aim to improve the initial rate of bone healing against the implant surface and then resorb within a short timeframe to allow the establishment of a bone to a metal contact.

The implant design has a great influence on the initial stability and subsequent function. The main design parameters are as follows:

- **Implant length**—implants are generally available in lengths from about 6 mm to as much as 20 mm. The most common lengths employed are 8–15 mm, which correspond quite closely to normal root lengths. However, it has a demographic variation. Westerners have about 20% longer teeth compared to Asians.

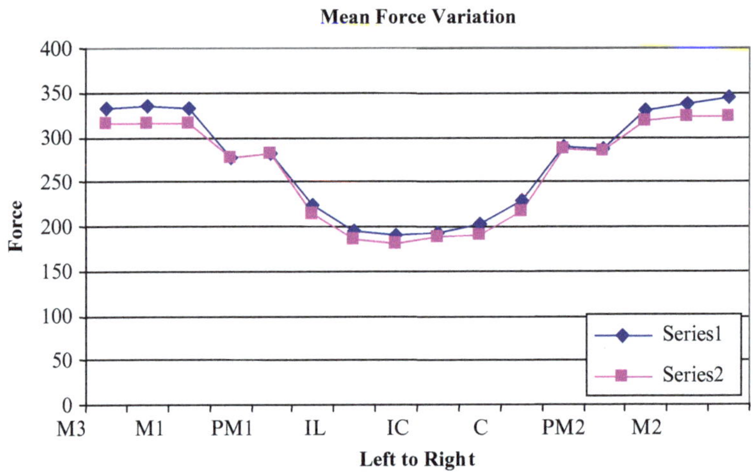

Fig. 5.9 The mean force variation in Newtons during bite in normal teeth (series 1) and the implanted teeth (our design using Titan 12 from Mishra Dhatu Nigam, India)

- **Implant diameter**—most implants are approximately 4 mm in diameter. At least 3.25 mm in diameter is required to ensure adequate implant strength. The implant diameter may be more important than the implant length in the distribution of loads to the surrounding bone. Implant diameters up to 6 mm are available, which are considerably stronger, but they are not widely used, because sufficient bone width is not available among patients. The most common size used in India is 4–4.5 mm. However, there is a wide variation due to the fusion of a wide range of ethnicities within India.

5.6 Biting Force

These forces were experimentally determined using a specially designed load cell by the author suitable for measuring biting forces. The data were obtained based on young students (20–22 years old), but the implanted teeth's biting forces were based on patients in the age range of 50–65 years. These data will give a basis for design and analysis of dental implants.

The biting force in human teeth varies over the entire range of 16 pairs of teeth and is nearly symmetrical over the midsaggital plane. The range of forces vary from 183–350 N, as determined by a group of the author's postgraduate students [6] (Fig. 5.9).

5.7 Implant Shape

Hollow cylinders, solid cylinders, hollow screws, or solid screws are common shapes of dental implants. The shapes are designed to maximize the potential area for osseointegration and provide good initial stability. Even minor alterations in the size, shape, and pitch of threads can enhance the latter property. Screw-shaped implants also offer good load distribution characteristics in function.

5.8 Surface Characteristics

The degree of surface roughness varies greatly among different systems. Surfaces that are machined, grit-blasted, etched, plasma-sprayed, and coated are available. The optimum surface morphology has yet to be defined, and some may perform better in certain circumstances. Increasing the surface roughness gives the potential to increase the surface contact with bone, but this may be at the expense of more ionic exchange and surface corrosion. Bacterial contamination of the implant surface will also be affected by the surface roughness if it becomes exposed within the mouth (Figs. 5.10 and 5.11).

The polished transmucosal neck is clearly demarcated from the plasma-sprayed body.

Figure 5.12 shows the plasma-sprayed surface at the same magnification as in Fig. 5.11a, b. The increase in surface area is considerable.

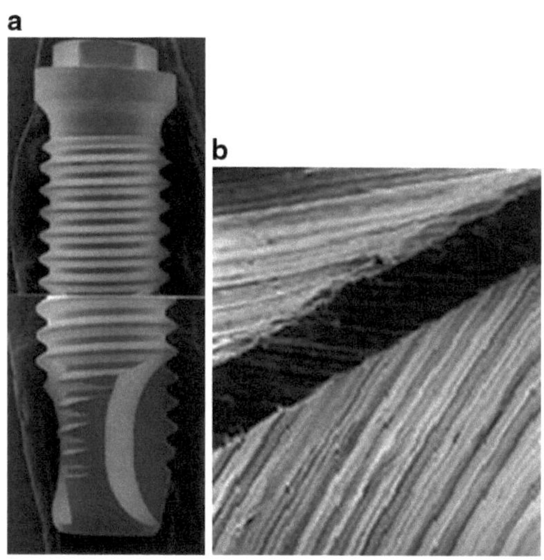

Fig. 5.10 (**a**) A scanning electron micrograph of a Brånemark/Nobel Biocare implant. It shows the basic thread design; (**b**) a scanning electron micrograph of a Brånemark/Nobel Biocare implant (a higher magnification view of the machined surface)

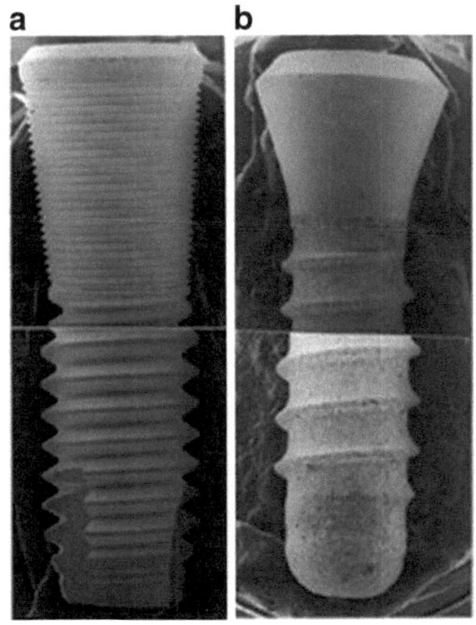

Fig. 5.11 (**a**) A scanning electron micrograph of an Astra ST implant. The conical neck has a micro thread and the apical part a coarser self tapping thread; (**b**) a scanning electron micrograph of an ITI Straumann solid screw implant. Adapted from the elaborate and profound work of Professor Brainemark of Implant Dentistry and Periodontology, Guy's Kings and St. Thomas' Medical and Dental School, London SE1 9RT [19]

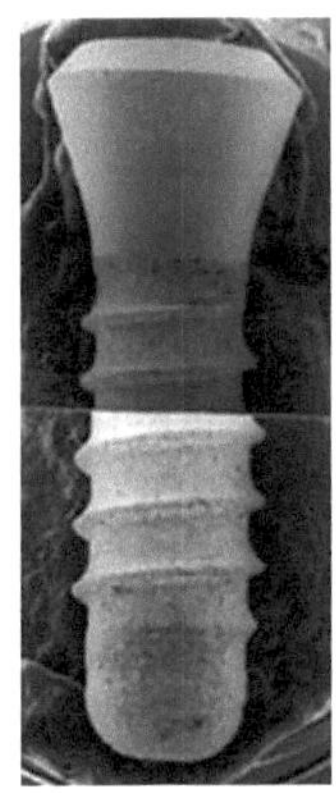

Fig. 5.12 A scanning electron micrograph of an ITI Straumann solid screw implant

5.9 Bone Factors

The stability of the implant at the time of placement is very important and is dependent upon bone quantity and quality as well as implant design. The edentulous ridge can be classified in terms of shape and bone quality. Following loss of a tooth, the alveolar bone resorbs in width and height. In extreme cases, bone resorption proceeds to a level that is beyond the normal extent of the alveolar process and well within the basal bone of the jaws.

The most favorable quality of jawbone for implant treatment is that it has a well-formed cortex and densely trabeculated medullary spaces with a good blood supply. Bone that is predominantly cortical may offer good initial stability at implant placement but is more easily damaged by overheating during the drilling process, especially with sites more than 10 mm in depth. At the other extreme, bone with a thin or absent cortical layer and sparse trabeculation offers very poor initial implant stability and fewer cells with a good osteogenic potential to promote osseointegration.

Success is highly dependent upon a surgical technique that avoids heating the bone. Slow drilling speeds, the use of successive incrementally larger sharp drills, and copious saline irrigation aim to keep the temperature below that at which bone tissue damage occurs (around 47 °C for 1 min). Further refinements include cooling the irrigant and using internally irrigated drills. Factors that compromise bone quality are infection, irradiation, and heavy smoking. The effects of the latter two are a result of a diminution of the vascular supply to the bone, which reduces the healing response, a feature that has been well described in the healing of fractures.

5.10 Loading Conditions

Following installation of an implant, it is important that it is not loaded during the early healing phase. Movement of the implant within the bone at this stage results in fibrous tissue encapsulation rather than osseointegration. This has been compared to the healing of a fracture, where stabilization of the bone fragments is very important to prevent non-union. In partially dentate subjects, it is desirable to provide temporary prostheses that are tooth-supported to avoid early implant loading. However, in patients who wear mucosally supported dentures, it is generally recommended that they should not be worn over the implant area for 1–2 weeks. This also helps to prevent breakdown of the soft tissue wound. Systems such as Brånemark implants have advised leaving implants unloaded beneath the mucosa for around 6 months in the maxilla and 3 months in the mandible, mainly because of differences in bone quality.

However, these are largely empirical guidelines, and bone quality and implant stability will vary greatly among individuals, jaws, and sites within jaws. Currently, there is no accurate measure that precisely determines the optimum period of healing before loading can commence. Bone quality can be assessed by measuring the cutting torque during preparation of the implant site. The stability of an implant and increasing bone-to-implant contact has been quantified using resonance frequency analysis. This newly developed noninvasive research tool measures the stiffness of the implant at the bone interface. In some circumstances, it has been shown that immediate loading is compatible with subsequent successful osseointegration, provided the bone quality is good and the functional forces can be adequately controlled.

The latter may involve placing an adequate number of implants and connecting them together as soon as possible with a rigid framework. However, these latter protocols should be considered experimental at the present time, and there is much data to support the more cautious approach advocated by Brånemark in ensuring a high level of predictable implant success. Some systems employ a single-stage approach in which the implant is installed so that it protrudes through the overlying mucosa (i.e., nonsubmerged), although avoidance of early loading is equally critical. Following the recommended healing period (around 3 months), abutments are connected to the implant to allow construction of the prosthesis. This protocol therefore avoids further surgery to uncover the implants. The loading of the implant-supported prosthesis is a further important consideration, which we will deal with in the following section.

5.10.1 Prosthetic Considerations

Carefully planned functional occlusal loading will result in maintenance of osseointegration and possibly increased bone-to-implant contact. In contrast, excessive loading may lead to bone loss and/or component failure. Clinical loading conditions are largely dependent upon the type of prosthetic reconstruction and the occlusal scheme.

5.10.2 The Type of Prosthetic Reconstruction

This can vary from a single tooth replacement in the partially dentate case to a full arch reconstruction in the edentulous individual. Implants that support overdentures may present particular problems with control of loading, as they may be largely mucosal-supported, entirely implant-supported, or a combination of the two.

5.10.3 The Occlusal Scheme

The lack of mobility in implant-supported fixed prostheses requires provision of shallow cuspal inclines and careful distribution of loads in lateral excursions. With single-tooth implant restorations, it is important to develop initial tooth contacts on the natural dentition and to avoid guidance in lateral excursions on the implant restoration. Loading will also depend upon the opposing dentition, which could be natural teeth, another implant-supported prosthesis, or a conventional removable prosthesis. Surprisingly high forces can be generated through removable prostheses.

5.11 The Number, Distribution, Orientation, and Design of Implants

The distribution of load to the supporting bone can be spread by increasing the number and dimensions (diameter, surface topography, length) of the implants. The spacing and three-dimensional arrangement of the individual implants will also be very important. The so-called tripod arrangement of three implants is recommended in situations of high load, such as replacement of molar teeth in the partially dentate individual.

Properties of Implant Connectors

Multiple implants are joined by a cast or milled framework. A rigid connector provides good splinting and distribution of loads between implants. It is equally important that the connector has a passive fit on the implant abutments so that loads are not set up within the prosthetic construction.

5.12 Placement of Implants into Extraction Sites

Generally, endosseous implants are placed immediately after tooth extraction. This process avoids potential narrowing of the alveolus from labial plate resorption that naturally occurs after tooth extraction and less time between extraction and final restoration. Successful immediate placement of an implant at the time of tooth extraction relies on excellent soft and hard tissue quality. The indications for the placement of implants immediately into the extraction sites are

1. Traumatic loss of teeth with a small amount of bone loss
2. Teeth lost because of gross decay without the presence of purulent exudates or cellulites
3. Inability to complete endodontic procedures
4. Presence of severe periodontal bone loss without purulent exudates
5. Adequate soft tissue health and quantity to obtain a complete crestal primary wound closure
6. Bone availability apical to the extraction site for stabilization of the implant
7. Appropriate location of extracted tooth for planned restoration. Although this process is widely used, there are several contraindications for immediate placement of implants into fresh extraction sites, including
   (a) Presence of purulent exudates at the time of extraction
   (b) Adjacent soft tissue cellulites and granulation tissue
   (c) Lack of adequate bone apical to the extraction site
   (d) Adverse location of the mandibular neurovascular bundle, maxillary sinus, or nasal cavity
   (e) Anatomic configuration of remaining bone or potential location of the implant preventing ideal prosthetics
   (f) Any clinical condition that prevents primary soft tissue wound closure.

If any of the aforementioned contraindications is present, delayed placement of the implant is indicated. The teeth are extracted, the socket is cleaned of soft tissue debris, and the site is allowed to heal for at least 6 weeks before placement of dental implants [7–9]. By delaying the implant placement, the surgeon avoids the potential infections and places the implant into healthier tissue [8].

If there is a large defect in the bone or if the tooth sockets are significantly larger than the diameter of the implant, one may delay implant placement until the extraction socket has filled in with bone. It may be treated with bone graft as well.

5.13 Clinical and Biomechanical Considerations for Fixed Tooth Replacement

Fixed or permanent tooth replacement is an advanced medical technology for the treatment of tooth loss. Before implantation, various clinical and biomechanical factors must be considered, as described next.

5.13.1 Bone Tissue Factors

The type of bone available determines the quality of support available. Maxillary and mandibular bone density and type have been classified by various authors [10] and include the following classification:

1. The entire jaw is composed of homogeneous compact bone.
2. A thick layer of compact bone surrounds a core of dense trabecular bone.
3. A thin layer of cortical bone surrounds a core of dense trabecular bone of favorable strength.
4. A thin layer of cortical bone surrounds a core of low-density trabecular bone.

Based on this classification, the most dense and favorable bone is generally found in the symphysis of the mandible. The posterior mandible and anterior and posterior maxilla generally have less favorable bone. The bone in these areas is generally composed of various combinations of cortical bone [11]. Therefore, alternative treatment plans should be considered that may change the site and number of implants used. An example of this is finding less desirable bone at the time of implant placement and increasing the number of implants to reduce the load on the individual implants [12].

5.13.2 Soft Tissue Factors

The quality and quantity of soft tissue at the site of implant placement must be evaluated. The soft tissue must be capable of withstanding the techniques used in completing the prosthesis and the rigors of maintaining the oral hygiene around the implants. Keratinized, fixed gingival tissues are the most desirable [7]. It is desirable to have a minimum of 1 mm of keratinized tissue around natural dentition.

Around implants, however, there should be a minimum of 2 mm of keratinized tissue [13]. Long-standing edentulous spaces may not have adequate quality or quantity of soft tissue. In these cases, it may be necessary to graft more desirable soft tissue to the area. This is particularly true in the posterior mandible, where movable mucosal tissue is often present.

The quantity of tissue at the implant site also deserves close attention. If implants are placed in thick soft tissue, excessive pockets may result around the abutments. This leads to the patient's having increased difficulty in complying with hygiene measures. The tissues should be thinned at the time of implant exposure and the placement of the temporary healing cuffs.

5.13.3 Number of Implants

The number of implants required for a fixed restoration depends on the length of the span, the location of the span in the arch, and whether the implants are to be placed in the maxilla or mandible [14]. In general, the following rules can be used to determine the number of implants required. Two implants can support a three-unit freestanding restoration in the mandible. One implant attached to a sound natural tooth can support a two-unit restoration in the mandible [12]. Multiples of these different types of support can be used for larger restorations. In the maxilla, every effort should be made to maximize the number of implants because of variances of the density of bone at different sites. A minimum of six implants should be used in the edentulous maxilla to support a totally implant-borne restoration.

5.13.4 Crown-to-Root Ratio

Guidelines have been developed to determine the load that can be placed on the natural dentition when supporting fixed restorations. Attempts have been made to correlate the surface area in contact with bone in implants and natural teeth. These have been imprecise due to the various morphologic designs used in the implant systems. It is also difficult to evaluate the differences in the manner in which bone "attaches" to the various surfaces used on implants.

Shorter implants have a higher rate of loss than longer ones of the same type [15, 16]. The diameter of the implant seems to play a lesser role than the length of the implant in assessing the implant's ability to sustain the added load of a fixed restoration. Therefore, the use of the longest implant possible should be the goal in selecting an implant. Restoration supported by multiple numbers of implants is far better than those supported by fewer implants.

5.13.5 *Implant Orientation*

The orientation of the implants in the edentulous space plays an important role in controlling the forces placed on the implant–bone interface. Optimum orientation provides for the forces on the implant to be directed along its axis. Any deviation from this optimum orientation results in a variety of unfavorable sequelae. To minimize the problems associated with implant orientation, it is recommended that the occlusal table be narrowed. This eliminates or minimizes the length of the lever arm of lateral force. Implants should not be placed in a straight line. Implants should be placed to maximize a tripodal support for the restoration [17]. Excessively long implant crowns should be avoided, and splinting of these should be considered if a long crown length cannot be avoided. The number of implants should be maximized to reduce the per-unit load of the individual implants. To facilitate drilling and avoid stress concentration, a minimum distance greater than $2d$ should be maintained, where d is the diameter of the screw.

5.14 Advantages of Permanent Tooth Replacement

Dental implants are a better solution to the problem of missing teeth. The ideal candidate for a dental implant is in good general and oral health. Adequate bone in the jaw is needed to support the implant, and the best candidates have healthy gum tissues that are free of periodontal disease.

The advantages of dental implants are they are aesthetically pleasing, they save teeth, they increase the patient's self-confidence, and they are reliable. The success rate of dental implants is highly predictable. They are considered an excellent option for tooth replacement.

5.15 Manufacture of Implants and Property Enhancement

The main manufacturing technique involves machining, using an automated screw-cutting lathe or thread rolling. The hexagonal or square slot in the head is produced by drilling and punching with a hexagonal punch. It was observed that the roughened surface of a screw root gives better osseointegration with bone. Therefore, many authors used sandblasting using alumina or hydroxyapatite powder of 20–80-μm size to enhance the surface area. Acid and alkali etching also give excellent results [18]. The pore size should be limited to 50–100 μm and roughness to 40–60-μm CLA value. The alumina powder should not lodge in the metal surface; ultrasonic cleaning helps to have a clean surface.

Microabrasive blasting can be used to roughen the smooth surface finish on the threaded portion of the implant. Roughening this area of the implant increases the surface area for the tissue to integrate, improving the bond. The ability to promote this tissue ingrowth is a critical component to the implant's success. We have used

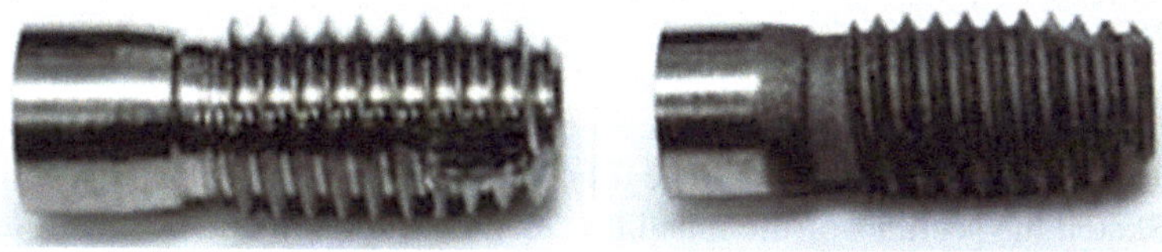

Fig. 5.13 Two threaded screws, one untreated and another treated with microblast, are shown

a different screw type, an open parabolic type, which helps to enter the jawbone with less cutting torque and force (Fig. 5.13).

References

1. Lemons J, Natiella J (1986) Biomaterials, biocompatibility and peri-implant considerations. Dent Clin N Am 30:4
2. Small IA (1975) Metal implants and the mandibular staple bone plate. J Oral Surg 33:571
3. Small IA (1978) Metal implants and the mandibular staple bone plate. J Oral Surg 36:604
4. Cranin AN, Denninson T (1970) The anterior vertical transosseous implant. In: Cranin AH (ed) Oral implantology. Charles C. Thomas, Springfield, IL, pp 200–208
5. Bosker H, Van Dijk L (1989) The transmandibular implant: a 12-year follow-up study. J Oral Maxillofac Surg 47:442–450
6. Biswas BK (2011) A study of the clinical efficacy of surface characteristics of root form 2-stage titanium dental implant. PhD thesis, Jadavpur University
7. Block MS, Kent JN (1990) Factors associated with soft and hard tissue compromise endosseous implants. J Oral Maxilliofac Surg 48:1153–1160
8. Yukna RA (1991) Clinical comparison of hydroxyapatite-coated titanium dental implants in fresh extraction sockets and healed sites. J Periodontol 62:468–472
9. Yukna RA (1992) Placement of hydroxyapatite-coated implants into fresh or recent extraction sites. Dent Clin N Am 36:97–116
10. Lekholni V, Zarb GA (1985) Patient selection and preparation. In: Brånemark P-I, Zarb G, Albrektsson TA (eds) Tissue integrated prostheses. Quintessence, Chicago, pp 199–209
11. Misch CE (1993) Contemporary implant dentistry. CV Mosby, St. Louis
12. Rangert B, Jemt T, Jorneus L (1989) Forces and moments on Brånemark implants. Int J Oral Maxillofac Implants 4:241–247
13. Artzi Z, Tal H, Moses O, Kozlovsky A (1993) Mucosal considerations for osseointegrated implants. J Prosthet Dent 70:427–432
14. Jensen O (1989) The classification of the osseointegrated implant. J Prosthet Dent 61:228–234
15. Bahat O (1993) Treatment planning and placement of implants in the posterior maxillae: Report of 732 consecutive Nobelpharma implants. Int J Oral Maxilliofac Implants 8:151–161
16. Saadoun AP, LeGall ML (1992) Clinical results and guidelines on Steri-oss endosseous implants. Int J Priodontics Restorative Dent 12:486–495
17. Chiche GJ, Pinault A, Weaver C, Finger I (1989) Adapting fixed prosthodontics principles to screw-retained restorations. Int J Prosthodont 2:317–322
18. Biswas BK (2010) Studies on screwroot dental implant with various surface treatment. PhD thesis, Jadavpur University, 2010
19. Professor of Implant Dentistry and Periodontology, Guy's Kings and St. Thomas Medical and Dental School, London SE1 9RT

Problems

1. Design a device to measure the maximum biting force for human subjects for all pairs of teeth. How will the force vary with age or gender and disease process? Think about whether it could be used as a device to check dental health.
2. Sketch the sections of the mandible and maxilla of an adult Bengali male with approximate sizes. You may consult a dental surgeon and study X-rays. Will this help you to find the diameters and length of the screw-root–type implant? If yes, try to find the approximate sizes. What type of thread profile will be suitable for a dental screw-root–type implant?
3. How does surface texture (roughness) or various bioactive coatings affect the screw root to anchor with bone? What might be the life expectancy of such constructs in human patients?
4. How do you plan to manufacture a screw-root–type dental implant and enhance its surface to improve osseointegration?

Glossary of Dental Terminology

This list is provided to help readers quickly identify the meaning of the terms used in this chapter and also elsewhere.

Abutment The teeth on either side of a missing tooth. Abutments are the part of the bridge used to support the replacement of the missing teeth (pontics)

Amalgam Silver filling. Amalgams are usually placed on the back teeth (posterior teeth)

Anesthesia Relieves the sensation of pain

Anterior teeth The front teeth (incisors and cupids)

Arch The upper or lower jaw

Back teeth See "posterior teeth"

Bicuspids The first and second bicuspids are the fourth and fifth teeth from the center of the mouth to the back of the mouth. These are the back teeth that are used for chewing; they only have two points (cusps)

Bitewing X-ray that shows the upper and lower teeth's biting surfaces on the same film. This X-ray shows the portion of the teeth above the gumline

Bridge A fixed appliance (prosthesis) that replaces missing teeth. A bridge is a series of crowns (abutments and pontics)

Bruxism Clenching or grinding of the teeth

Calculus The sticky film on teeth (plaque) that has hardened. Also known as tartar

Canal	The narrow chamber inside the root of the tooth that contains the nerve and blood vessels
Caries	Correct technical term for decay
Cleaning	See "prophylaxis"
Complete series	See "full mouth X-rays"
Composite filling	Tooth-colored filling. Insurance companies usually only allow them on the front teeth (anterior teeth). When composites are done on the back teeth (posterior teeth), the insurance company usually pays for them as an amalgam. Composites are also known as resin fillings
Crown	Full coverage for a tooth
Curettage	Surgical scraping of bacteria from the soft tissue. This is a periodontal procedure and is usually performed one quadrant at a time
Cusps	The high points on the chewing surfaces of the back teeth (posterior teeth)
Cuspids	The third tooth from the center of the mouth to the back of the mouth. These are the front teeth that have one rounded or pointed edge used for biting. Also known as canines
Deciduous teeth	See "primary teeth"
Denture	A removable appliance (prosthesis) that replaces all of the teeth in either the upper or lower jaw
Diagnostic	Procedures performed by the dentist to identify what's going on in the mouth. The most common procedures are the exam and X-ray, which Trojan classifies as preventive
Edentulous	All the teeth are missing in either the upper or lower arch
Endo	See "endodontics"
Endodontics (Endo)	The treatment of diseases or injuries that affect the root tip or nerve of the tooth. The most common procedure is a root canal
Fluoride	Topical application of a gel or liquid that prevents decay
F.M.X.	Full mouth X-rays
Front teeth	See "anterior teeth"
Full mouth X-rays (F.M.X.)	X-rays showing all the teeth. Includes 14 periapicals and 2 or 4 bitewings. Also known as a complete series
General anesthesia	Relieves the sensation of pain in the whole body. General anesthesia renders a patient unconscious
Gingiva	The gums

Impaction	An unerupted or partially erupted tooth that will not fully erupt because it is obstructed by another tooth, bone, or soft tissue
Implant	A post that is implanted in the bone. A crown, bridge, or denture is then placed over the implant
Incisors	The central and lateral incisors are the first and second teeth from the center of the mouth to the back of the mouth. These are the front teeth with the flat edges for biting
Inlay	A laboratory-processed restoration made of metal, acrylic, or porcelain. This filling does not involve the high points of the tooth (cusps)
Local anesthesia	Relieves the sensation of pain in a localized area
Mandible	The lower jaw
Maxilla	The upper jaw
Molars	The first, second, and third molars are the sixth, seventh, and eighth teeth from the center of the mouth to the back of the mouth. The back teeth with the large chewing surface on top. They have four points (cusps)
Nightguard	A removable acrylic appliance to minimize the effects of grinding the teeth (bruxism) or joint problems (TMJ). Usually worn at night to prevent the grinding of teeth or to relieve joint pain. Also known as an occlusal guard
Occlusal	The chewing surfaces of the back teeth
Occlusal guard	See "nightguard"
Onlay	A laboratory-processed restoration made of metal, porcelain, or acrylic that replaces one or more of the highest points of the tooth (cusps)
Oral surgery (O.S.)	Surgery of the mouth
Ortho	See "orthodontics"
Orthodontics (Ortho)	Straightening of the teeth
O.S.	See "oral surgery"
P.A.	Common dental language for a periapical. See "periapical"
Palate	Roof of the mouth
Panorex	An X-ray taken outside the mouth that shows all the teeth on one film
Partial denture	A removable appliance (prosthesis) that replaces some of the teeth in either the upper or lower jaw. See illustration below
Pedo	See "pedodontics"
Pedodontics (Pedo)	The treatment of children's teeth
Periapical	An X-ray that shows the whole tooth (above and below the gumline). Also known as a single film or P.A.
Perio	See "periodontics"

Perio charting	Measures the depth that the gums have detached from the side of the tooth, forming a pocket (perio pocket)
Periodontal maintenance	Cleaning of the teeth following periodontal treatment, includes perio charting. Procedure code 4910. Also known as a perio prophy or perio recall
Periodontics (Perio)	The treatment of diseases of the gum or bone (supporting structure)
Perio pocket	The pocket that forms when the gums detach from the side of the tooth
Perio prophy	See "periodontal maintenance"
Perio recall	See "periodontal maintenance"
Permanent first and second molars	The adult first and second molars, the sixth and seventh teeth from the center of the mouth to the back of the mouth. Does not include the third molar (wisdom tooth)
Permanent molars	The adult first, second, and third molars
Permanent dentition	See "permanent teeth"
Permanent teeth	The adult teeth. Also known as the permanent dentition
Plaque	A sticky film on the teeth. If it is not removed by brushing, it can harden into calculus
Pontic	The part of a bridge that replaces the missing teeth
Posterior teeth	The back teeth (bicuspids and molars)
Preventive	Procedure performed to prevent decay and gum disease. The procedures that Trojan classifies as preventive are the exam, X-rays, and prophylaxis. The exam and X-rays are also known as diagnostic procedures
Primary dentition	See "primary teeth"
Primary teeth	The baby teeth. Also known as the primary dentition. The baby teeth are replaced by the adult teeth (permanent teeth)
Prophy	See "prophylaxis"
Prophylaxis	Cleaning the teeth. Also known as a prophy
Prosthetics	A fixed or removable appliance to replace missing teeth. Example: bridges, dentures, and partials. Sometimes single crowns are considered prosthetics
Prosthodontics	Dealing with the replacement of missing teeth

Quadrant	One of the four equal sections of the mouth. The upper right, upper left, lower right, or lower left
Root canal therapy (R.C.T.)	The nerve of the tooth is removed from the canal inside the root and replaced with a filling material
Root planing	Deep cleaning of the teeth to remove calculus below the gumline. This is not a prophylaxis. This is a periodontal procedure and is usually performed one quadrant at a time
Sealant	Clear application of acrylic placed over the biting surface of the tooth to prevent decay
Single film	See "periapical"
Tartar	See "calculus"
Temporomandibular joint (TMJ)	Temporo (temporal bone), mandibular (lower jaw). This is the connecting hinge between the lower jaw and base of the skull.
Third molar	See "wisdom tooth"
TMJ	Temporomandibular joint
Virgin teeth	Teeth that have no decay or fillings
Wisdom tooth	The third molar, the eighth tooth from the center of mouth to the back of the mouth. Wisdom teeth are often impacted (obstructed from erupting) and have to be extracted

Chapter 6
The Design of a Bone Fracture-Fixation Device

6.1 Introduction

In these days of rapid industrialization and globalization, skeletal fractures due to traffic and railroad accidents, indoor accidents due to highly polished modern floor surfaces, and competitive sports-related injuries are increasing at a rapid rate. There are two principal methodologies for repairing fractured skeletal tissues. One is external fixation, and the other is internal fixation.

6.2 External Fixation

An **external fixator** is a surgical treatment used to set **bone fractures** in which a cast would not allow proper alignment of the fracture. In this kind of reduction, holes are drilled into uninjured areas of bones around the fracture, and special screws or pins are screwed into the holes. Outside the body, a rod or a curved piece of metal with special ball-and-socket joints joins the screws to make a rigid support. The fracture can be set in the proper anatomical configuration by adjusting the ball-and-socket joints. Since the screws pierce the skin, proper cleaning to prevent infection at the site of surgery must be done.

The external fixator is installed in an operating room, normally under **general anesthesia**. Removing the external frame and screws usually requires special wrenches and can be done without anesthesia as an office procedure. External fixation is usually used when internal fixation is contraindicated—often to treat open fractures, or as a temporary solution.

External fixation is also used in limb lengthening. For example, younger people with shorter limbs can have legs lengthened to achieve a desired height. In most cases, the thigh bone (femur) is cut diagonally in a surgical procedure under anesthesia. External fixator pins or wires are placed on each side of the "manmade fracture" and the external metal apparatus is used to very gradually pull the two parts of

S. Pal, *Design of Artificial Human Joints & Organs*,
DOI 10.1007/978-1-4614-6255-2_6, © Springer Science+Business Media New York 2014

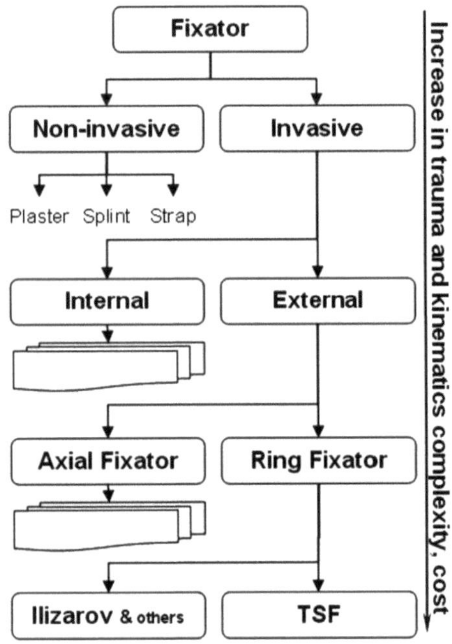

Fig. 6.1 Types of orthopedic fracture-fixation devices

the femur apart millimeter by millimeter, day by day, and week by week. Bone is extremely sensitive tissue and will gradually grow into the small gap created by this "distraction" technique. Such a process can take several months.

In most cases, it may be necessary for the external fixator to be in place for many weeks or even months. Most fractures heal in 6–12 weeks. However, in complicated fractures and when there are problems with the healing of the fracture, this may take a longer time. It is known that bearing weight through a fracture by walking on it, for example, with the added support of the external fixator frame actually helps fractures to heal. Loading the fracture site judiciously is essential for rapid healing. Sometimes a pulsed electromagnetic field of a definite magnitude, repetition rate, and duration is used to heal nonunions and infected fractures.

The bone fracture-healing method, time, quality and geometry of union are mainly dependent on fracture stabilization. An improper stability at the fracture site can lead to a deformity, delayed healing, or bone loss. Selecting the method of stabilization depends on a range of factors, such as the severity and complexity of damage to soft tissues and the bone, anatomical access restrictions, and the conception of the patient.

There is a wide range of commercially available orthopedic fixators on the market today, and each has its advantages and disadvantages. Figure 6.1 presents a flowchart of the types of fixation used to date. As one progresses down the chart, the cost of treatment, the complexity of the kinematics of fixation, and the severity of the injury increase.

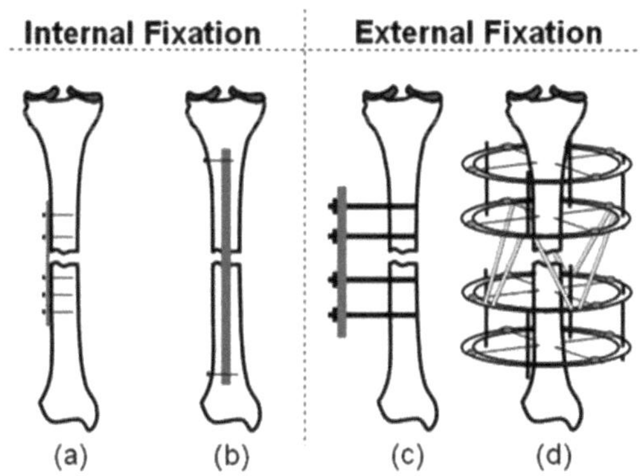

Fig. 6.2 Examples of schema of invasive fracture-fixation methods (internal and external). (**a**) Plate and screws: (**b**) intramedullary nail fixed with screws inside the bone; (**c**) unilateral bar connected to bone via half-pins; (**d**) TSF ring fixator with two accessory rings and connection to bone via eight fine wires

First, all fixation methods can be classified into two groups: invasive and noninvasive. Noninvasive stabilization methods are suitable for treating relatively simple bone fractures. They provide the least control over the mechanical environment of the fracture, because they have no direct contact with the skeletal system. Positional control of the broken bone segments is performed via surrounding tissues and therefore is neither very accurate nor versatile. Noninvasive stabilization was one of the first methods used in medicine. An advanced version of this type of stabilization has used a plaster cast. The plaster cast successfully mimicked the shape of the bones and its surrounding tissues. If a long plaster cast was used, it was possible to avoid unwanted angulations; however, there was very poor control over the length of the bone, axial rotation, and position of the fractured bone segments. The advantage of noninvasive fixators is that they do not cause any direct damage to the anatomy; they are easy to assemble and use. They also avoid infection. However, there is a limit to control of the fracture due to the soft tissue's properties and presence. The common type of healing for fractures stabilized with noninvasive fixation devices is an external bridging callus. The Taylor spatial frame (TSF) is a modern multiplanar external fixator that combines the ease of application and computer accuracy in the reduction of fractures.

Unlike noninvasive fixators, invasive fixators penetrate the soft tissue in order to establish a direct connection with the skeletal system. They can be grouped into internal and external types (Fig. 6.2).

Examples of internal fixators would be intramedullary to the limb anatomy for the healing period and act as a splint that shares the load with the bone. The advantages of internal fixators are excellent control over the position of the bone segments, early stability/rigidity, and early usage of joints and muscles. Stress shielding, commonly experienced due to high plate stiffness, may lead to delayed union and poor bone formation. The highly stiff (value of the modulus of elasticity) plate may also act as a "stress raiser," causing a new fracture at the end of the plate. Primary cortical healing is often the mechanism by which a fracture heals when plates are used.

Hidaka and Gustilo [1] have shown that removing the plates introduces risks of refracture of the healed bone. The present author developed a composite plate material using ultra-high-molecular-weight polyethylene (UHMWPE) powder and alumina ceramics up to 30% by weight and a heat-compression technique to mould the plates (doctoral thesis of S. Roy, [2]). Intramedullary nails (IMN) have an unquestionable place in the management of fractures of the femoral shaft. However, this is only true for the femur due to the anatomy of the blood supply to the shaft. For other bones, the IMN with or without reaming can interfere with the blood supply, which may negatively influence healing. Furthermore, it is not possible to perform bone transport, shortening, lengthening, and postoperative deformity correction unless a special IMN is used at a specialized treatment center. The common healing mechanisms when the IMN is used are external bridging and later medullary callus formation.

The other group of the invasive fixators is external. These minimally invade the human anatomy during treatment periods. The majority of the external fixator's structure (exoskeleton) is located outside the human anatomy. The exoskeleton is connected to the bone segments via fine wires and half/full pins. Axial external fixators (uniaxial/biaxial/monolateral), whose geometry is normally parallel to the axis of the bone, are connected using pins. Advantages of the axial fixator are its simple structure and simple kinematics. However, this type of fixation has a low overall bending stiffness, which is significant since the main axis of the fixator is offset from the load axis of the bone. In addition, it is very complicated (if possible at all) to perform deformity correction in more than one plane using such fixators. The axial fixators are simple in structure and suitable mainly for stabilizing fractured bone segments where low loads are exhibited during treatment. It was observed by Khalily et al. [3] that the stiffness of axial fixators decreases with increasing load. Since this type of fixator allows micromotion, the typical bone healing occurs by the formation of an external bridging callus.

In recent years, the focus has been on ring fixators, as they are highly versatile, allowing postoperative adjustments. Contrary to axial fixators, ring fixators become stiffer with increasing load. The vertical axis of the ring fixator is aligned with the bone load axis, minimizing the unwanted bending effects observed in axial fixators. Subject to frame design, it is possible to change the mechanical properties of the frame during the treatment period. Since the exoskeleton of the fixator is located a few inches (more than 5–6 cm) away from the anatomy, access to the skin and soft tissues is maintained, allowing access to fresh air and postoperative treatment of any damage or infection. Frames can be applied with minimal blood loss or soft tissue damage, due to the small diameters of half-pins and fine wires. In turn, this provides pain relief and early mobility.

The main disadvantages of the ring fixators are size, form, weight, pin tract infections, lack of means of assessing the fracture stability with the fixator in situ, and high cost. A few types of ring fixators have been available commercially to date. An Ilizarov fixator [4, 5] is one of the more popular ring fixators. It was pioneered by Prof. Ilizarov in the 1950s in the former USSR and has been used over the last 60 years in Europe and the U.S. The Taylor spatial frame (TSF) [6] was a later introduction.

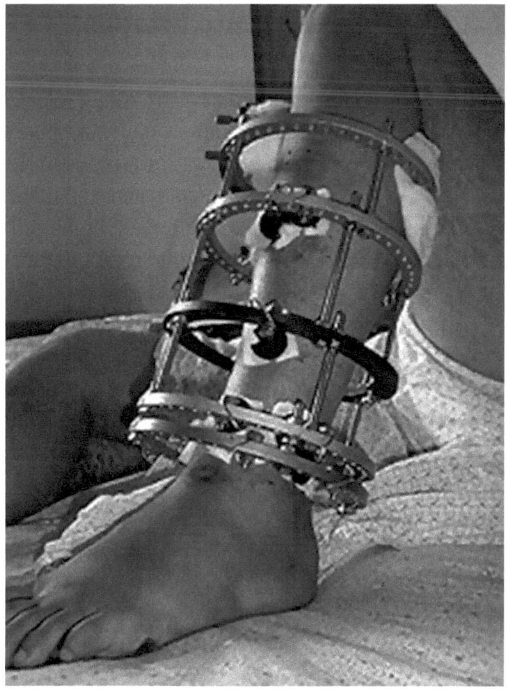

Fig. 6.3 Ilizarov apparatus in place over a broken tibia

6.2.1 The Ilizarov Apparatus

The Ilizarov device is a specialized form of an external fixator, a **circular fixator**, modular in construction. Stainless steel rings are fixed to the bone via stainless heavy-gauge wire (called "pins" or Kirschner wires). The rings are connected to each other with threaded rods attached through adjustable nuts. The circular construction and tensioned wires of the Ilizarov apparatus provide far more structural support than the traditional monolateral fixator system. This allows early weight bearing (Fig. 6.3).

The top rings of the Ilizarov device (fixed to the healthy bone by the tensioned wire) allow force to be transferred through the external frame (the vertical metal rods), bypassing the fracture site. Force is then transferred back to the healthy bone through the bottom ring and the tensioned wires. This allows the Ilizarov apparatus to act as a sort of bridge, both immobilizing the fracture site and relieving it of stress, while allowing for the movement of the entire limb and partial weight bearing. Middle rings (and tensioned wires) hold the bone fragments in place and give greater structural support to the apparatus and limb. However, the critical load-bearing rings are the top and bottom rings, which transfer the force from healthy bone down to healthy bone, bypassing the fracture site.

6.3 Internal Fracture Fixation

In internal fracture fixation, the fixating elements are used in vivo after surgical intervention. The fixators, such as nails and plates, are implanted into the bones at the fracture sites. When members of the American Academy of Orthopedic Surgeons were asked to list the most significant advances in treatment during the twentieth century, the development of internal fixation ranked high on the list. Internal fixation allows shorter hospital stays, enables individuals to return to function earlier, and reduces the incidence of nonunion (improper healing) and malunion (healing in improper position).

A broken bone must be carefully fixed in position and supported until it is strong enough to bear weight. Until the last century, physicians relied on casts and splints to support the bone from outside the body (external fixation). But the development of sterile surgery reduced the risk of infection so that doctors could work directly with the bone and could implant devices in the body. New materials such as stainless steel, cobalt, and titanium were not only strong and durable, but also had the flexibility necessary to support the bone. These materials are also compatible with the body and rarely cause an allergic reaction or implant failure. The only problems are their density and high modulus of elasticity.

The most common types of internal fixation are wires, plates, rods, pins, nails, and screws used inside the body to support the bone directly.

6.3.1 Wires

Surgical wires are used to attach large fragments of bone, like the greater trochanter, which is often detached during total hip replacement. Wires are also used to provide additional stability in long-oblique or spiral fractures of long bones that have already been stabilized by other means. Similar approaches based on the use of wires have also been employed to restore stability in the lower cervical spine region and in the lumbar segment.

Twisting and knotting are unavoidable when fastening wires to bone; however, they reduce the strength of the wire by 25% or more due to stress concentration. This can be partially overcome by using a higher-diameter wire, since strength increases directly proportionally to diameter squared. The deformed regions of the wire are more prone to corrosion than the undeformed regions are because of the higher strain energy absorbsion. To reduce this problem and improve ease of handling during surgery, most wires are annealed to increase their ductility and workability.

6.3.2 Braided Multifilament Wire

Braided multifilament wire is an attractive alternative because it has similar tensile strength as a monofilament wire of equal diameter but is more flexible and has better fatigue strength. It was observed that bone often grows into the inlays of the braided

Fig. 6.4 Kirschner wire with a 2.5-mm-diameter used to hold bone fragments together

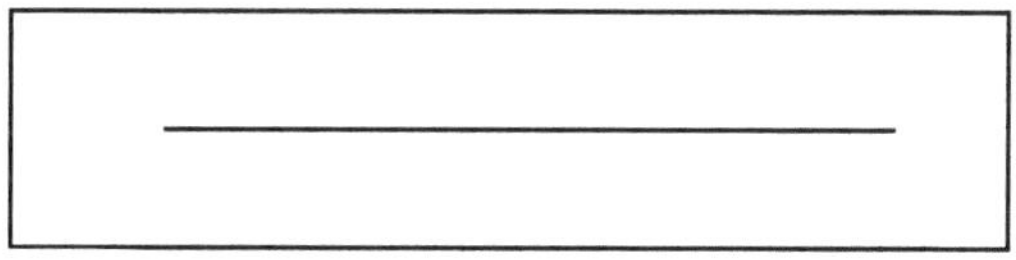

Fig. 6.5 Trocar pin

wire, making it difficult to remove, as it prevents the wire from sliding when pulled. When a wire is used with other metallic implants, then direct contact should be avoided or materials should be matched to prevent galvanic corrosion (Fig. 6.4).

6.3.3 Pins

Straight wires with a pointed end are called Steinmann pins; however, if the pin diameter is less than 2.38 mm, it is called Kirschner wire. These pins are widely used, primarily to hold fragments of bones together provisionally or permanently and to guide large screws during insertion. To facilitate implantation, the pins have different tip designs that have been optimized for different types of bone. The trochar tip, which has three cutting faces, is not efficient in cutting; it is often used for cortical bone like nail, which is hammered into the bone, displacing the material due to high pressure generated at the tip. Most pins are made of 316L stainless steel; however, recently, biodegradable pins made of polylactic or polyglycolic acid have been used for the treatment of minimally loaded fractures (Fig. 6.5).

During traction, we will have to ensure that the pin does not bend. It is supported in the tibia and being pulled by the string passing over the frictionless pulley, and dead weights are hanged.

6.3.4 Screws

Screws are the most widely used devices for the fixation of bone fragments. There are two types of bone screws: (1) cortical bone screws, which have smaller pitch threads, and (2) cancellous screws, which have large pitch threads, to get more thread-to-bone contact. Screws may have V-threads. The cortical screws are of two types, self-tapping and non-self-tapping (Fig. 6.6). The self-tapping screws have cutting flutes that cut thread in the pilot drill hole during insertion; in contrast, the non-self-tapping screws require a tapped drill hole for fixation.

The holding power of screws can be affected by the size of the pilot drill hole (which is equal to the internal diameter of the screw), the depth of the screw thread,

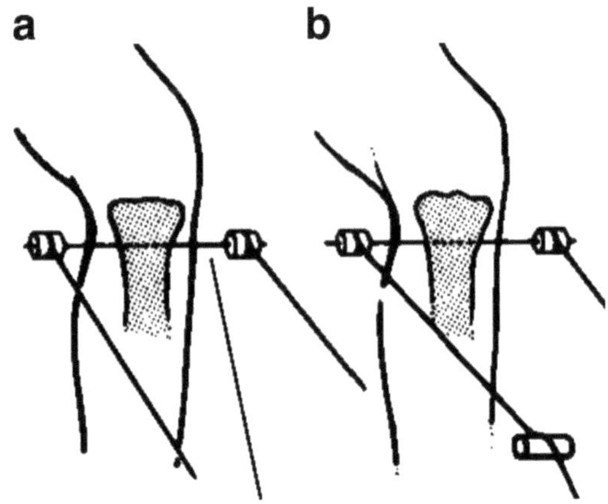

Fig. 6.6 Pins are also used to apply traction to a broken bone for conservative treatment

the outside diameter of the screw, and the quality of the bone. Therefore, the selection of the screw type should be based on the assessment of the quality of the bone at the time of insertion. Under identical conditions, self-tapping screws provide a slightly greater holding power than non-self-tapping screws.

6.3.4.1 Mechanical Properties of Screws

Screw pull-out strength varies with the time elapsed after insertion in vivo, and it depends on the growth of bone into the screw threads and/or resorption of the surrounding bone. The bone immediately adjacent to the screw often undergoes *necrosis* initially, but if the screw is firmly fixed, when the bone revascularizes, permanent secure fixation may be achieved. This is particularly true for titanium alloy screws or screws with a roughened thread surface, with which bone ongrowth results in an increase in the screw removal torque. When the screw is subject to micro or macro movement, the contacting bone is replaced by a fibrous membrane, the purchase is diminished, and the screw loosens. The pilot drill hole should be made carefully and profusely irrigated so that the temperature does not exceed 45 °C. It has been shown experimentally that the bone will be charred if the temperature is more than 55 °C and lasts more than few minutes (Fig. 6.7).

6.3.4.2 Applications of Screws

The two principal applications of bone screws are (1) as interfragmentary fixation devices to “lag” or fasten bone fragments together, or (2) to attach a metallic plate to bone. Interfragmentary fixation is used in most fractures involving cancellous bone and in oblique fractures in cortical bone. In order to lag the fracture fragment, the head of the screw must engage the cortex on the side of insertion without gripping the bone, while the threads engage cancellous bone and/or the cortex on the

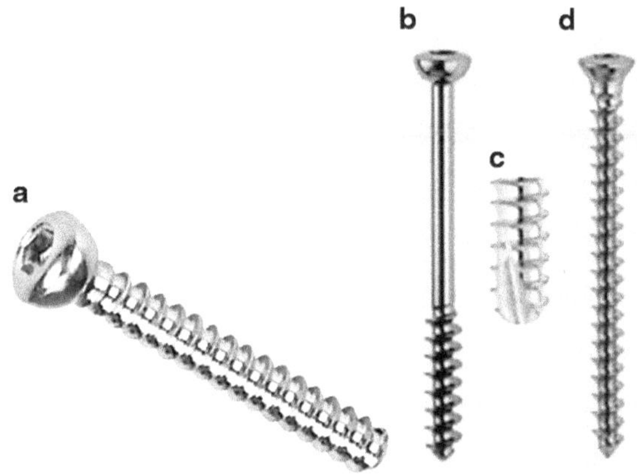

Fig. 6.7 Bone screws, for cortical (**a**) and cancellous bone (**b**, **d**). The smaller part (**c**) indicates the self-tapping screw, which will not need tapping of the bone before screwing in the compact bone

opposite side. When screws are employed for bone plate fixation, the bone screw threads must engage both cortices. Screws (ordinary or compressive type) are also used for the fixation of spinal fractures.

6.3.4.3 Design of Screws

Screws are designed on the basis of application. Their diameter ranges between 3 and 6.5 mm. The thread profile is usually a V-type for anchorage with a thread angle of 60°. The screw diameter and pitch are related. For cancellous bone, the thread pitch is larger and the angle much lower, for less damage to the bone. The screws are to be designed based on the core diameter subjected to torsional moment and tension force. The head diameter and hexagonal socket head are standardized. Charts are available from manufacturers regarding these details. A partial table from INOR-India group has been reproduced here as an example (Table 6.1).

6.3.5 Plates

Plates are available in a wide variety of shapes and are intended to facilitate the fixation of bone fragments. They range from the very rigid, designed to produce primary bone healing, to the relatively flexible, intended to facilitate physiological loading of bone.

The rigidity and strength of a plate in bending depend on the cross-sectional size (mainly thickness) and material from which it is made. Consequently, the weakest region in the plate is the screw hole, especially if the screw hole is left empty, due to a reduction in the cross-sectional area in this region. The effect of the material on the rigidity of the plate is defined by the elastic modulus of the material for bending and by the shear modulus for twisting.

Table 6.1 Cortical screw chart

Cortical XL screws 2.7 mm Diameter		
		Description
Cat. No.	Item code	24 TPI (To be ordered in multiples of 10 only)
0006X	027-24-006	Length: 6 mm
0006X	027-24-008	Length: 8 mm
0006X	027-24-010	Length: 10 mm
0006X	027-24-012	Length: 12 mm
0006X	027-24-014	Length: 14 mm

Thus, given the same dimensions, a titanium alloy plate will be less rigid than a stainless steel plate, since the elastic moduli of the alloys are 110 GPa and 200 GPa, respectively.

Stiff plates often shield the underlying bone from the physiological loads necessary for its healthful existence. Similarly, flat plates closely applied to the bone prevent blood vessels from nourishing the outer layers of the bone. For these reasons, the current clinical trend is to use more flexible plates to allow micromotion and low-contact plates (LCP) to allow restoration of vascularity to the bone.

The underlying goals with the use of plates are

1. To improve the fracture healing rate
2. To reduce the loss of bone mass in the region shielded by the plate
3. To reduce the incidence of refractures, which may occur following plate removal.

The interaction between the bone and the plate is extremely important, since the two are combined into a composite structure. The stability of the plate–bone composite and the service life of the plate depend upon accurate fracture reduction. The plate is most resistant in tension; therefore, in fractures of long bones, the plate is placed along the side of the bone that is typically loaded in tension. Having excellent apposition of the bone fragments, as well as developing compression between them, is critical in maintaining the stability of the fixation and preventing the plate from repetitive bending and fatigue failure.

Compression between the fracture fragments can be achieved with a special type of plate called a **dynamic compression plate (DCP)**. This plate has elliptical-shaped screw holes with the long axis oriented parallel to that of the plate. The screw hole has a sliding ramp to the long axis of the plate that conforms to the shape of the screw head. As the screw is driven home, the plate goes to tension and the fracture site in compression.

Bone plates are often contoured in the operating room to conform to an irregular bone shape and thus to achieve maximum contact of the fracture fragments. However, excessive bending decreases the service life of the plate. The most common failure modes of a bone plate-screw fixation are screw loosening and plate failure. The latter typically occurs through a screw hole, due to fatigue and/or crevice corrosion. The shape of the holes of the dynamic compression plate allows inclination of the screws in a transverse direction of +7° and in a longitudinal direction of 25°.

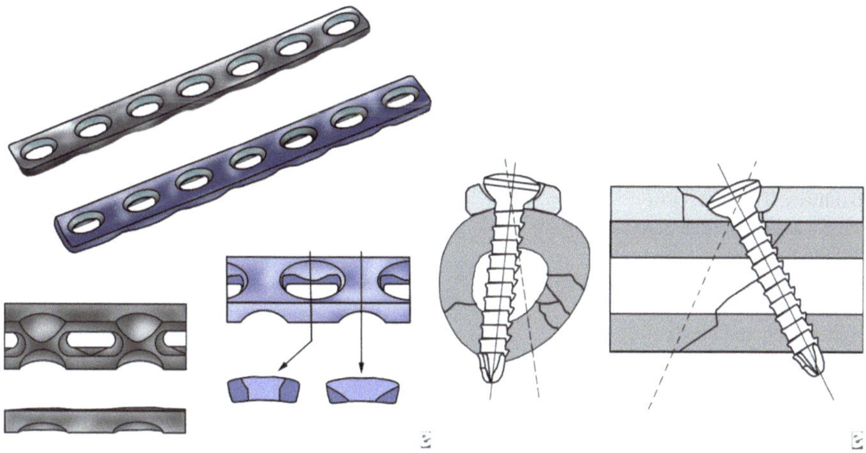

Fig. 6.8 Principle of a dynamic compression plate for fracture fixation: During tightening, the screw head will slide down on a ramp in a plate screw hole. This will displace the plate away from the fracture and compress the fracture fragments

In the vicinity of the joints, where the diameter of long bones is wider, the cortex thinner, and cancellous bones abundant, plates are often used as a buttress or as a retaining wall. A buttress plate applies force to the bone perpendicular to the surface of the plate and prevents shearing or sliding at the fracture site. Buttress plates are designed to fit specific anatomic locations and often incorporate other methods of fixation besides cortical or cancellous screws, for example, a large lag screw or an I-beam.

For the fusion of vertebral bodies following diskectomy, spinal plates are used along with bone grafts. These plates are secured to the vertebral bodies using screws. Similar approaches have been employed to restore stability in the thoracolumbar and cervical spine regions as well (Figs. 6.8, 6.9, 6.10).

6.3.6 *Intramedullary Nails*

Intramedullary devices (IM nails) are used as internal struts to stabilize long bone fractures. Intramedullary nails are also used for the fixation of femoral neck or intertrochanteric bone fractures; however, this application requires the addition of screws. A variety of designs is available, going from solid to hollow cylindrical, with shapes such as cloverleaf, diamond, and C (slotted cylinders).

Compared to plates, IM nails are better positioned to resist multidirectional bending since they are located in the center of the bone. However, their torsional resistance is less than that of the plate. Therefore, when designing or selecting an

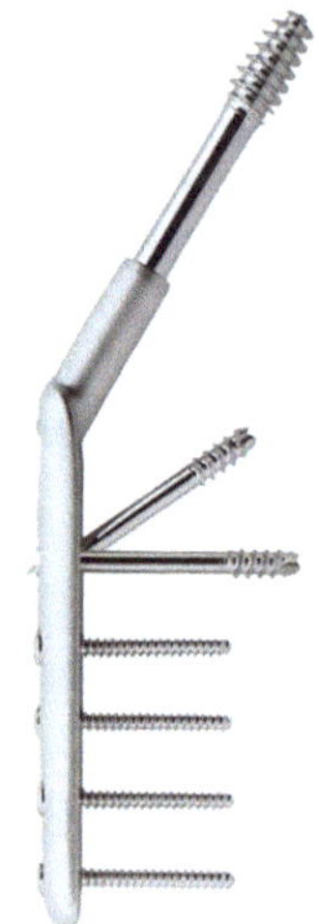

Fig. 6.9 Assembled dynamic hip screw (DHS). The lag screw for the femoral head is within the sleeve of the side plate. The side plate shows the fixation screws that are inserted into the femoral diaphysis. The two upper screws are cancellous screws, and the four lower screws are cortical screws. Note that the cancellous screws have a higher pitch and depth of thread for better purchase in spongy bone

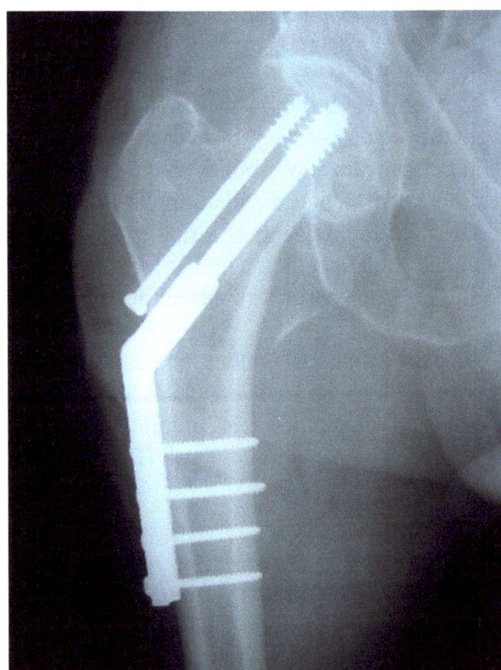

Fig. 6.10 Dynamic hip screw transfixing an intertrochanteric fracture of the right femur shown in a post-op X-ray. A cannulated cancellous screw that serves as a lag screw has also been inserted to increase compression and to control rotation of the femoral head. Note that the lesser trochanter has been fractured off the femur and has been displaced medially

intramedullary nail, a high polar moment of inertia is desirable to improve torsional rigidity and strength. The following formula may be useful:

$$T / J = \tau / r = G\theta / L,$$

where T = torque, J = polar moment of Inertia of the section of the nail, G = modulus of rigidity, θ = angle of twist, and L = length of the element, r = radius.

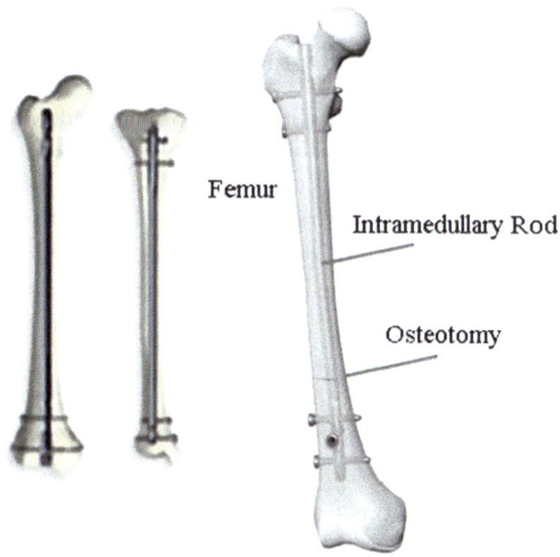

Fig. 6.11 Examples of intermedullary nails used in tibia and femur bone

The torsional rigidity of an IM nail is proportional to the elastic modulus and the moment of inertia. For nails with a circular cross section, the torsional stiffness is proportional to the fourth power of the nail's radius. The wall thickness of the nail also affects the stiffness. A slotted, open-section nail is more flexible in torsion and bending, allowing for easy insertion into a curved medullary canal—for example, that of the femur. However, in bending, a slot is asymmetrical with respect to rigidity and strength. For example, a slotted nail is strongest when bending is applied so that the slot is near the neutral plane; the nail is weakest when oriented so that the slot is under tension (Fig. 6.11).

In addition to the need to resist bending and torsion, it is vital for an IM nail to have a large contact area with the internal cortex of the bone to permit torsional loads to be transmitted and resisted by shear stress. Two different concepts are used to develop shear stress: (1) a three-point, high-pressure contact, achieved with the insertion of curved pins; and (2) a positive interlocking between the nail and intramedullary canal, to produce a unified structure. Positive interlocking can be enhanced by reaming the intramedullary canal. Reaming permits a larger, longer, nail–bone contact area and allows the use of a larger nail with increased rigidity and strength.

The addition of screws through the bone and nail, proximal and distal to the fracture, known as *interlocking*, increases the torsional stability and prevents shortening of the bone, especially in unstable fractures. The IM nail, which has not been interlocked, allows interfragmentary compressive force, due to its low resistance to axial load. Another advantage of the intramedullary nails is that they do not require opening the fracture site, since they can be inserted through a small skin incision, typically located in one extreme of the bone. The insertion of an IM nail, especially one that requires reaming of the medullary canal, destroys the intramedullary vessels, which supply two thirds of the blood supply to the cortex. However, this is usually not of clinical significance because revascularization occurs rapidly.

6.3.7 Use of Machine Elements in External Fixators

Pins, screws, and rods are also used to construct external fixators, such as frames and rings. Although they are outside the body, the screws and pins go through the skin and muscle to connect to the bone. In this way, they differ from casts and splints, which rely solely on external support. There may be some inflammation or, less commonly, infection associated with the use of external fixators. Normally, these can be managed with wound care and/or oral antibiotics.

6.4 Other Considerations

Sterile conditions and advances in surgical techniques reduce, but do not remove, the risk of infection when internal fixation is used. The severity of the fracture, its location, and the medical status of the patient must all be considered.

In addition, no technique is foolproof. The fracture may not heal properly, the plate or rod may break or deform, or the patient may have an allergic reaction to the implant. Although some media attention has focused on the possibility that cancer could develop near a long-term implant, there is little evidence documenting an actual cancer risk and much evidence against that possibility. Orthopedic surgeons are continuing their research to develop improved methods for treating fractures.

6.5 Materials Involved in Bone–Plate Design

Usually, conventional high-stiffness stainless steel (SS) has been employed for long-bone fracture fixation. However, the big difference in modulus between the plate and bone as well as the compressive stresses occurring between the plate and the bone (due to overtightening of screws) disturb the vascularity of the bone underneath the plate, cause bone resorbtion underneath the plate, and reduce its strength as a long-term effect.

In recent years, there have been considerable awareness and discussion about the need for using less stiff plates to improve fracture healing and prevent bone weakening due to stress shielding. It is not entirely correct to say that bone plates with a high stiffness (or Young's modulus, E) cause excessive stress shielding, because stiffness is characterized by the product of E and the moment of inertia I of the plate's cross section. Hence, the plate geometry also has a bearing on the stiffness and thereby on the stress shielding of the bone. However, for a uniform plate geometry, plates with a lower E will offer less stress shielding than plates with a higher Young's modulus.

Metal alloys, bioceramics, titanium alloys, pure titanium, composite materials, and polymers (nonresorbable and bioresorbable) were used in fabricating fracture plates. Each of the above materials can broadly be categorized as (1) bioinert, (2) porous, (3) bioactive, and (4) bioresorbable. In general, bioinert material is selected for bone plates because bioactive material becomes bonded with the bone (along with the soft tissues) and causes problems if plate removal or corrective surgery is required.

The bioceramic materials that are bioinert (such as Al_2O_3, ZrO_2) possess a Young's modulus (*E*) in the range of **400 ± 20 GPa**, in contrast to that of hydroxyapatite. While the properties of ceramics (such as high hardness, chemical inertness, oxidation resistance, high strength, high melting points, and low fracture toughness) are suited to the requirements of the bone plate, their brittleness and high *E* result in stress shielding of the bone, thus limiting their use for bone plates.

Metallic alloys, such as cobalt-base alloys (e.g., CoCrW, CoCrMo), have an *E* of about **250 ± 10 GPa** along with wear, corrosion, and heat resistance. However, they are not suitable for usage, because of their poor fabricability and high cost. Stainless steel (e.g., 316L) is one of the most preferred biomaterials for bone plates because of its mechanical properties (*E* = **200 ± 20 GPa**, ductility, etc.), corrosion resistance, bioinert state, and cost-effectiveness in comparison with other biocompatible metals. Titanium alloys (e.g., Ti-6Al-7Nb, Ti-6Al-4V), with an *E* of **110 ± 10 GPa**, are especially preferred for bone screws because of their increased corrosion resistance and improved ductility. However, although titanium alloys offer improved strength (with less ductility) compared to pure titanium, they are not preferred for plate implants because of difficulty in their contouring (as required for pelvic and mandibular plates). Titanium alloys are preferred, however, for intramedullary rods, spinal clamps, self-drilling bone screws, and other implants because of their high strength and low *E*. Tables 6.2 and 6.3 give the details about biomaterials' properties and modes of failure.

Pure titanium metal (CP-titanium) is also one of the most widely chosen materials for bone plates because of its excellent biocompatibility and corrosion resistance. The ductility of titanium is lower compared to SS because of its hexagonal crystal structure. This makes contouring of titanium plates difficult compared to stainless steel plates. Titanium plates also offer less stress shielding to bone (for the same geometries) after healing, because their *E* is 68 GPa compared to the 200 GPa of SS. However, they are not as amenable to contouring as SS plates.

6.6 Composite Materials

Composite materials, such as carbon fiber-reinforced polymers (CFRP), which consist of a polymer matrix and fiber, are combined to achieve the requisite high strength and adequate *E* value. The polymer matrix materials can be broadly classified as resorbable (e.g., polysorb, biosyn) and nonresorbable [such as polyether ether ketone (PEEK) or ultra-high-molecular-weight polyethylene (UHMWPE)]. Polymers per se do not have the strength and stiffness required for bone plates; hence, polymers reinforced by fibers are employed for the bone-plate application or used as scaffolds in the preparation of bone grafts. Composite materials used for bone plates mainly

Table 6.2 Biomaterials used in fracture repair

Materials	Properties	Application
Stainless steel (AISI-316L)	Low cost, easy to fabricate	Surgical wire, (annealed) pin, plate, screw, IM nail
Ti alloy (Ti-6Al-4V)	High cost Low density and modulus Excellent osseointegration	Surgical wire Plate, screws, IM nails
CoCrMo alloys (wrought)	High cost High density and modulus Difficult to fabricate	Surgical wire IM nails
Polylactic acid	Resorbable Low strength	Pin, screw
Nylon	Nonresorbable plastic	Cerclage band

Table 6.3 Failure modes of fracture-fixation device

Failure mode	Failure location	Reasons for failure
Overload	Bone fracture site Implant screw hole Screw thread	Small-sized implant Unstable reduction Early weight bearing
Fatigue	Bone fracture site Implant screw hole Screw thread	Early weight bearing Small-sized implant Unstable reduction, fracture nonunion
Corrosion	Screw head-plate hole Bent area	Mismatch of implant alloys Overtightening screw Overbent Scratches during insertion Motion
Loosening	Screw	Wrong choice of screw type, osteoporotic bone

consist of a thermoplastic polymer matrix [such as PEEK or polymethyl methacrylate (PMMA), etc.] and fibers such as glass or carbon. The disadvantage of using composite material arises if the implant fails, when revision surgery is warranted. This is because of the risk of fiber breakage and the subsequent penetration of small fiber particles into the bone tissue, causing irritation and inflammation.

The increased use of bioresorbable polymers (i.e., polymers that degrade *in vivo* into nonharmful byproducts) in recent years poses the problem of their loss of strength while bone healing is in progress. Bone-plate fracture fixation should sustain loads for 1.5–2 years, a goal that has yet to be achieved with resorbable materials. Hence, a new class of resorbable materials needs to be developed, having adequate mechanical properties and a resorption time increased by 1–2 years.

In view of the above discussion, polymers and calcium phosphates are osteoinductive and resorbable; they cannot behave as load-sharing members and fail in *in vivo* loading conditions. For a reinforced fractured bone, it is important to initially have a plate with sufficient stiffness to prevent tensile stresses at the fracture interface, while allowing the bone away from the fracture site to be stressed under loading conditions (to prevent loss of bone strength). An optimal plate needs to be designed that caters to the above-mentioned objectives.

Based on these considerations, the use of stiffness-graded materials (SGMs) may be recommended for bone plates. SGMs are characterized by a smooth and continuous change of the mechanical properties from one characteristic surface to the other. Stiffness-graded material is a relatively new concept in bone plates in order to decrease stress shielding (this concept is well documented for dental implants). Controlled segregation, controlled blending, vapor deposition, plasma spraying, electrophoretic deposition, controlled powder mixing, slip-casting, sedimentation forming, centrifugal forming, laser cladding, metal infiltration, controlled volatilization, and self-propagating high-temperature synthesis are some manufacturing techniques that have evolved in the fabrication of SGMs. The current production of SGMs is hampered by the current manufacturing process technology.

Axial compressive load is more prominent in long bones. However, it does not endanger bone healing by opening the fracture gap, and it contributes to more interfragmentary compression at the fracture interface. On the other hand, load eccentricity from the center of the bone plate and the intrinsic curvature of long bones cause bending moments to be applied to the fracture-fixed bone. Bending moments will induce both tension and compression stresses across the fracture interface and open up the fracture, leading to a reduction in the stability of the fixation.

6.7 Bioactive Fixation Using Bioactive Materials

A bioactive material is one that elicits a specific biological response at the interface of the material, which results in the formation of a bond between the tissues and the material. This concept has now been expanded to include a large number of bioactive materials with a wide range of rates of bonding and thicknesses of interfacial bonding layers. Examples of such materials include bioactive glasses (e.g., Bioglass®), bioactive glass ceramics (e.g., Ceravital®, A/W glass ceramic, machinable glass ceramics), dense hydroxyapatite (e.g., Durapatite® or Calcitite®), bioactive composites such as polyethyleneBioglass®, polysulfone-Bioglass®, and polyethylene-hydroxyapatite (Hapex®) mixtures. All of these bioactive materials form an interfacial bond with bone. However, the time dependence of bonding, the strength of the bond, the mechanism of bonding, and the thickness of the bonding zone differ for the various materials (Table 6.4).

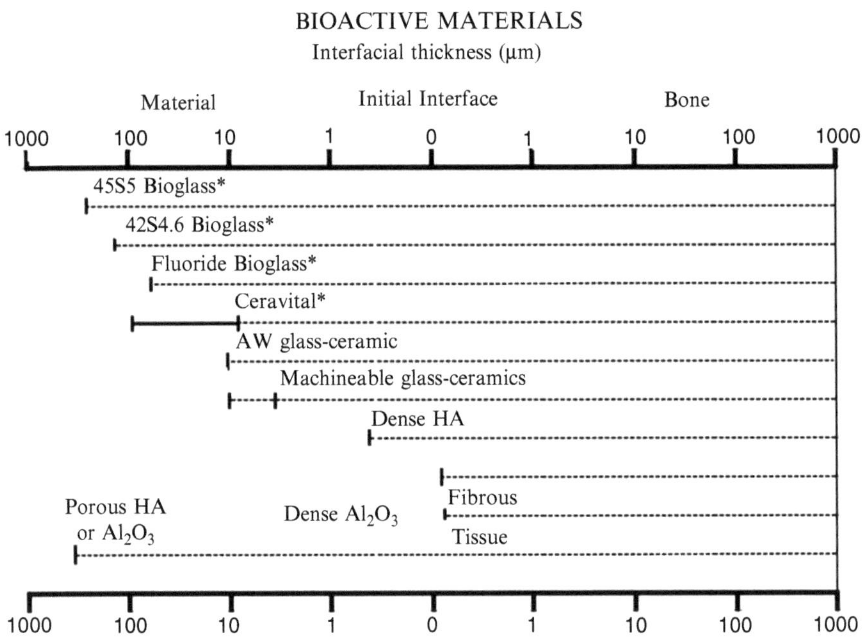

Table 6.4 Types of bioceramics—tissue attachment and bioceramic classification

Types of bioceramics	Type of attachment	Example
1	Dense, nonporous, nearly inert ceramics attach by bone growth into surface irregularities by cementing the device into the tissues, or by press-fitting into a defect (termed morphological fixation).	Al_2O_3 (single crystal and polycrystalline)
2	For porous inert implants, bone ingrowth occurs, which mechanically attaches the bone to the material (termed biological fixation).	Al_2O_3 (porous polycrystalline), hydroxyapatite-coated porous metals
3	Dense, nonporous, surface-reactive ceramics, glasses, and glass ceramics attach directly by chemical bonding with the bone (termed bioactive fixation).	Bioactive glasses, bioactive glass-ceramics, hydroxyapatite
4	Dense, nonporous (or porous), resorbable ceramics are designed to be slowly replaced by bone.	Calcium sulphate (plaster of Paris), tricalcium phosphate, calcium phosphate salts

6.8 Compositions

Special compositions of glasses, ceramics, glass ceramics, and composites develop a mechanically strong bond to bone. These materials have become known as "bioactive ceramics." Some even more specialized compositions of bioactive glasses will bond to soft tissues as well as bone. A common characteristic of bioactive glasses and bioactive ceramics is a time-dependent, kinetic modification of the surface that occurs upon implantation. The surface forms a biologically active hydroxyl carbonate apatite (HCA) layer that provides the bonding interface with tissues. The HCA phase that forms on bioactive implants is equivalent chemically and structurally to the mineral phase in bone. It is the biological equivalence of the HCA layer that forms on the bioactive implant surface, which is responsible for interfacial bonding.

6.9 Strong Interfacial Bond with Bone

Materials that are bioactive develop an adherent interface with tissues that resists substantial mechanical forces. In many cases, the interfacial strength of adhesion is equivalent to or greater than the cohesive strength of the implant material or the tissue bonded to the bioactive implant. There were three key compositional features to these glasses that distinguished them from traditional soda-lime-silica glasses: (1) less than 60 mol% SiO_2, (2) high Na_2O and CaO content, and (3) high CaO/P_2O_5 ratio. These compositional features make the surface highly reactive when exposed to an aqueous medium.

Many bioactive silica glasses are based upon the formula "45S5," signifying 45 wt.% SiO_2 (S = the network former) and 5:1 molar ratio of Ca to P. Glasses with substantially lower molar ratios of Ca to P (in the form of CaO and P_2O_5) do not bond to bone. However, substitutions in the 45S5 formula of 5–15 wt.% B_2O_3 for SiO_2 or 12.5 wt.% CaF_2 for CaO or ceraming the various bioactive glass compositions to form glass ceramics have no measurable effect on the ability of the material to form a bone bond (Table 6.5).

Suggested Study for Fracture Fixation

1. Hidaka S, Gustilo RB (1984) Refracture of bones of the forearm after plate removal. J Bone Joint Surg Am 66(8):1241–3
2. Roy S, Bag S (2005) For polymer ceramic composites, the two PhD theses of Sukumar Roy and Sandip Bag offered by Jadavpur University, Kolkata , 2005 and 2007, respectively, may be consulted
3. Khalily C, Voor MJ, Seligson D (1998) Fracture site motion with Ilizarov and "hybrid" external fixation. J Orthop Trauma 12(1):21–6
4. Ilizarov GA (1989) The tension-stress effect on the genesis and growth of tissues: Part II. The influence of the rate and frequency of distraction. Clin Orthop Relat Res 239:263–85

Table 6.5 Composition of bioactive glass and ceramics

Component	4SSS Bioglass*	4SSS 4F Bioglass*	4581SSS Bioglass*	52S4.6 Bioglass*	56S4.3 Bioglass*	KGC Coravital	KGS Coravital	KGy213 Coravital	A/W Glass–ceramic	MB Glass–ceramic	S45P7
SiO_2	45	45	30	52	55	46.2	46	38	34.2	19–52	45
P_2O_5	6	6	6	6	6				16.3	4–24	7
CaO	24.5	14.7	24.5	21	19.5	20.2	33	31	44.9	9–3	22
$Ca(PO_3)_2$						25.5	16	13.5			
CaF_2		9.8							0.5		
MgO						2.9			4.6	5–15	
MgF_2											
Na_2O	24.5	24.5	24.5	21	19.5	4.8	5	4		3–5	24
K_2O						0.4				3–5	
Al_2O_3								7		12–33	
B_2O_3			15								2
Ta_2O_3/TiO_2								6.5			
Structure	Glass and glassceramic	Glass	Glass	Glass		Glass ceramic	Glass ceramic	Glass ceramic	Glass ceramic	Glass ceramic	

*Bioglass

5. Ilizarov GA (1989) The tension-stress effect on the genesis and growth of tissues. Part I. The influence of stability of fixation and soft-tissue preservation. Clin Orthop Relat Res 238:249–81
6. Taylor HS, Taylor JC (1997) Orthopaedic fixation device, Patent No: 5,702,389. 1997, Smith & Nephew Richards, Inc.: USA. 1–24
7. Hillard PJ, Harrison AJ, Atkins RM (1998) The yielding of tensioned fine wires in the Ilizarov frame. Proc Instn Mech Eng H 212:37–47

Problems

1. List and sketch the bone fracture-fixation devices, indicating their simplicity and complexity and their effectiveness in healing fractures. What materials are suitable for such devices? Make a table listing their physical properties, and compare them with those of bone.
2. What is bioactive fixation of bone fractures? Explain the mechanism for such fracture fixation, indicating the interfacial mechanism of bone tissue conduction and induction. Also, indicate how the coating on metallic implants can be used for anchorage with bone.
3. Study the fracture-fixation screws for joining cortical bone and cancellous bone. What is the significant difference between the two types, and what is the reason for this difference? How do you propose to design such screws?
4. There is an oblique fracture in a femoral cortex almost at a 45° angle to its axis. At what angle must the cortical screws be fixed? Should it be perpendicular to the axis or perpendicular to the fracture line from the viewpoint of good compression at the fracture site? What minimum number of screws will be necessary to stabilize such a fracture?
5. Compare the efficacy of an external ring fixator and an IM nail and plate and screw fixator from the viewpoints of the duration of healing and patient compliance and cost.

Chapter 7
The Shoulder Joint and Its Artificial Replacement

7.1 Introduction to Joint Replacement

Human upper- and lower-body joints are vulnerable to disease processes, such as various types of arthritis, or dislocation or damage due to trauma in road accidents or household falls. Quite often, conservative treatment repairs such damaged body parts and the patient goes back to his or her normal life. But it is not uncommon to have restricted painful movement, leading to disability and loss of function to a great extent.

In those cases, artificially reconstructing the joint with metal, ceramic, and polymer components is the only effective alternative available. Usually, the metal components, which are quite similar to the joint, are to be prepared and surgically fixed to the joint.

The joints that are usually replaced in the upper body are as follows:

1. Shoulder joint
2. Elbow joint
3. Wrist joint
4. Finger joint.

Similarly, for the lower body, the joints that are load-bearing are

1. Hip joint
2. Knee joint
3. Ankle joint.

If we study the anatomy of these joints, they are all diarthodial synovial joints having a pair of articulating surfaces lined with soft cartilage and enclosed in a membranous housing lubricated by synovial fluid. These joints are almost frictionless; their coefficient of friction is less than 0.001, which no manmade joint can achieve. To design the replacement of such joints, we need to know the kinematics, meaning the range of movements. These include flexion, extension, and rotation in different axes and planes; this information can be obtained from an anatomy

S. Pal, *Design of Artificial Human Joints & Organs*,

DOI 10.1007/978-1-4614-6255-2_7,

textbook, as we have to replicate such movement. Then we need to find the joint's reaction forces during the activities of the joint under both static and dynamic conditions. Usually, major muscles that act at the joint during a particular situation need to be identified. The muscle cross section can be obtained from a CT scan. The forces may be assumed to be proportional to the cross section, as nature always tries to optimize material use. Details regarding such force analysis could be found in *The Textbook of Biomechanics* by this author (Viva Book, New Delhi, 2009).

Now the congruent joint parts are to be shaped. Following the principle of bearing design in engineering, one part is made of harder material, like metal alloys or ceramic, and the other part is made of polymer, e.g., ultra-high-molecular-weight polyethylene (UHMWPE). Usually, the cartilage-lined parts are diseased due to osteoarthritis, rheumatoid arthritis, or traumatic arthritis. These parts are surgically removed, and usually the bony cavities are shaped using a rasp or other tools to insert the prosthetic part. The metal alloy components may be fixed to the bone with bone cement (polymethyl methacrylate) or may be porocoated with bioactive material like Bioglass or hydroxyapatite. Sometimes the prosthesis is push-fit in the bony canal. Each system has its individual suitability and application and also may be patient-specific.

7.2 The Shoulder Joint and Its Artificial Replacement

7.2.1 Introduction

The shoulder joint is the most efficient and complex joint of the human body, as it has the largest range of motion compared to the other major joints in the body. The shoulder is made up of three bones: the scapula (shoulder blade), the humerus (upper arm bone), and the clavicle (collarbone), which functions as a movable but stable base for the motions of the humerus. The rotator cuff connects the humerus to the scapula. The rotator cuff is formed by the tendons of four muscles: the supraspinatus, infraspinatus, teres minor, and subscapularis. Tendons attach muscles to bones. Muscles move bones by pulling on the tendons. The rotator cuff helps raise and rotate the arm. As the arm is raised, the rotator cuff also keeps the humerus tightly in the socket. A part of the scapula, called the glenoid, makes up the socket of the shoulder. The glenoid is very shallow and flat. The part of the scapula that connects to the shoulder is called the acromion. A bursa is located between the acromion and the rotator cuff tendons. A bursa is a lubricated sac of tissue that cuts down on the friction between two moving parts. Bursae are located all over the body where tissues must rub against each other. In this case, the bursa protects the acromion and the rotator cuff from grinding against each other. The humeral head of the shoulder is the ball portion of the joint. The humeral head has several blood vessels, which enter at the base of the articular cartilage. Articular cartilage is the smooth, white material that covers the ends of bones in most joints. Articular cartilage provides a

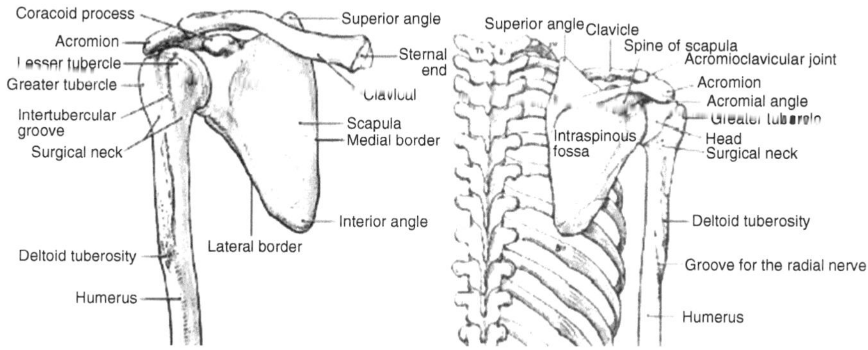

Fig. 7.1 The shoulder joint and the bones at the joint: a simplified view

slick, rubbery surface that allows the bones to glide over each other as they move. Cartilage also acts as a shock absorber (Fig. 7.1).

The shoulder joint is surrounded by a watertight sac called the joint capsule. The joint capsule holds fluids that lubricate the joint. The walls of the joint capsule are made up of ligaments. Ligaments are connective tissues that attach bones to bones. The joint capsule has a considerable amount of slack, loose tissue, so that the shoulder is unrestricted as it moves through its large range of motion.

7.2.2 The Different Joints

Movement of the shoulder joint takes place at the following joints:

1. Manubrioclavicular joint
2. Acromioclavicular joint
3. Scapulothoracic joint
4. Glenohumeral joint
5. Sternoclavicular joint
6. Subdeltoid joint.

The three synovial joints are the sternoclavicular joint (joining the sternum part of the thorax to the clavicle), the acromioclavicular joint (joining the clavicle to the acromion of the scapula), and the glenohumeral joint (connecting the glenoid cavity of the scapula and the head of the humerus). The shoulder joint has different planes of movement during different parts of motion. To allow normal shoulder abduction, both elevation and forward movement of the clavicle, swinging of the scapula takes place at the manubrioclavicular joint. During movement of the arm, the medial border of the scapula glides over the dorsal side of the thoracic cage, pressed onto it by the combined action of the muscles. This connection gives rise to the fourth joint, the scapulothoracic gliding plane (STGP) (Figs. 7.2 and 7.3).

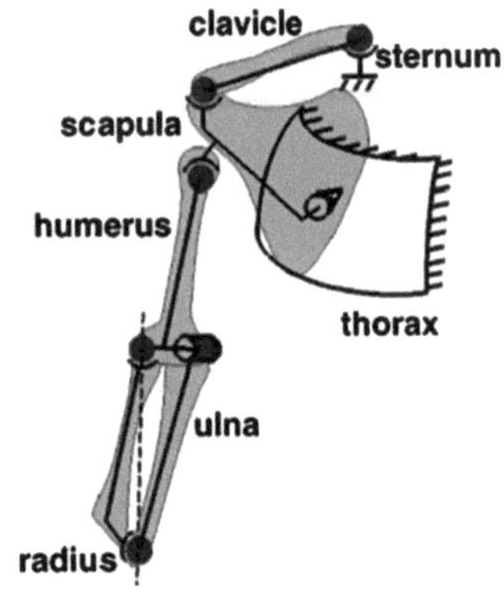

Fig. 7.2 Arm skeleton model

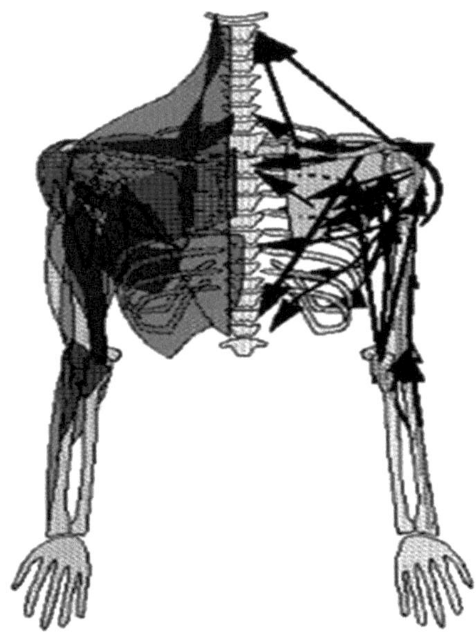

Fig. 7.3 Arm musculature model (courtesy of W. Maurel et al.)

7.2.3 Musculature

The shoulder joint exhibits the most complex motion among all the joints in the human body. For every movement, the cooperation of several muscles is necessary. To perform the movements, the upper limb is equipped with at least 22 muscle actuators, some of which even divide in several bundles attached onto different bones [1]. They can be divided in several groups according to the bones they move and the degrees of freedom (DOF) they control. Dvir and Berme noticed that most muscles acting on the scapula insert close to its medial border [2]. This concerns the levator scapulae, the rhomboids, and the middle and lower parts of the trapezius. These muscles make the scapula a strong base for performing the arm movements.

The rotator cuff refers to the group of muscles that covers the humeral head and controls some of its rotations. These are the subscapularis/teres major, as opposed to the infraspinatus/teres minor for controlling the axial rotations, and the supraspinatus/deltoideus, which handle the abduction.

The other actuators of the humerus are the latissimus dorsi and pectoralis major, which cooperate in its adduction, but they oppose each other in flexion/extension and axial rotation (Grant 1991). Two prime antagonist groups of muscles control the flexion/extension movements of the forearm: the brachialis and biceps brachii for the flexion, as opposed to the anconeus and triceps brachii for the extension. When the brachialis is inactive, the biceps brachii also contributes toward controlling the supination movement of the forearm, together with the brachioradialis, as opposed to the pronator teres, which controls the pronation (Chao 1978). As muscles never work in isolation, natural movements always involve the motions of all the bones. For a complete analysis, it is necessary to consider the motion of the mechanism as a whole: Almost all investigations on the shoulder girdle motion focus on the quantification of the scapulohumeral rhythm during elevation. This rhythm describes the way in which the humerus elevation is composed of rotations in the glenohumeral joint and scapulothoracic gliding plane (Högfors 1991). All the shoulder bone rotations are involved when, for example, the arm performs circumduction movements. Based on the muscle fiber distribution within a muscle, each of these muscles was represented by one to six muscle lines of action of force between the origin and the insertion [3].

7.2.4 Ligaments

There are three extracapsular ligaments in the shoulder girdle: the costoclavicular ligament, limiting the range of motion of the sternoclavicular (SC) joint, and the conoid and trapezoid ligaments, acting on the acromioclavicular (AC) joint.

7.2.5 Glenohumeral Joint

The glenohumeral joint or the scapulohumeral joint is an articulation between the humerus and the scapula, the articular surfaces involved being the head of the humerus and the glenoid fossa of the scapula. It is a true anatomical joint of the ball-and-socket type. In this joint, the humeral head constitutes nearly a hemisphere, which is inclined medially, upward and slightly backward. The socket is the cavity of the glenoid fossa of the scapula, which is shallow and points forward and laterally.

As the humeral head and the glenoid have the shapes of a sphere and the glenoid articulates with a very small area of the humerus, it has a large range of motion.

A brief description of the glenohumeral joint is given next.

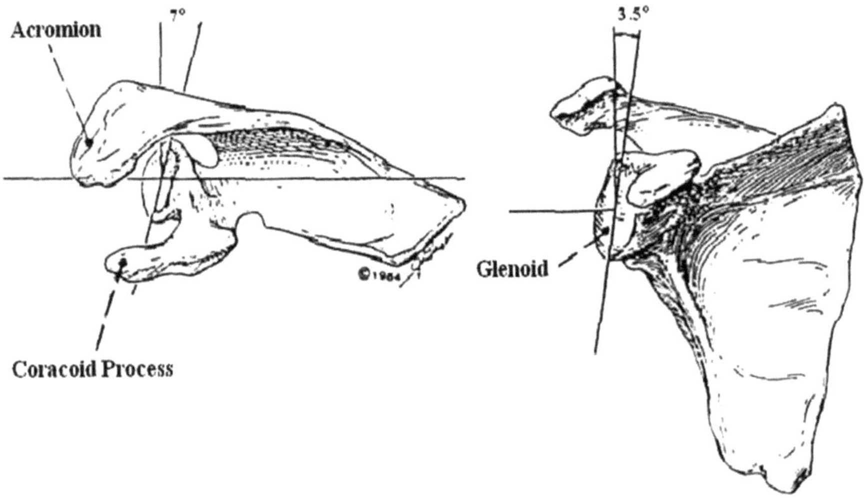

Fig. 7.4 The right scapula including the glenoid [4]

7.2.6 Glenoid

The glenoid fossa is the portion of the scapula that articulates with the head of the humerus, forming the ball-and-socket joint of the shoulder girdle. The articular surface of the glenoid is very small. In the coronal plane (a vertical plane from the front of the body to the back of the body), the articular surfaces of the body comprises 75° [4]. The glenoid has a slight upward tilt of 5° referable to the medial border of the scapula and is retroverted a mean of approximately 7°, with individual variation. The diameter of the glenoid is approximately 56.6 mm (range 43.2–73.8 mm) (Van der Helm et al. 1989). The articular surface of the glenoid can be described as having the shape of a sphere (Van der Helm et al. 1989), as in Fig. 7.4.

Saha (1971) determined there are three types of glenohumeral joints, as in Fig. 7.5:

1. The glenoid has a larger diameter than the humeral head.
2. The glenoid and the humeral head are concentric.
3. The glenoid has a smaller diameter than the humeral head.

7.2.7 Humerus

The articular surface of the humerus exists of approximately one third of the surface of a sphere with an arc of about 120° [4]. This articular surface called the humeral head has an angle with the humeral shaft of 135° (range 132–138°) and is retroverted approximately 32° (range 27–37°) referable to the condylar line of the distal

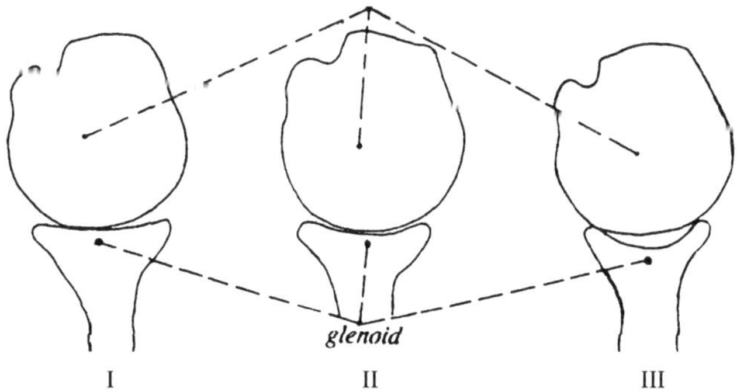

Fig. 7.5 Three types of glenohumeral joints (courtesy of Saha 1971)

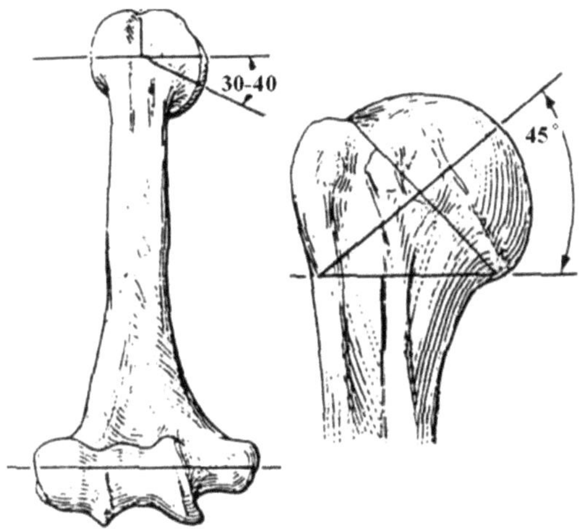

Fig. 7.6 Frontal view of the right humerus and the articular surfaces of the humeral head [4]

humerus as in Fig. 7.6 [5]. The diameter of the humeral head has an average of 51.4 mm (range 44.0–62.0 mm) (Van der Helm et al. 1989).

7.2.8 Kinematics of the Glenohumeral Joint

The great mobility of the shoulder joint comes from that fact that it has several axes of rotation, as follows:

1. An anterior-posterior axis, about which the abduction and adduction take place.
2. A transverse axis, which allows flexion and extension movements.
3. A vertical axis, about which flexion and extension can take place, with the arm abducted to 90°.
4. The longitudinal axis of the humerus, allowing internal and external rotation. The vertical axis and the humeral longitudinal axis coincide when the arm is in the anatomical reference position, i.e., hanging vertically from the body.

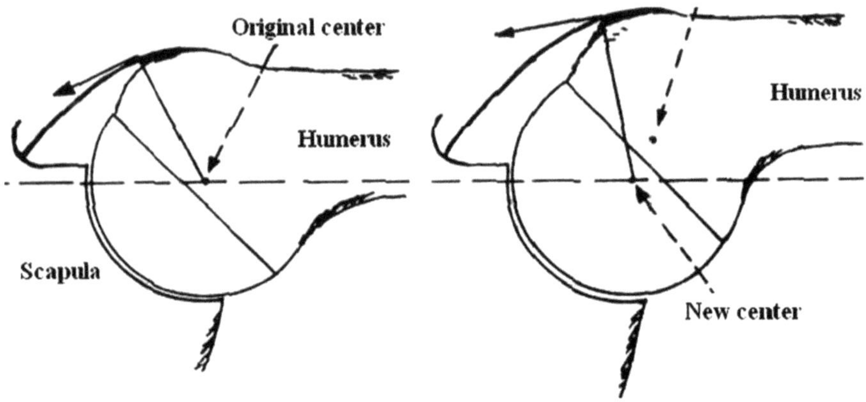

Fig. 7.7 Displacement of the geometrical center of the glenoid in relation to the humerus at rest

7.2.9 Geometrical Center and Center of Rotation

The geometrical center of the glenohumeral joint lies at the center of the glenoid; the center of rotation is dependent on the total motion of the arm, as shown in Fig. 7.7. The muscles that provide motion in the shoulder have lever arms. The geometrical center is important because it is an important parameter in the design of the shoulder prosthesis: It determines the size and direction of the lever arms of the muscles. There are two geometrical centers, one relative to the humerus and another relative to the glenoid. If either of the diameters does not match, then there will be a change in the moment pattern during motion of the arm.

7.3 Humeral Translation Relative to the Scapula

Translation in the glenohumeral joint is the displacement of the geometrical center of the humeral head with regard to the scapula, as in Fig. 7.8. During passive perpendicular movement, a translation of a few mm can occur in relation to the glenoid. This space is due to the soft tissue, which permits the humeral head to translate when the muscles are not active. The healthy rotator cuff muscles will pull the humeral head in the glenoid and make translation in a perfect ball-and-socket joint very unlikely, if not impossible (Van der Helm et al. 1989).

If, however, a large translation is possible, large movements of the humeral head in relation to the glenoid would be highly restricted by the surrounding healthy musculature and soft tissue. The position of the geometrical center of the humeral head would then move in relation to the scapula and be dependent on the diameters of the humeral head and the glenoid. If the two spheres are concentric, only rotation would be possible. If the diameter of the glenoid was larger than that of the humeral head, then the geometrical center of the humerus would still lie on the plane. If the

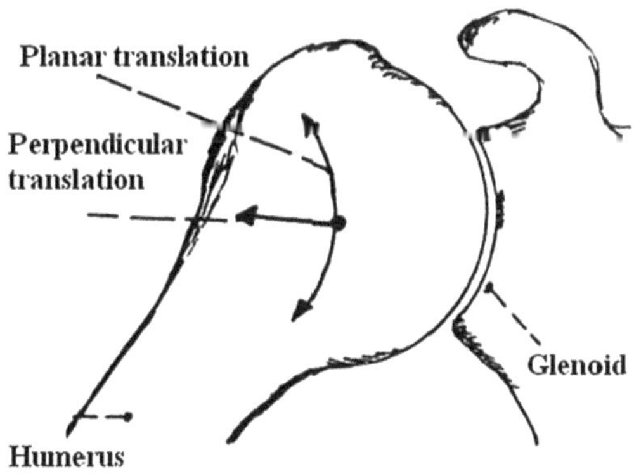

Fig. 7.8 Planar translation versus perpendicular translation of the humeral head in relation to the glenoid

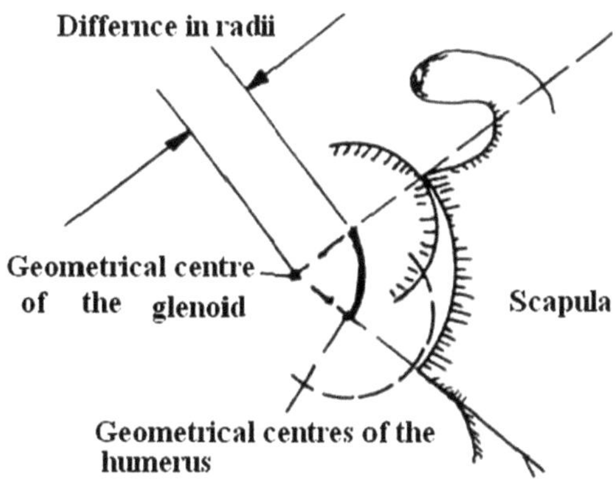

Fig. 7.9 The outlined plane for movement of the geometrical center

geometrical center of the humerus goes outside the area, it would probably a cause of subluxation or even a dislocation (Fig. 7.9).

7.4 Degrees of Freedom (DOF)

Assuming translations negligible compared to rotations, all joints, except the scapulothoracic joint, are usually assumed to be ball-and-socket joints, having more or less 3 rotational degrees of freedom (DOF). The scapulothoracic joint is a special case since it does not properly involve articular structures between the scapula and thorax. However, due to its surrounding muscles, the scapula is usually considered constrained to glide on the thorax (Dvir 78). This reduces the number of DOF of the scapulothoracic joint to 4. Considering all joints independently, the number of DOF of the upper limb would amount to 22. However, as they are organized in closed chains, the number of DOF of the upper arm reduces to 12 [1] (Fig. 7.10).

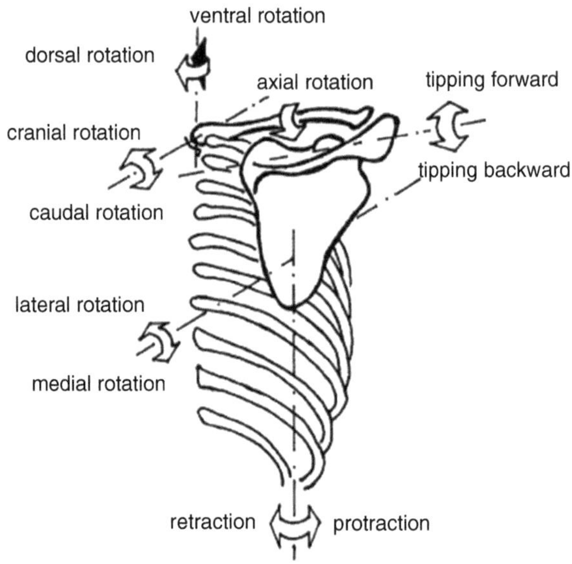

Fig. 7.10 Shoulder rotations (courtesy of Pronk 1991)

7.5 Range of Motion

The combined motions of the glenoid along with the scapula are called the scapulohumeral rhythm. At the onset of arm elevation, the participation of the scapula is highly variable. Through the first 60° of flexion and 30° of abduction, the scapula hardly moves at all (Inman 1994), after which a continuous and synchronous movement between the scapula and the arm was found. For people with defects at the shoulder, the scapulohumeral rhythm may be highly variable (Reitveld 1986).

The different types of movements that take place in the shoulder girdle are given below along with the range of each kind of movements.

Abduction and adduction: This is the motion of the arm in which the moving body part (here the arm) is moving in the coronal plane; the average range of motion is about 180° of abduction and 75° of adduction, as in Fig. 7.11.

Flexion and extension (or elevation and backward elevation, respectively): This is the motion of the arm in a vertical plane parallel to the saggital plane; the average range of motion is about 180° for flexion and 60° for extension, shown in Fig. 7.11.

Horizontal flexion and horizontal extension: This is the motion of the arm in the transversal plane; the average range of motion is about 130° for horizontal flexion and 60° for horizontal extension, as in Fig. 7.11.

Exorotation and endorotation (or outward rotation and inward rotation): This is the motion of the arm with the lower arm at a 90° angle with the upper arm around the shaft of the upper arm; the average range of motion is about 60° for exorotation and 80° for endorotation, as in Fig. 7.11.

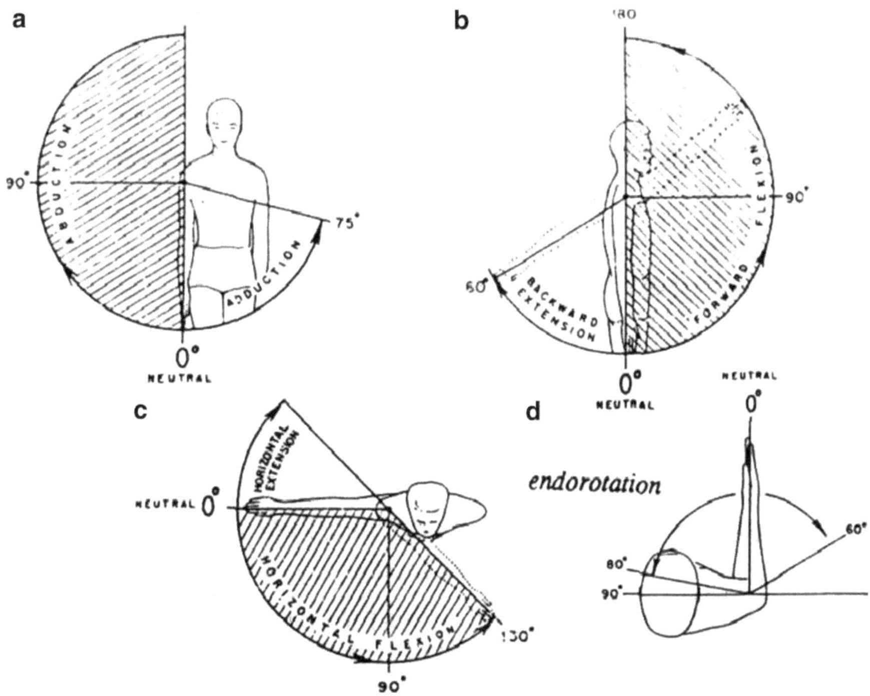

Fig. 7.11 (**a**) Abduction; (**b**) flexion/extension; (**c**) horizontal flexion/extension; (**d**) exorotation/endorotation

A "normal" range of motion does not exist, as it depends upon age, gender, weight, etc. Thus, an *average* range of motion exists.

In 1993, an extensive description of the finite-element model of the shoulder mechanism was published by Van der Helm. In part, the model includes the following descriptions:

Glenohumeral joint (GH): Both articular surfaces of the GH joint were modeled as spheres with identical radii (Van der Helm et al. 1991). Then the GH joint behaves as a spherical joint with a rotation center fixed with respect to the scapula. Because of the large range of motion of the GH joint, the joint capsule is lax and it was assumed not to transmit any forces within the normal physiological range of motion. Consequently, the vector resulting from muscle forces and external forces at the humerus must intersect the articular surface of the glenoid cavity; otherwise, it cannot be counterbalanced by any joint reaction force and the joint will dislocate [3]. Pointing the resultant force vector to the glenoid cavity was added as a constraint in the optimization procedure. Van der Helm [6, 7] analyzed the shoulder joint elaborately and calculated the forces in three directions, as listed in Table 7.1.

In the analysis of the kinematic and dynamic behavior of the shoulder mechanism, Van der Helm, gave a detailed description of the muscular forces. He stated that the *deltoid muscle* has the largest physiological cross-sectional area (PCSA) of the muscles of the shoulder mechanism and exerts by far the largest moments around

Table 7.1 Glenohumeral joint reaction forces (*N*) corresponding to humeral abduction angles [6, 7]

Load case	Abduction angle (°)	F_X	F_Y	F_Z	Resultant force (N)
1	0	−6.06	−16.41	8.07	371.14
2	30	164.46	14.03	−16.14	165.84
3	60	323.74	−36.88	−3.68	352.85
4	90	383.71	−77.28	34.62	392.95
5	120	314.03	−137.96	45.56	346.01
6	150	137.74	−134.43	11.86	192.83
7	180	39.78	−72.51	−3.73	82.79

the GH joint. Through the combined lateral rotation and tipping backward of the scapula, together with axial rotation of the humerus, the moment arm of the posterior part becomes smaller. Above 90° anteflexion, the scapular part of the deltoid muscle has the larger moment arm and becomes active. The biarticular part of the clavicular part of the deltoid muscle more or less has the same pattern of activity as the scapular part. Peak levels of activity are at 90° abduction and 60° anteflexion.

7.6 Total Shoulder Replacement (Glenohumeral Arthroplasty)

A shoulder replacement is a commonly performed surgery for arthritis and joint diseases that has been successfully performed for more than 40 years as of this writing in 2011.

The first shoulder replacement was performed in 1893 by a French surgeon, Péan, for a tuberculosis infection. However, it was not until the 1970s that shoulder replacements were routinely performed. Initially, implant designs were highly constrained devices that did not accurately restore shoulder biomechanics and ultimately failed. Modern total shoulder implants allow for more motion and less constraint. A shoulder replacement involves replacing the end of the arm bone (humerus) and the socket (glenoid) with metal and plastic parts that then act as a new shoulder joint. A plastic (polyethylene) glenoid component is cemented to the glenoid. The parts are made of cobalt chrome, titanium, and/or polyethylene.

7.6.1 Reasons for Shoulder Arthroplasty

The most common reason for undergoing shoulder replacement surgery is *osteoarthritis*. Osteoarthritis is caused by the degeneration of the joint over time, through wear and tear. Osteoarthritis can occur without any injury to the shoulder, but that is uncommon. Because the shoulder is not a weight-bearing joint, it does not suffer as much wear and tear as other joints. Osteoarthritis is more common in the hip and knee. Most of the time osteoarthritis occurs many years after an injury to the shoulder. A shoulder dislocation can result in an unstable shoulder. The extra movement or repeated dislocation of the unstable joint causes damage to the articular cartilage and other joint tissues.

Over time, this damage leads to osteoarthritis. Osteoarthritis is not the only type of arthritis that affects the shoulder joint. Systemic diseases, such as *rheumatoid arthritis*, may affect any joint in the body. Certain types of shoulder fractures can injure the blood vessels of the humeral head. The fracture may heal, but the blood vessels don't. When the blood vessels are damaged, the humeral head no longer has any blood supply. This condition leads to a condition called necrosis. In necrosis, parts of the joint surface actually die. Eventually, this necrosis can lead to arthritis. Shoulder joint replacement should be carried out when fractures affect the humeral head. Like any arthritic condition, osteoarthritis of the shoulder may respond to anti-inflammatory medications such as aspirin or ibuprofen. Acetaminophen may also be prescribed to ease the pain. Some of the newer medications such as *glucosamine* and *chondroitin sulfate* are more commonly prescribed today. They seem to be effective in helping reduce the pain of osteoarthritis in all joints. Physical or occupational therapy may be suggested to help patients regain as much of the motion and strength in their shoulder as possible before they undergo surgery. An injection of cortisone into the shoulder joint may give temporary relief. Cortisone is a powerful anti-inflammatory medication that can ease inflammation and reduce pain, possibly for several months.

7.7 Types of Shoulder Prostheses

There are mainly three types of shoulder arthroplasty, as shown in the Table 7.2, namely, the **hemi-arthroplasty**, the **total arthroplasty,** and the **reverse shoulder arthroplasty**. A hemi-arthroplasty involves replacing the humeral head and not replacing the glenoid (socket), which might be the best option if the glenoid does not have any arthritis or if there is some concern that the glenoid component might fail if it is replaced. Hemi-arthroplasty of the shoulder has given excellent results for glenohumeral arthritis in patients with massive irreparable rotator cuff tears [8]. A total shoulder involves replacing the humeral head and the glenoid. A total shoulder might be the best option if the glenoid is damaged but sufficient bone and rotator cuff remain to ensure that the glenoid component will last. A total shoulder is contraindicated if the rotator cuff is not intact. A reverse shoulder arthroplasty involves replacing both the humeral head and the glenoid, but the ball and socket are reversed to improve the muscle function. Because the center of rotation is translated medially, the deltoid muscle has a longer moment arm and can generate more force. The deleterious effects of translating the center of rotation are a decreased range of

Table 7.2 Types of shoulder arthroplasty

Hemi-arthroplasty	Total shoulder arthroplasty	Miscellaneous
Conventional	Fixed fulcrum	Excisional arthroplasty
Bipolar	Unconstrained	Allograft/alloprostheses
Surface replacement	Semiconstrained/fully constrained	Interpositional arthroplasty

motion and increased impingement. This increased impingement causes scapular notching and can undermine the glenoid component.

There are two major types of artificial shoulder replacements: a **cemented prosthesis** and an **uncemented prosthesis**. A cemented prosthesis is held in place by bone cement that fixes the metal to the bone. An uncemented prosthesis has a fine mesh of holes on the surface. Bone grows into the mesh. Over time, this anchors the prosthesis to the bone.

The motion of the shoulder joint is the combined motions of the different joints and the muscles constituting the shoulder girdle. Based on this, there can be another type of classification: (1) the fully constrained type; (2) the unconstrained or non-constrained type; (3) the semiconstrained type.

7.8 Fully Constrained Type

Initially, the total shoulder arthroplasty (TSA) was of the constrained type in which the humeral and glenoid components were coupled around a fixed center of rotation, as in Fig 7.12. This surgery was used in patients with rotator cuff deficiencies and helped in preventing superior migration of the humeral components. In this kind of design, the majority of the loads are borne by the prosthesis alone; hence, the propensity for loosening of the prosthesis and disused osteoporosis of bones can occur. With this type of prosthesis, there is almost no translation between the humeral and glenoid components of the prosthesis, which was achieved by a fixed ball-and-socket joint that allows rotational movement but prevents translation. Thus, it is not possible for this type of prosthesis to dislocate or subluxate. This type of prosthesis was used by Péan [9]. Translation between the components was not possible, as the humeral component was locked to the glenoid component via two metal loops and a rubber ball with grooves. In those cases in which the rotator cuff can be repaired, a semiconstrained prosthesis should be used [10]. This contains a fixed center of rotation, so the possibility of fracture of the scapula increases postoperatively as this

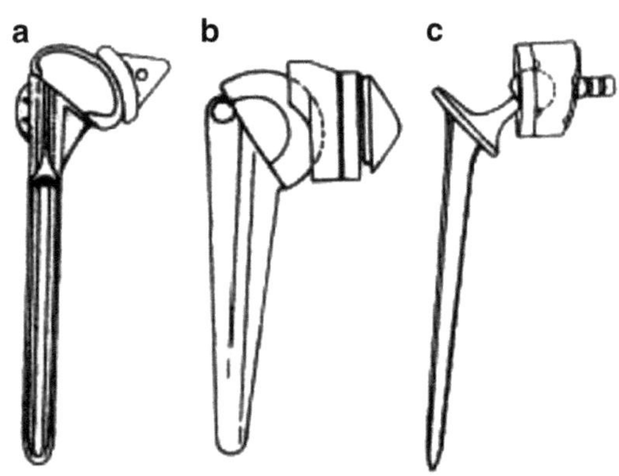

Fig. 7.12 (**a**) Nonconstrained; (**b**) semiconstrained; (**c**) fully constrained

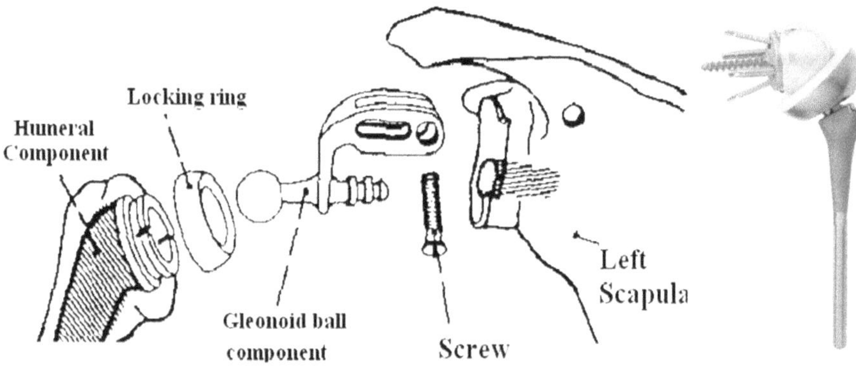

Fig. 7.13 Schema of Kolbel reversed prosthesis (from Kolbel et al. 1986) and a photo

is caused by the arm not being to translate due to a fixed center of rotation. These types of prostheses are not generally used nowadays.

Thus, the constrained type can further be classified into two types: the anatomical type and the reversed type.

In the **anatomical type**, the ball is attached to the humerus and the socket to the scapula. There are several constrained prostheses of the anatomical type [Bickel prosthesis (Cofield 1977), Buechel's "floating socket prosthesis" (1977), Michael Reese prosthesis (Post et al. 1978), Zippel prosthesis]. This prosthesis consists of a metal glenoid cup, two polyethylene pieces that lock around the metal humeral ball, a tightened metal locking ring with screw thread, a humeral stem, two screws, and a central metal glenoid stem. A perfect concentricity would cause large frictions and should therefore be prevented. The prosthesis has a fixed fulcrum (fixed center of rotation).

In the **reversed type**, the ball is attached to the scapula and the socket to the humerus. The constrained reversed type of prosthesis is not very well known. Kolbel (1986) designed a prosthesis in which the ball is attached to the scapula and the socket to the humerus (Fig. 7.13). The metal ball is attached to the scapula via a stem; this stem has rims for better fixation and a side piece or arm with two grips that go around the spina scapulae. The glenoid stem is cemented. The humeral component exists of a polyethylene stem with a socket; this is also cemented. The geometrical center relative to the humerus has remained almost exactly at the same position, and relative to the scapula it is displaced slightly laterally.

The reverse ball-and-socket design has had early success in restoring shoulder function and relieving pain in patients with a rotator cuff–deficient shoulder [1, 5, 9–13]. Shoulder prostheses that were derived directly from hip replacements, with a constrained ball and socket, have failed. Hemi-arthroplasty, the current standard of care for this condition, offers only "limited goals" for functional improvement and only modest improvements in pain [4, 14–16]. Likewise, bipolar implants, used in an effort to improve stability, have produced results not significantly better than hemi-arthroplasty [17, 18].

In the reversed design, the forces in the joint are directed through the center of the glenosphere, converting the centrifugal (outward) forces into centripetal (inward) forces. This in turn creates inherent stability in the reversed design because of the congruency of the humeral socket and glenosphere.

7.9 Nonconstrained Type

The largest experience to date is with the Neer II system. Neer advocates cementing the glenoid component, which may be either all polyethylene or metal-backed (Zadeh et al. 1998). On the humeral side, the prosthesis may be fixed with cement or inserted in a press-fit fashion. These types of prostheses do not have any physical link between the humeral and glenoid components and depend totally on the surrounding musculotendinous cuff for stability, as in Fig. 7.14. The glenoid component conforms to the humeral head. The components try to recreate the natural anatomy and relationship. This minimizes less stress concentration at the interface and allows early rehabilitation. This is the type of prosthesis used in most cases. An unconstrained prosthesis replaces the glenohumeral joint, with the stability provided by the soft tissue and a functioning rotator cuff. This means it must be possible to save the rotator cuff to a reasonable standard. Some surgeons use this prosthesis independent of the status of the rotator cuff and soft tissue around the joint, because it resembles the natural geometry of the joint best. The unconstrained prosthesis usually is a prosthesis that resembles the natural glenohumeral joint anatomically well. These types of prostheses also have a translation possibility, and the range of such translation lies between the constrained and nonconstrained types, as in Fig. 7.14. The semiconstrained prosthesis provides more stability than the unconstrained prosthesis and is used in cases of a malfunctioning rotator cuff and/or soft tissue. The semiconstrained prosthesis differs from the constrained prosthesis in that it is possible for the humeral head to dislocate. The difference between a semiconstrained and unconstrained prosthesis is a bit unclear. The glenoid component is more constrained than the unconstrained prosthesis in that the glenoid component has a larger articular surface that grips the humeral component more firmly. This

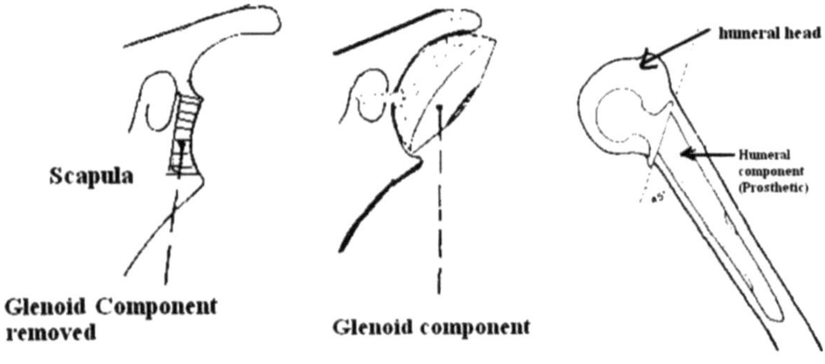

Fig. 7.14 Semiconstrained type, letournal Lagrange from Francobal, Howmedica int. (Laurence 1991)

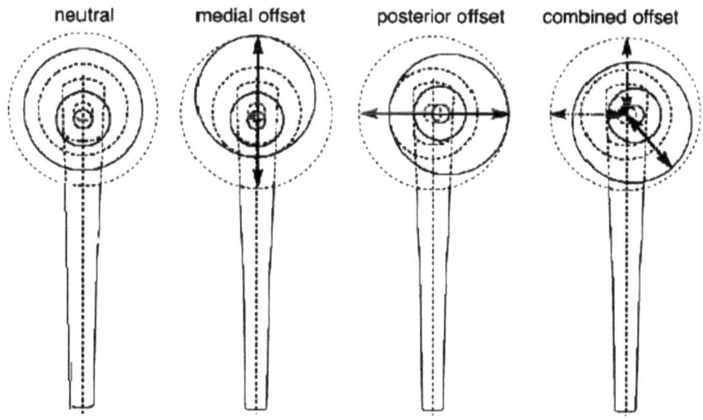

Fig. 7.15 Various offsets of glenoid and humeral radii

can be achieved, for example, by an upper lip on the glenoid to prevent upward subluxation or dislocation. If we look at the constraintness index of the semiconstrained prosthesis, a value up to 1/2 can be found. Either component leaves one direction open for translation of the other component.

Most semiconstrained prostheses differ from the unconstrained ones only by, for example, an extra upper lip on the glenoid or any other form of extra constraintness on the glenoid component. The most commonly used semiconstrained prosthesis is the anatomical Stanmore total shoulder prosthesis [10].

The difference between the semiconstrained and constrained is that in this semiconstrained prosthesis, a subluxation or dislocation is possible. If the forces across the prosthesis are too large, the humeral component can snap out of the glenoid component before a fracture occurs. In the constrained types, this is not possible and the prosthesis or bone either deforms or breaks. This prosthesis maintains the geometrical center relative to the glenoid but displaces that relative to the humerus upward.

The geometry of the prostheses determines shoulder function. First, the combined offset and the retro-torsion of the head are adjusted. Then, the correct version of the glenoid must be restored in order to avoid instability and excessive rim loading and to distribute the pressures evenly. As opposed to the subchondral bone plate, the glenoid and humeral cartilage surfaces are closely conforming. Congruence, a measure of the conformity between two surfaces, is defined as the difference in the radii of curvature of the humeral head and the glenoid as in Fig. 7.15. The closer this difference is to zero, the more congruent the joint will be. Their matching articular surfaces can be considered a stabilizing factor and an improvement of the transmission of forces to the subchondral bone. Radial match improves joint stability and muscle efficiency. Reduction of the glenoid size and its special geometry allow a better range of motion and prevent glenohumeral impingement. Abutment of the humerus on the glenoid rim limits abduction and rotation. This can lead to high shear forces, causing early glenoid loosening. The glenoid implant was designed to

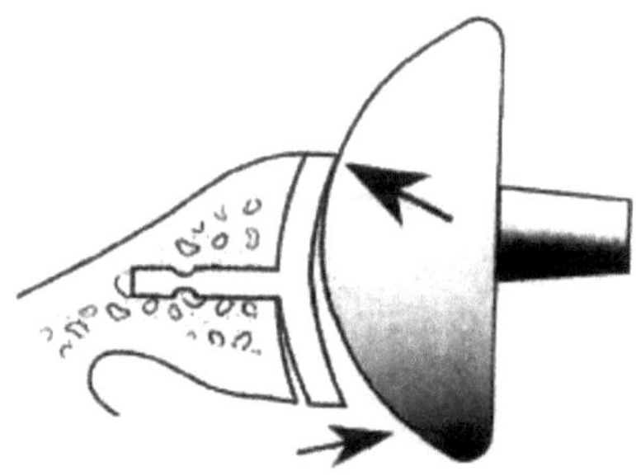

Fig. 7.16 The "rocking horse" glenoid

minimize the risk of impingement through reduction of size, an oblique rim, and conformity of the radius. To avoid overstuffing of the joint, the glenoid implant was designed to be as thin as possible.

The convex design is for minimal bone resection and the macro structure for an even cement mantle at the bony surface. Two pegs are located far apart from one another to achieve stable fixation in the cement and to reduce the "rocking horse" effect as shown in Fig. 7.16.

Humeral head translation and eccentric glenoid loading are present in normal shoulders, but the effect becomes more marked in shoulder instability amid rotator cuff dysfunction (Harryman et al. 1995).

In 1955, Neer introduced his prosthesis for the replacement of a fractured humeral head. The hemi-arthroplasty performed by Neer involved an articular replacement of the humeral head by a metal-stemmed humeral head; a glenoid was not used. Later, in the 1970s, a total shoulder prosthesis was developed following the concept of total shoulder replacement, replacing both the humeral head and the glenoid by artificial surfaces, as in Fig. 7.17. The Neer humeral component consists of a stem and a humeral head. The glenoid component has a keel to attach it to the scapula, and the radius of curvature of its articular surface is similar to that of the humeral component. The glenoid component is available in a high-density polyethylene or a metal-backed polyethylene design. The geometrical center relative to the humerus remains almost at the same position, and that relative to the scapula is laterally displaced a distance equal to the thickness of the glenoid component.

The second is the **bio-modular total shoulder prosthesis** (Neer) as in Fig. 7.18. This is a prosthesis that has several humeral components of different stem sizes on which several humeral heads of different diameters can be attached, depending upon the natural diameter of the humeral head. The humeral component is coated and press-fit. The glenoid component is available in a polyethylene version with or without a metal backing and has a rim and screws for attachment. It has one diameter for all head sizes, which results in the conformity index varying per prosthesis. The geometrical center in relation to the humerus is maintained, but that in relation to the scapula is again displaced laterally by a distance equal to the thickness of the glenoid component.

The modular systems have become more popular nowadays. The Neer II type of prosthesis is now available in both modular and conventional forms. The modular

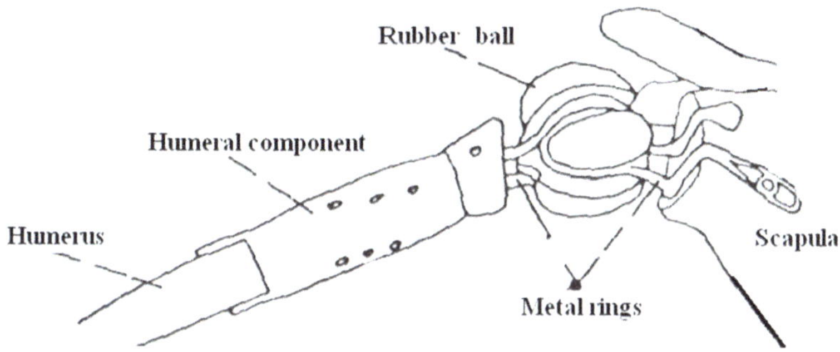

Fig. 7.17 The earliest Neer I prosthesis

Fig. 7.18 The bimodular shoulder prosthesis

systems are available in a variety of humeral head sizes (diameters) and humeral stem lengths. The flexibility afforded by these modular systems makes the adjustment of soft tissue tension, rotator cuff closure, implant fixation, and stability easier to achieve (Flatow 1995, [19]). At times, revision surgery is also made easier by retaining a well-fixed humeral stem (Fenlin et al. 1994). The more complex designs are the Aequalis® (Forth Medical Ltd., Newbury, Berkshire, UK) (Fig. 7.19a), Nottingham® (Biomet Ltd., Bridgend, South Glamorgan, UK) (Fig. 7.19b), and Randeli® (Lime-Ltd. Medical Systems, Casiacco, Italy) systems, which allow the humeral head to be offset in various directions.

The Biangular® system (Biomet Ltd., Newbury, Berkshire, UK) allows the anteversion of the humeral component to be adjusted individually. The Delta (Grammont®) (Medinov, Routine, France) shoulder system has options for having the ball-and-socket part of the prosthesis in either anatomical or reversed orientation, as in Fig. 7.20. This comes with a hydroxyapatite-coated stainless steel humeral stem. The Eska® modular (Eska Implants GmbH & Co., Lubeck, Germany), shown in Fig. 7.21, and RPS® shoulder systems (Lima-Ltd. Medical Systems, Casiacco, Italy) accommodate various lengths of upper humeral bone loss by using interchangeable lengths of metal spacers.

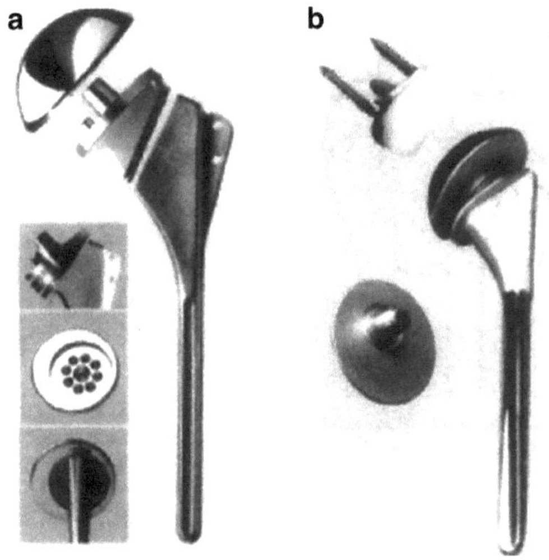

Fig. 7.19 (**a**) The Aequalis system; (**b**) the Nottingham system

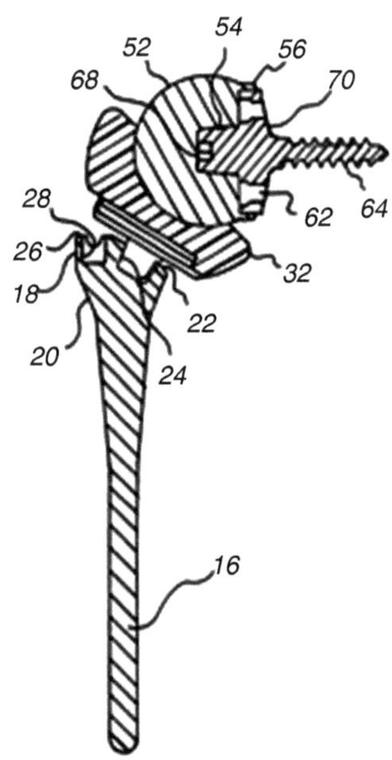

Fig. 7.20 The Delta (Grammont) prosthesis

The disadvantages of complex modular systems are increased cost, complexity, and potential technical errors at surgery. Increased wear debris production is more likely with a greater number of implant interfaces. Furthermore, component dissociation is a well-recognized complication (Fenlin et al. 1994). Another factor to take into account is that, for a given head, the amount of articulating surface area

Fig. 7.21 The Eska prosthesis

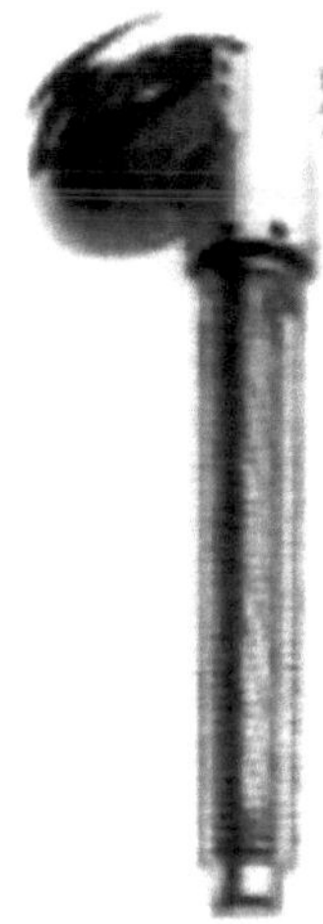

available for the humeral component is larger in a one-piece implant as compared with the equivalent modular type. This is due to the space taken by the Morse taper and the humeral component collar in the modular version. Therefore, the maximum range of movement is potentially lower in a modular system as compared with the equivalent one-piece type as in Fig. 7.22 (Fenlin et al. 1994). The dissociation of modular humeral heads has been further investigated by Blevins and co-workers (Blevins et al. 1997). They proposed that the most likely cause of this complication was contamination of the Morse taper by blood. They made the observation that as little as 0.4 ml of fluid can prevent proper seating of the taper, as in Fig. 7.23.

The **cup prosthesis** is a type of unconstrained prosthesis that has a cup instead of a humeral stem and head. Only Jonnson et al. and Steffee and Moore (1984) have used the cup prosthesis. The one used by Jonnsson et al. (1986, 1990) consists only of a humeral component. The humeral component is a hemispheric cup. The cup has both circumferential and radial grooves on the inside to provide fixation and to permit extrusion of excess cement. The main disadvantage is the geometrical center, which is situated on the plane through the cutting of the cup. This means the geometrical center is at least displaced upward since the natural head is only one third of a sphere.

7.10 Fixation Procedures

There are two methods of fixation of prostheses: cemented or cementless. PMMA (polymethyl methacrylate) is widely used for fixation. But the use of PMMA has been reported to be less reliable, particularly in the presence of instability or a

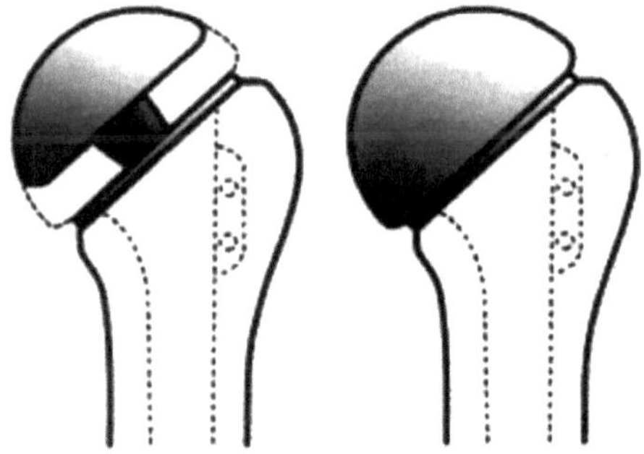

Fig. 7.22 The articulating surface is reduced

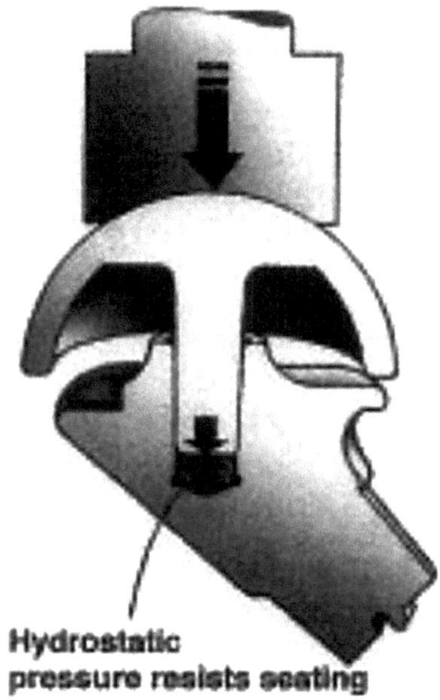

Fig. 7.23 Morse taper in modular system as compared with the equivalent one-piece version

massive tear of the rotator cuff (Barett et al. 1987: [14, 20]). Sometimes both the total polyethylene and the metal-backed components fixed with cement demonstrated radiographic lucent lines at the cement–bone interface of the glenoid component (Amstutz et al., Cofield and Daly 1992). The problem of progressive radiographic changes has prompted the development of implants with porous surfaces having capabilities of bone ingrowth to achieve secure fixation with bone [21].

7.11 Causes of Failure

The failure scenarios for total hip arthroplasty, introduced by Huiskes [11], are also useful, more or less, to analyze failure mechanisms of other reconstructed joints like the TSA. Although the failure of an implant is mainly due to biological causes, the initiation of the failure process may be due to mechanical events. Two dominant failure scenarios of the cemented reconstruction can be identified [11]:

1. According to the particulate reaction failure, the **materials or interfaces are too weak** to sustain the effect of long-term, dynamic loads applied on the implants. The likelihood of mechanical failure depends on the stress induced in a material versus the strength of the material. The repetitive nature of the external loads generates high dynamic stresses in the materials and at the interfaces. As a result, mechanical damage, typically micro cracks, is gradually accumulated in the cement. These micro cracks reduce the strength of the cement and its bonds at the layer of the interface between implant and bone, eventually causing failure. With uncemented prostheses, loosening may occur due to the failure of the implant–bone interface as well as the polyethylene–metal interface. The polyethylene cup may be dissociated from the metal backing, which may still maintain a secure fixation with bone, thus resulting in failure of the prosthesis. The eventual gross loosening of the implant may be due to cement–bone interface loosening, failure (cracking) of the cement due to excessive stresses, and relative motions between the materials.
2. According to the particulate reaction scenario: The **cement–bone interface** gradually disintegrates due to the migration of wear particles. These tiny particles may be polyethylene debris, abraded from the cup of the glenoid prosthesis, cement particles abraded from the cement mantle, or metal debris burnished from the implant. The polyethylene wears debris in particular is considered to be the major threat to any arthroplasty. These wear debris cause particulate reactions by macrophages, osteolysis, soft tissue interposition, and, finally, gross loosening.

7.11.1 Loosening of the Component

Symptomatic loosening of the glenoid and humeral components after a total shoulder replacement arthroplasty is common and represents nearly one third of all complications that are associated with this operation (Averill 1980).

7.11.1.1 Loosening of the Glenoid Component

There is a close relationship among the symptomatic loosening of the glenoid component, glenohumeral instability, and irreparable tears of the rotator cuff (Barret 1987).

7.11.1.2 Loosening of the Humeral Component

Difficulties with fixation of the glenoid component account for most of the complications related to aseptic loosening of total shoulder prostheses.

7.11.2 Anterior Instability

Anterior instability is most commonly associated with malrotation of the humeral component, dysfunction of the anterior part of the deltoid, or disruption of the sutured subscapularis tendon (Fenlin 1975).

7.11.3 Superior Instability

The amount of proximal humeral migration was independent of the size of the defect of the rotator cuff but was positively associated with a torn rotator cuff and poor pre-operative function. Although superior migration of the humeral head is a well-recognized complication of shoulder hemi-arthroplasty, it does not appear to be directly related to the development of discomfort in the shoulder or to impending failure after this operation. The potential complication of loosening of the glenoid component as a result of progressive superior migration of the humeral component after total shoulder arthroplasty is more of a problem. In this setting, the humeral head articulates with the superior portion of the glenoid component and results in eccentrically applied glenoid compressive forces, increased stress at the bone–cement interface, and eventual loosening of the glenoid component within the glenoid fossa (Franklin et al. 1988).

7.11.4 Posterior Instability

Asymmetrical wear of the posterior aspect of the glenoid is characteristic of long-standing osteoarthritis, and a failure to recognize substantial posterior erosion of the glenoid may lead to placement of the glenoid component in excessive retroversion at the time of the operation, resulting in an increased propensity for posterior instability (Cofield 1990).

7.11.5 Inferior Instability

Inferior instability after shoulder arthroplasty usually occurs as a complication of treatment of acute fractures of the proximal aspect of the humerus but has also been noted after total shoulder replacement for prosthetic revision, chronic fracture, and previous osteosynthesis. This optimizes function by minimizing the tendency for inferior instability and subsequent weakness during elevation.

7.11.6 Rotator Cuff Tears

Postoperative tearing of the rotator cuff is the third-most-frequent complication of total shoulder arthroplasty.

7.11.7 *Intraoperative Fractures*

For the most part, intraoperative fractures of the humerus or glenoid are the result of operative errors, many of which are avoidable. These errors include inadvertent reaming, overzealous impaction, or manipulation of the upper extremity during exposure of the glenoid (Bonutti 1992).

7.11.8 *Infection*

Infection after total shoulder arthroplasty is a rare but potentially devastating complication. In most cases, an increased susceptibility to infection is associated with host-related risk factors, such as diabetes mellitus, rheumatoid arthritis, systemic lupus erythematosus, remote sites of infection, and previous operations on the shoulder. Additionally, extrinsic factors such as immunosuppressive chemotherapy, systemic administration of corticosteroids, and multiple injections of steroids have been noted as risk factors.

7.11.9 *Neural Injuries*

Most of these injuries involved a neurapraxia, and nonoperative treatment yielded good results.

References

1. Högfors G, Sigholm P, Herberts P (1987) Prosthetic replacement of the shoulder joint requires familiarity with its anatomy and biomechanics. Maintenance of the origin and insertion of the deltoid muscle is critical to the final outcome of this procedure. Biomechanical model of the human shoulder—I. Elements. J Biomech 20:157–166, 1987
2. Dvir Z, Berme N (1978) The shoulder complex in elevation of the arm: a mechanism approach. J Biomech 11:219–225
3. Van der Helm FCT, Veenbaas R (1991) Modeling the mechanical effect of muscles with large attachment sites: application to the shoulder mechanism. J Biomech 24(12):1151–1163
4. Morrey BF, An KN (1990) Biomechanics of the shoulder. In: J. Shoulder Elbow Surg, Bristol. pp. 207–245
5. Maki S, Gruen T (1976) Anthropometric study of the glenohumeral joint. 22nd annual ORS, New Orleans, LA, January 28–30, 1976: p.173
6. Van der Helm FCT (1994) Analysis of the kinematic and dynamic behaviour of the shoulder mechanism. J Biomech 27:527–550
7. Van der Helm FCT (1994) A finite element musculoskeletal model of the shoulder mechanism. J Biomech 27:551–569
8. Wirth MA (1996) Part I: periprosthetic fractures of the upper extremity. In: Rockwood CA, Green DP, Bucholz RW, Heckman JD (eds) Fractures, vol 1, 4th edn. Lippincott, Philadelphia, pp 540–576

9. Lugli T (1978) Artificial shoulder joint by Péan (1893). The facts of an exceptional intervention and the prosthetic method. Clin Orthop Relat Res 133:215–218
10. Lettin AW, Copeland SA, Scales JT (1982) The Stanmore total shoulder replacement. J Bone Joint Surg Br 64(1):47–51
11. Huiskes R (1993) Failed innovation in total hip replacement. Acta Orhtop Scand 64:699–716
12. R.S. Khandpur, Handbook of biomedical engineering, 6th edn.
13. MIMICS, Theory Manual, V. 10.0, MIMICS® Inc.
14. Franklin JL, Barrett WP, Jackins SE, Matsen FA III (1988) Glenoid loosening in total shoulder arthroplasty. Association with rotator cuff deficiency. J Arthroplasty 3:39–46
15. Post M (1987) Constrained arthroplasty of the shoulder. Orthop Clin N Am 18:455–462
16. Post M, Jablon M (1983) Constrained total shoulder arthroplasty. Long-term follow-up observations. Clin Orthop 173:109–116
17. Post M, Haskell SS, Jablon M (1980) Total shoulder replacement with a constrained prosthesis. J Bone Joint Surg Am 62:327–335
18. Post M, Jablon M, Miller H, Singh M (1979) Constrained total shoulder joint replacement: a critical review. Clin Orthop 144:135–150
19. Dines DM, Warren RF (1994) Modular shoulder hemiarthroplasty for accurate fractures: surgical considerations. Clin Orthop 307:18–26, Semin arthroplasty 1995; 6:233–244
20. Weiss A-PC, Adams MA, Moore JR, Weiland AJ (1990) Unconstrained shoulder arthroplasty. A five year follow-up study. Clin Orthop Relat Res 257:86–90
21. Gupta S, Van der Helm FCT, Van Keulen F (2004) Stress analysis of cemented glenoid prostheses in total shoulder arthroplasty. J Biomech 37(11):1777–1786

Problems

1. The shoulder joint is a wonderful joint and has a wide range of movement but is very unstable—explain the statement with quantitative data.
2. Describe the various types of shoulder prostheses that were developed over the years. What is the fixation modality of such a joint? Indicate the materials. Give your opinion about the Indian effort.
3. What are the probable modes of loosening or failure of the shoulder joint?
4. How will you approach customized shoulder joint replacement? A modular prosthesis is desired.

Chapter 8
The Elbow Joint and Its Artificial Replacement

8.1 Introduction

The **elbow** joint [1] is a special hinge joint in the middle of the arm. The three bones that form the elbow joint are the humerus of the upper arm and the radius and ulna of the forearm [1]. The bony prominence at the very tip of the elbow is the olecranon process of the ulna, and the inner aspect of the elbow is called the antecubital fossa.

8.1.1 Movements

The different **movements** possible at the elbow joint include

- Flexion (touch the shoulder with the fingertips of the same side)
- Extension (straighten your upper limb)
- Supination (palm facing upward)
- Pronation (palm facing downward).

Flexion and extension occur mainly at the ulnohumeral joint. Supination and pronation occur at the radioulnar joint. The elbow joint comprises three different portions: the joint between the ulna and humerus, the joint between the head of the radius and the humerus, and the proximal radioulnar articulation, described below. All these articular surfaces are enveloped by a common synovial membrane, and the movements of the whole joint should be studied together. The combination of the movements of flexion and extension of the forearm with those of pronation and supination of the hand, which is ensured by the two movements being performed at the same joint, is essential to the accuracy of the various minute movements of the hand.

The portion of the joint between the ulna and humerus is a simple hinge joint and allows movements of flexion and extension only. Due to the obliquity of the trochlea of the humerus, this movement does not take place in the anteroposterior plane of the body of the humerus. When the forearm is extended and supinated, the axes of the arm and forearm are not in the same line; the arm forms an obtuse angle with the

S. Pal, *Design of Artificial Human Joints & Organs*,
DOI 10.1007/978-1-4614-6255-2_8, © Springer Science+Business Media New York 2014

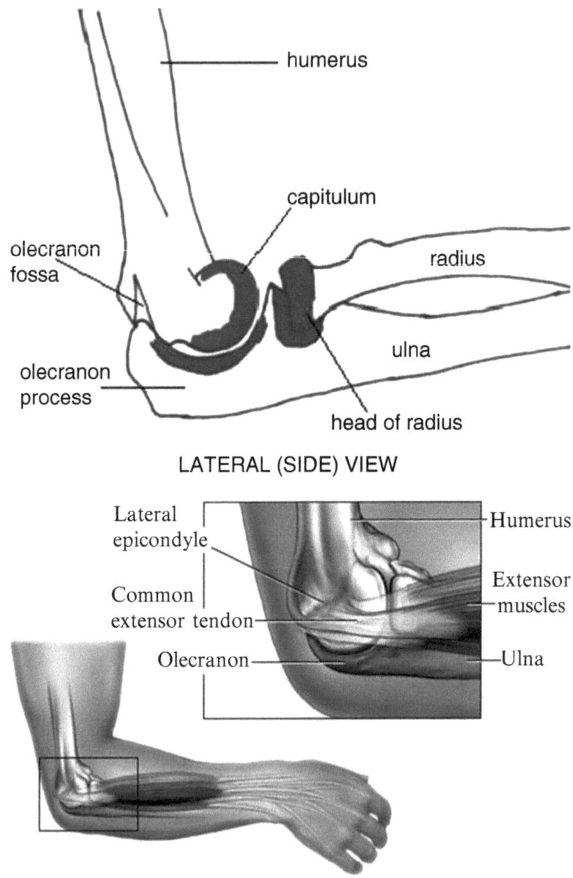

Fig. 8.1 Different views of the elbow joint, indicating bones, muscles, and tendons

forearm, the hand and forearm being directed lateralward (outward). During flexion, however, the forearm and the hand tend to approach the middle line of the body, which enables the hand to be easily carried to the face or mouth. The accurate adaptation of the trochlea of the humerus, with its prominences and depressions, to the semilunar notch of the ulna prevents any lateral movement. *Flexion* is produced by the action of the biceps brachii and brachialis, assisted by the brachioradialis and the muscles arising from the medial condyle of the humerus. *Extension* is produced by the triceps brachii and anconeus, assisted by the extensors of the wrist, the extensor digitorum communis, and the extensor digiti quinti proprius (Fig. 8.1).

Two main movements are possible at the elbow:

1. The hingelike bending and straightening of the dynamic (flexion and extension) joint between the humerus and the ulna, about 80–85°.
2. The complex action of turning the forearm over (pronation or supination) occurs at the articulation between the radius and the ulna (this movement also occurs at the wrist joint).

The real hinge joint moves in only one plane, like a door hinge, but this is not a pure hinge joint. In the anatomical position (with the forearm supine), the radius and ulna lie parallel to each other. During pronation, the ulna remains fixed, and the radius rolls around it at both the wrist and the elbow joints. In the prone position, the radius and ulna appear crossed.

Most of the force through the elbow joint is transferred between the humerus and the ulna. Very little force is transmitted between the humerus and the radius. (By contrast, at the wrist joint, most of the force is transferred between the radius and the carpus, with the ulna taking very little part in the wrist joint.)

8.2 Muscles, Arteries, and Nerves

The muscles in relation with the joint are

- In *front*, the brachialis, the brachioradialis
- *Behind*, the triceps brachii and anconeus
- *Laterally*, the supinator, and the common tendon of origin of the extensor muscles
- *Medially*, the common tendon of origin of the flexor muscles, and the flexor carpiulnaris.

The arteries supplying the joint adontre derived from the anastomosis between the profunda and the superior and inferior ulnar collateral branches of the brachial, with the anterior, posterior, and interosseous recurrent branches of the ulnar, and the recurrent branch of the radial. These vessels form a complete anastomotic network around the joint. The nerves of the joint are a twig from the ulnar nerve, as it passes between the medial condyle and the olecranon, a filament from the musculocutaneous nerve, and two from the median nerve.

8.3 Parts of the Joint

The elbow joint comprises three different portions. All these articular surfaces are enveloped by a common synovial membrane. The movements of the entire joint should be studied together; details are given in Table 8.1. The combination of the movements of flexion and extension of the forearm with those of pronation and supination of the hand, which is ensured by the two being performed at the same joint, is essential to the accuracy of the various minute movements of the hand during music and dance performances with various postures (mudra). The hand is only directly articulated to the distal surface of the radius, and the ulnar notch on the lower end of the radius travels around the lower end of the ulna. The ulna is excluded from the wrist joint by the articular disk. Thus, rotation

Table 8.1 Details of the three portions of the elbow joint

Joint	From	To	Description
Humeroulnar joint	Trochlear notch of the ulna	Trochlea of humerus	A simple hinge joint, allowing movements of flexion and extension only to the extent of 80–85°
Humeroradial joint	Head of the radius	Capitulum of the humerus	An arthrodial joint
Proximal radioulnar joint	Head of the radius	Radial notch of the ulna	In any position of flexion or extension, the radius, carrying the hand, can be rotated. This movement includes pronation and supination nearly 90° or above

of the head of the radius around an axis passing through the center of the radial head of the humerus imparts circular movement to the hand through a very considerable arc.

8.4 Ligaments

The trochlea of the humerus is received into the semilunar notch of the ulna, and the capitulum of the humerus articulates with the fovea on the head of the radius. The articular surfaces are connected together by a capsule, which is thickened medially and laterally and, to a lesser extent, in front and behind. These thickened portions are usually described as distinct ligaments.

The major ligaments are the ulnar collateral ligament, radial collateral ligament, and annular ligament.

8.5 Synovial Membrane

The synovial membrane extends from the margin of the articular surface of the humerus and lines the coronoid, radial, and olecranon fossa on that bone. It is reflected over the deep surface of the capsule and forms a pouch between the radial notch, the deep surface of the annular ligament, and the circumference of the head of the radius. Projecting between the radius and ulna into the cavity is a crescentic fold of synovial membrane, suggesting the division of the joint into two; one is the humeroradial, the other is the humeroulnar.

Between the capsule and the synovial membrane are three masses of fat:

- The largest, over the olecranon fossa, is pressed into the fossa by the triceps brachii during the flexion.
- The second is over the coronoid fossa.
- The third, over the radial fossa, is pressed by the brachialis into its respective fossa during extension.

8.6 Carrying Angle

In a normal radiograph, the picture of the straightened arm shows the carrying angle of the elbow. With the arms extended at the sides and the palms facing forward, the forearm and hands are normally slightly away from the body. This is the normal "carrying angle" of the elbow, which is 5–15°. This angle permits the forearms to clear the hips in swinging movements during walking, and it is important when carrying objects. Women, on average, have smaller shoulders and wider hips than men, which may necessitate a greater carrying angle. There is, however, extensive overlap in the carrying angle between individual men and women, and a gender bias has not been consistently observed in scientific studies [2–4].

The angle is greater in the active limb than in the nondominant limb of both sexes [5, 6], suggesting that natural forces acting on the elbow modify the carrying angle. Developmental [7], aging, and possibly racial influences add further to the variability of this carrying angle. People with a more extreme carrying angle may be more likely to pronate the forearm when holding objects in the hand to keep the elbow closer to the body.

8.7 Diseases of the Elbow

8.7.1 Tendonitis

The types of disease most commonly seen at the elbow are due to injury, usually overuse injuries: tennis elbow and golfer's elbow.

Golfer's elbow involves the tendon of the common flexor origin, which originates at the medial epicondyle of the humerus (the "inside" of the elbow). Tennis elbow is the equivalent injury but at the common extensor origin (the lateral epicondyle of the humerus).

8.7.2 Fractures

As we know, there are three bones at the elbow joint, and any combination of these bones may be involved in a fracture of the elbow. Patients who are able to fully extend their arm at the elbow are unlikely to have a fracture (98% certainty) and an X-ray is not required as long as an olecranon fracture is ruled out [8].

8.7.3 Arthritis

Elbow arthritis is usually seen in individuals with rheumatoid arthritis or after fractures that involve the joint itself. When the damage to the joint is severe, arthroplasty or elbow joint replacement needs to be considered [9].

8.8 Elbow Replacement

In elbow replacement surgery, the painful surfaces of the damaged elbow are replaced with artificial elbow parts usually made of stainless steel (AISI-316L) or titanium alloy. There are two components, a humeral part and an ulnar part. One part fits into the humerus (upper arm), and the other part fits into the ulna (forearm), not the radius. The two parts are then connected and held together by a pin. The resulting hinge allows the joint to flex and extend. The pin will be a double shear, and its diameter is within 8–12 mm depending on the size of the joint, which can be determined from the X-ray of the joint. The maximum force on the joint can be estimated from the weight of the lower limb and any weight being carried by the hand in flexed position at 90°. The joint reaction force may be taken as 6–10 times the weight of the forearm, including the hand, which is approximately 2.1–2.2% of the body weight the hand may carry a 4–5-kg load. The cross section of the stems is usually made of a tapered triangular or rectangular section in the humerus, and a rounded square section may be used for the ulnar part. Usually, bone cement is used for fixation of the implant, and it can make up for any minor size difference. Figure 8.2 shows two such prostheses in assorted forms and assembled and fixed in the bony cavity of the two bones.

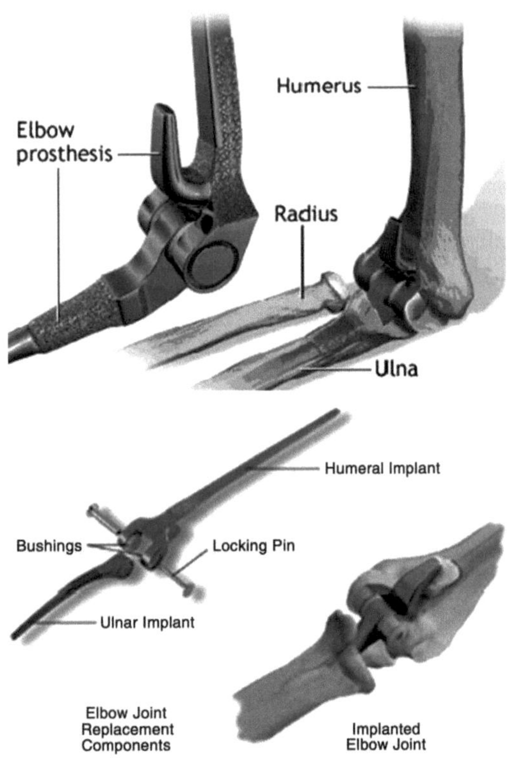

Fig. 8.2 Two sets of elbow prostheses in separation and in installed forms (Western design)

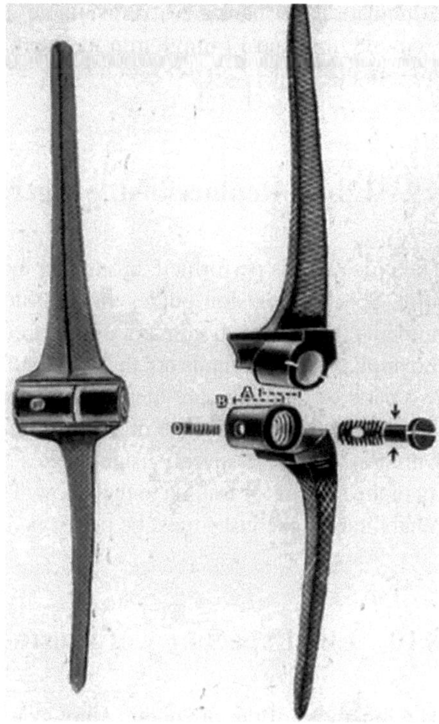

Fig. 8.3 "Sloppy hinge" elbow joint prosthesis showing the assembled and exploded views. Notice the cross sections of the two parts and crisscrossing of the surfaces

Dr. Bakshi (JBJS 1991), orthopedic surgeon in Kolkata, developed an affordable "sloppy hinge" elbow prosthesis over years of research and clinical experimentation while working at Calcutta University.

The upper part is the humeral component, and the lower one is the ulnar part. The surfaces of the stems are knurled (crisscrossed surface), and the cross sections are triangular and square, respectively, in the two components. The two eye ends, when placed axially aligned, are assembled by a threaded screw onto the ulnar part. The screw, shown at the right, has a lower-diameter shoulder. This screw fixed the two components, and unscrewing is prevented by a small screw, as shown in Fig. 8.3. The steams are serrated at the surface to enhance fixation with bone cement. The triangular section of the upper part starts from $15 \times 10 \times 10$ mm and gradually reduces to 4 mm at the tip. The linking screw provides a smooth bearing surface, which is smaller in diameter than the hole in the humeral component (Fig. 8.3). This provides a gap between the bearing surfaces and only partial articular contact during elbow movement; there is 7–10° of laxity, which allows some varus-valgus movement but still limits axial rotation. This particular feature gives it the name "sloppy hinge" and tries to mimic the natural movement of the elbow.

The **humeral** stem is triangular and the ulnar stem quadrangular in cross section; their lengths are 95 mm and 75 mm, respectively, and their curvatures conform to those of the medullary canals. They have cross-cut knurled surfaces. Five different combinations of shank sizes are available: 13 mm/7 mm, 14 mm/7 mm, 15 mm/8 mm, 16 mm/8 mm, and 17 mm/8 mm for the humeral and ulnar stem sizes at the starting point in the joint.

8.9 Elbow Replacement Surgery

The procedure is performed through an incision over the elbow that will expose the joint. Special precision guides and instruments will be used to cut the ends of the humerus (upper arm bone) and ulna (forearm bone) and prepare the bone to accept the implant. The implants are then inserted and fixed in place with bone cement. The two parts of the hinge are then brought together and secured with a pin. When the surgeon is satisfied with the fit and function, the incision will be closed and covered with dressings. The surgery usually takes 1–3 h, although this depends on the severity of the arthritis or trauma to the elbow. The patient is released after a few days, and rehabilitation exercises must be performed, which are important for recovery.

8.10 Life Expectancy of Prosthetic Elbow Joint

The longevity of the prosthetic elbow varies from patient to patient. It depends on many factors, such as a patient's physical condition and activity level, as well as the accuracy of the implant placement during surgery. It is useful to keep in mind that prosthetic joints are not as strong or durable as a natural, healthy joint, and there is no guarantee that a prosthetic joint will last the rest of a patient's life.

Today, total elbow replacement is becoming a common and predictable procedure. Many patients enjoy relief from pain and improved function, compared to their status before surgery. As a result, some patients may have unrealistic expectations about what the prosthetic elbow can do and how much activity it can withstand. As with any mechanical joint, the components move against each other. Natural fluid in the joint space, called synovial fluid, helps to lubricate the implants, just as it lubricates the bones and cartilage in a natural joint. Still, the prosthetic components do wear as they roll and slide against each other during movement. As with car tires or brake pads, the rate of wear depends partly on how the elbow joint is used. Activities that produce a lot of stress on the joint implants, as may be the case with more active patients, may reduce the service life of the prosthesis. Implant loosening and wear on the components can lead to the necessity for revision surgery to replace the worn parts, or all of the parts. The surgeon will be in the best position to discuss these issues with the patient, taking into account the patient's particular clinical circumstances, the type of implants used, and the patient's postsurgical lifestyle. Heavy work, like hammering, impact, boxing, or lifting more than a few kilograms during 90° flexion, should be avoided. Direct load carrying should be kept under 5 kg, if not less. Elderly people should avoid high load carrying in their daily activities.

References

1. eMedicine/Stedman Medical Dictionary
2. Steel F, Tomlinson J (1958) The carrying angle in man. J Anatomy 92(2):315–317, PMID 13525245. http://www.pubmedcentral.nih.gov/articlerender.fcgi?artid=1249704
3. Van Roy P et al (2005) Arthro-kinematics of the elbow: study of the carrying angle. Ergonomics 48(11–14):1645–1656, PMID 16338730
4. Zampagni M et al (2008) Estimating the elbow carrying angle with an electrogoniometer: acquisition of data and reliability of measurements. Orthopedics 31(4):370, PMID 19292279
5. Paraskevas G et al (2004) Study of the carrying angle of the human elbow joint in full extension: a morphometric analysis. Surg Radiol Anat 26(1):19–23, PMID 14648036
6. Yilmaz E et al (2005) Variation of carrying angle with age, sex, and special reference to side. Orthopedics 28(11):1360–1363, PMID 16295195
7. Tukenmez M et al (2004) Measurement of the carrying angle of the elbow in 2,000 children at ages six and fourteen years. Acta Orthop Traumatol Turc 38(4):274–276, PMID 15618770
8. Appelboam A et al (2008) Elbow extension test to rule out elbow fracture: multicentre, prospective validation and observational study of diagnostic accuracy in adults and children. BMJ 337:a2428. doi:10.1136/bmj.a2428
9. Warme WJ, Matsen FA III (eds) (2010) Total elbow joint replacement for elbow arthritis: surgery with a dependable, time-tested prosthesis can lessen pain and improve function in elbows, especially those with rheumatoid arthritis

Problems

1. Study the elbow joint anatomy to find the kinematics and kinetics of the joint. Find the range of movement of the joint. How many articulating joints are there in the elbow joint? Try to estimate the forces in the joints.
2. The elbow joint is not a perfect hinge joint, but it is more than that. If an arthritic joint is to be replaced, how may a sloppy hinge joint prosthesis help? Check the sloppy hinge elbow joint designed by Dr. D. Bakshi, orthopedic surgeon in Kolkata. Is there any drawback to the joint from the viewpoint of friction in the articulating surface? How you can improve its frictional characteristics? The stems were cut crisscrossed for better fixation. Can you improve upon it? It has triangular and square cross sections in its humeral and ulnar components; why not conical or cylindrical? Explain from the point of view of effective function.
3. Explain a step-by-step procedure to design an elbow prosthesis for a 65-year-old lady suffering from painful osteoarthritis. How will you proceed? Make a nice sketch with the approximate main dimensions.

Chapter 9
The Wrist Joint and Its Artificial Replacement

Joint replacement surgery in the wrist is less common than other joint replacements, but it can be an option if the patient has severe painful arthritis that does not respond to other treatments. Three very illustrative views of the wrist joint are shown in Figs. 9.1a–c, showing the anatomy, the bones' articulating surfaces, and the nomenclature of each bone. A wrist X-ray is shown in Fig. 9.1c.

9.1 Anatomy

The wrist is a more complicated joint than the shoulder and the elbow. On the hand side of the wrist, there are two rows of bones at the base of the hand. There are four bones in the first row. The bones in these rows are called the **carpals**. The long thin bones of the hand radiate out from one row of carpals and form the basis of the fingers and thumb.

The radius and the ulna are the two long bones of the forearm that form a joint with the first row of carpals.

The ends of the bones are covered with an elastic tissue cartilage. Cartilage creates a slick surface that enables the bones to move smoothly when they move against each other. The friction is very low, and synovial fluid lubricates the joints as usual.

9.2 Rheumatoid Arthritis of the Hand and Wrist

If the cartilage is worn away or damaged by injury, infection, or disease, the bones themselves will rub against each other, wearing out the ends of the bones. This causes a painful, arthritic condition (Fig. 9.2). Osteoarthritis, the most common form of arthritis, results from a gradual wearing away of the cartilage covering on the bones (Papp., S.R. et al., 2006).

S. Pal, *Design of Artificial Human Joints & Organs*,
DOI 10.1007/978-1-4614-6255-2_9, © Springer Science+Business Media New York 2014

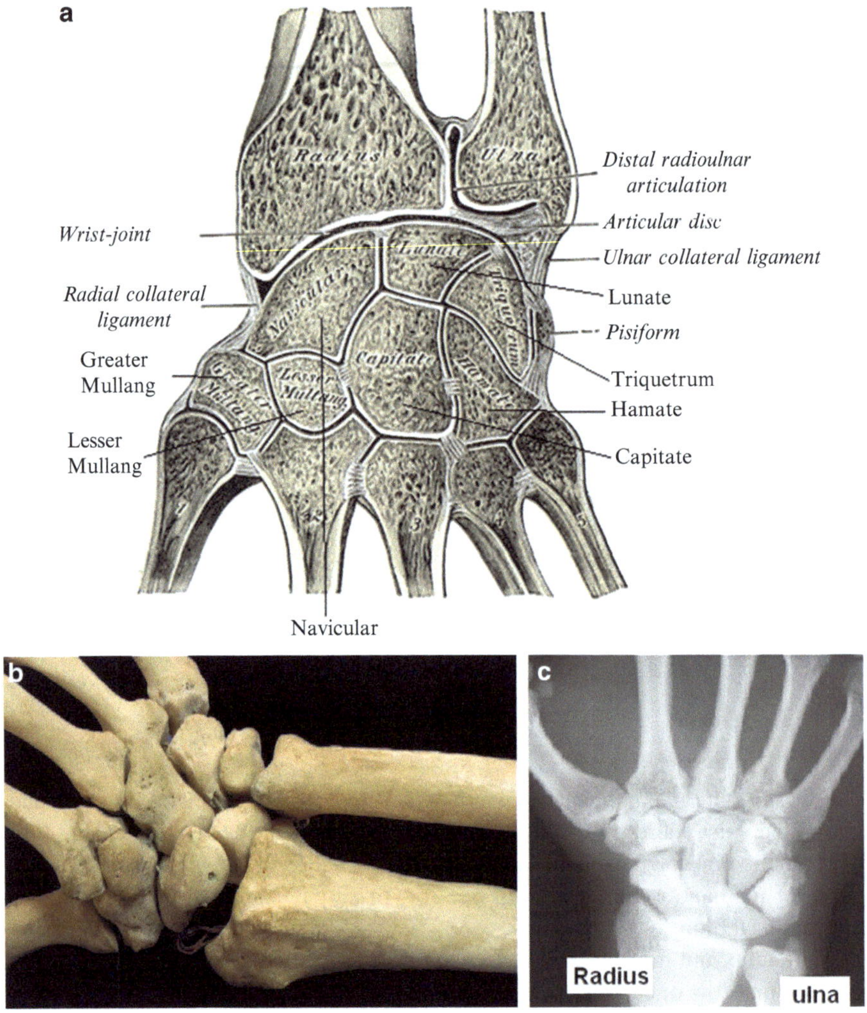

Fig. 9.1 (**a**) Wrist joint features; (**b**) the bones; and (**c**) the X-ray view

Rheumatoid arthritis is a chronic inflammatory disease of the joints that results in pain, stiffness and swelling. Rheumatoid arthritis usually affects several joints on both the right and left sides of the body. Both forms of arthritis may affect the strength of the fingers and hand, making it difficult to grip or pinch. The typical candidate for wrist replacement surgery has severe arthritis but does not need to use the wrist to meet heavy demands in daily use. The primary reasons for wrist replacement surgery are to relieve pain and to maintain function of the hand and wrist.

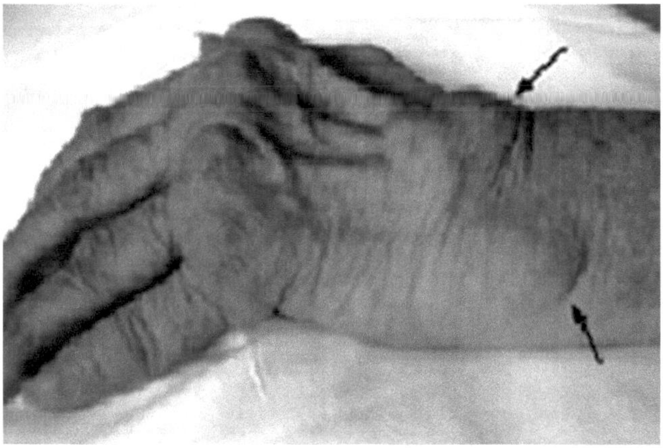

Fig. 9.2 Arthritis of hand bones

9.3 Treatment

Wrist replacement surgery may help retain or recover wrist movements. It may also improve the ability to perform daily living activities, especially if there is arthritis in the elbow and shoulder. During any total joint replacement, the worn-out ends of the bones are removed and replaced by an artificial joint (prosthesis) (Figs. 9.3 and 9.4).

In some cases, fusing the bones of the wrist together will reduce or eliminate pain and improve grip strength. However, if the bones are fused together, the wrist will not be able to bend, and that will limit several activities.

9.3.1 Forces Generated in the Wrist Joint

Chadwick and Nicol [2] developed a three-dimensional, mathematical model of the elbow and wrist joints, including 15 muscle units, 3 ligaments, and 4 joint forces. The device measures radial forces divided into six components and forces of up to 250 N per segment. Ten normal volunteers were asked to complete four tasks representing occupational activities; their grip force was monitored during the activities. Together with kinematic information from the six-camera Vicon data, the moment effects of these loads at the joints were calculated. These external moments are assumed to be balanced by the internal moments, generated by the muscles, passive soft tissue, and bone contact. The effectiveness of the body's internal structures in generating joint moments was assessed by studying the geometry of a simplified model of the structures, where information about the lines of action and moment arms of muscles, tendons, and ligaments is contained. The assumption of equilibrium between these external and internal joint moments allows the formulation of a set of equations from which muscle and joint forces can be calculated. A two-stage, linear optimization routine minimizing the overall muscle stress and the sum of the

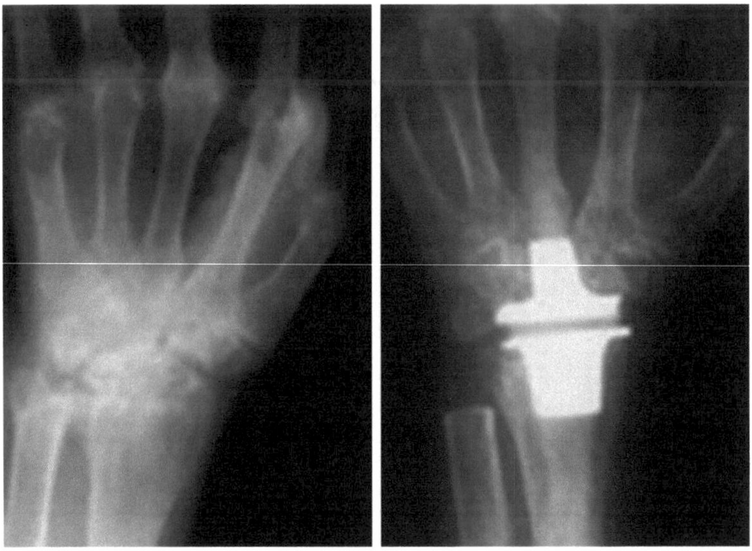

Fig. 9.3 X-ray of a wrist with severe rheumatoid arthritis throughout the wrist before (*left*) and after (*right*) replacement of the wrist joint with an implant [reproduced from Carlson JR, Simmons BP (1998) Total wrist arthroplasty. J Am Acad Orthop Surg 6:308–315, for teaching and training]

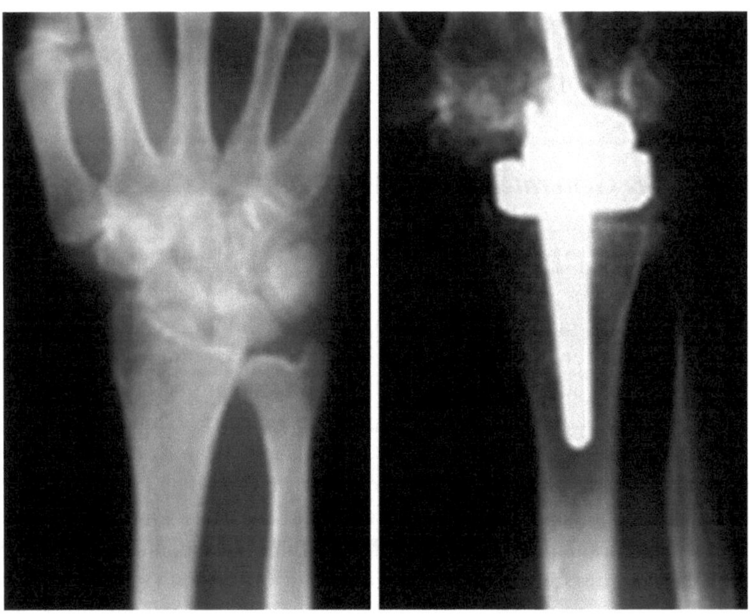

Fig. 9.4 X-ray of a wrist with osteoarthritis before (*left*) and after (*right*) wrist replacement

joint forces has been used to overcome the force-sharing problem. Humeroulnar forces of up to **1,600 N**, humeroradial forces of up to **800 N**, and wrist joint forces of up to **2,800 N** were found for moderate-level activity. These data will help us to design an artificial wrist joint.

9.4 Implants: Components of a Wrist Arthroplasty

There are several different designs for wrist arthroplasty. Most have two components, one for each side of the joint. These components are made of metal alloy. A high-quality polyethylene is used as a spacer between the two metal components. Newer implant designs have tried to replicate the anatomy of the wrist. One component inserts into the radius of the forearm, the wider bone at the wrist. The portion of this component that faces into the wrist joint has a curve that fits the second component in the wrist side. The component that inserts into the hand bone (the carpal component) has a flat surface that faces the first component (Fig. 9.5).

It inserts into a carpal bone through one long stem and one or two shorter stems. A plastic spacer fits into the components in the joint area. Spacers come in different sizes so they can match the hand. A spacer is normally flat on one

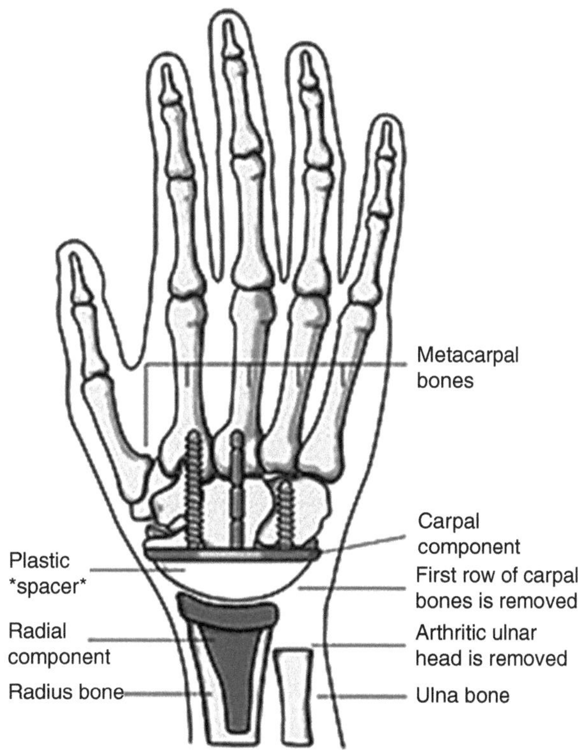

Fig. 9.5 The radial part is fixed with bone cement in the radius, and the carpal components are fixed as shown in small bones

side and rounded on the other. This design enables it to fit into the carpal component while it rocks on the radial component, creating a more natural wrist motion. The steam of the radial component will be accommodated in the marrow cavity, and the size can be determined from an X-ray. It should be gradually tapered and have a rounded rectangular section. The spacer should be made of ultra-high-molecular-weight polyethylene (UHMWPE) and shaped as part of a sphere articulating on the radial part and fixed with the second. The row of carpal bones are fixed with screws through a round plate.

9.5 Surgery

Wrist joint replacement can be done as an outpatient procedure, unlike a hip or knee replacement. Wrist replacement surgery is often combined with other procedures to correct deformities or disorders in the tendons, nerves, and small joints of the fingers and thumb.

An incision is made first on the back of the wrist, as shown in Fig. 9.6. The damaged ends of the lower arm bones are removed, and the first row of carpal bones may also be removed. The radial component of the prosthesis is inserted into the center of the radius bone on the outside of the lower arm. It is held in place and fixed with bone cement.

Depending on the component design, the carpal component is then inserted into the center hand bone (third metacarpal) or screwed into the remaining row of carpal bones. Bone cement may be used to hold the component in place. The carpal bones

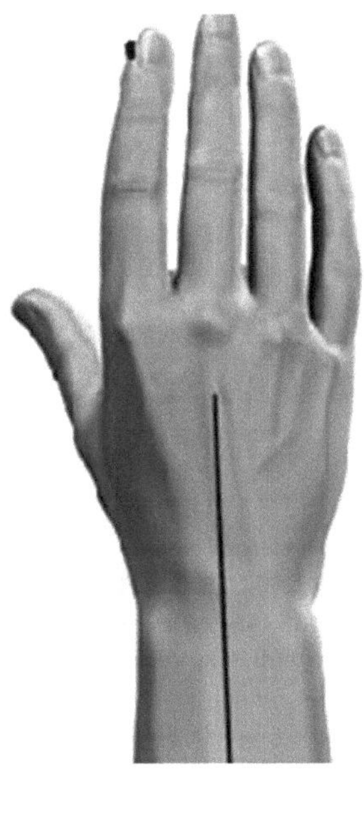

Fig. 9.6 The incision in wrist replacement surgery

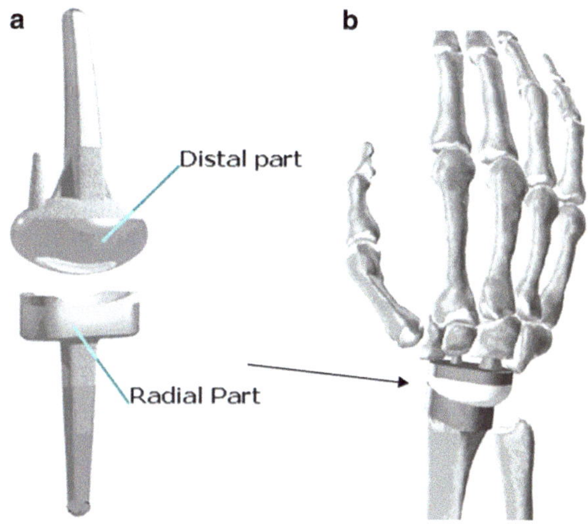

Fig. 9.7 Wrist arthroplasty: (**a**) the two components; (**b**) the carpal and radial components fixed to bone and radius. The components are fixed with bone cement

may be linked or fused together to better secure this component. An appropriately sized spacer is used between the metal components.

9.6 Postsurgical Care

A plaster cast needs to be worn for the first several weeks. When the cast is removed, a protective splint will need to be worn for the next 6–8 weeks. Although pain relief is immediate, gradual exercises will have to be performed for several weeks to restore movement and, eventually, to increase power and endurance. Wrist arthroplasty can improve motion to about 50 % of normal.

The physical demands from the joint will have limitations. The use of vibrating and jolt-producing tools, e.g., a hammer or pneumatic tool, is to be avoided. The amount of weight lifted should also be limited. A fall on the outstretched hand may break the prosthesis, and so highly demanding sports should be avoided.

On average, a wrist replacement can be expected to last 10–15 years with careful use. As with all implants, long-term follow-up is advised. Generally, follow-up every year or every 2 years will identify any developing conditions or problems (Fig. 9.7).

References

1. Papp SR, Athwal GS, Pichora DR (2006) The rheumatoid wrist. J Am Acad Orthop Surg 14:65–77
2. Chadwick EKG, Nicol AC (2000) Elbow and wrist joint contact forces during occupational pick and place activities. J Biomech 33(5):591–600

Problems

1. The wrist joint is a very complex joint, and humans can do a wide range of activities with it. Think of the cricket bowlers fast, and spinners and batsmen as well. However, this joint replacement is not very common in India.

   How do you propose to design an artificial wrist joint based on its anatomy? Do you think you will achieve all the movements with your design? Consult an orthopedic surgeon to help you in your design, especially for surgical insertion of the components.
2. Study Fig. 9.7, and discuss how you will find the sizes of the carpal and radial components and what materials you propose to use. Consider the fixation modality as well. Can you suggest a simplified joint design? Could you make a cost estimate of the artificial wrist joint?

Chapter 10
The Finger Joint and Its Artificial Replacement

10.1 Introduction

The finger anatomy includes the bones, joints, nerves, and blood vessels of the fingers. Each finger is made of **three bones** called the phalanges. Starting from the fingertip to the knuckle, they are

- Distal phalanx
- Middle phalanx
- Proximal phalanx.

The thumb has no middle phalanx. Each finger also has **three joints**:

- The distal interphalangeal (DIP) joint between the distal phalanx and middle phalanx.
- The proximal interphalangeal (PIP) joint between the middle phalanx and proximal phalanx.
- The metacarpophalangeal (MP) joint between the head of metacarpal and the proximal phalanx (Fig. 10.1).

The **movements** possible at the different joints are

- Flexion and extension at all three joints
- Abduction and adduction (side-to-side movement) at the metacarpophalangeal joint.

Flexion at the DIP joint is produced by the flexor digitorum profundus muscle. Flexion at the PIP joint is produced by the flexor digitorum superficialis muscle. Abduction and adduction (side-to-side movement) at the metacarpophalangeal joint are produced by the palmer and dorsal interossei muscles (Figs. 10.2, 10.3, and 10.4).

S. Pal, *Design of Artificial Human Joints & Organs*,
DOI 10.1007/978-1-4614-6255-2_10,

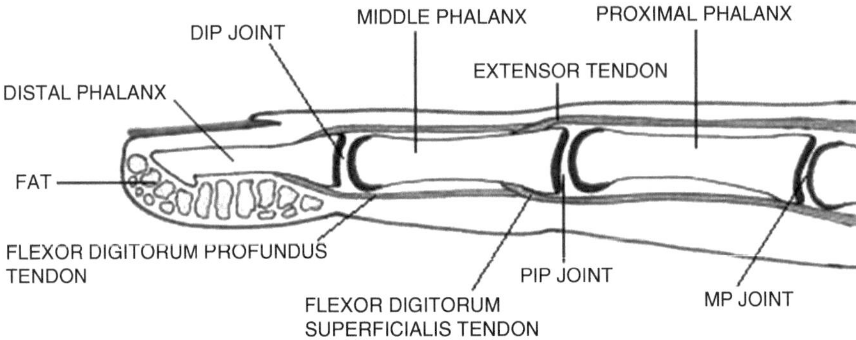

Fig. 10.1 Anatomy of finger joints

Fig. 10.2 This photograph shows abduction movement of the fingers. This movement is produced by the dorsal interossei muscles

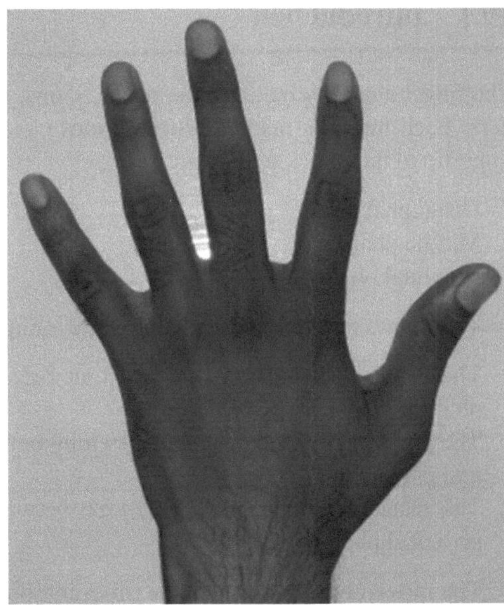

10.2 Diseases of the Finger Joint

Arthritis of the finger joints may occur due to many reasons, and arthritic joints can make it difficult to do daily activities due to pain and deformity. Unbearable pain or progressive deformity from arthritis may lead to surgical treatment. The common types of joint diseases of the finger are as follows: **osteoarthritis, rheumatoid arthritis, others (systemic lupus erythematosus, psoriatic arthritis, scleroderma, gout, pseudogout)** (Fig. 10.5).

The finger and thumb joints are covered on the ends with **articular cartilage**. This white, shiny material has a rubbery consistency. The function of articular

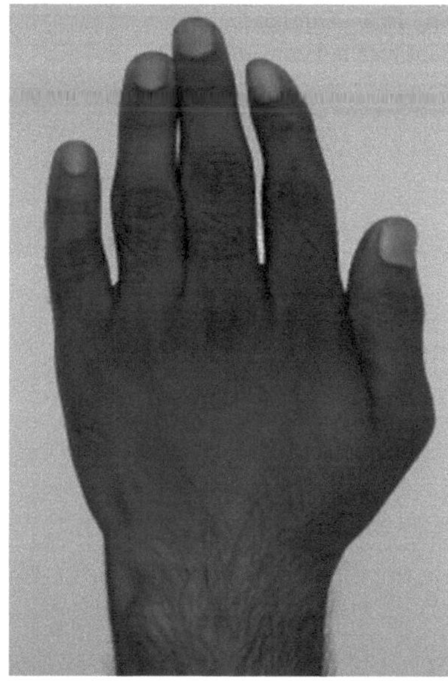

Fig. 10.3 This photograph shows adduction movement that is caused by the palmer interossei muscles. Extension of all the joints is caused by the extensor digitorum muscle. Each finger has a pair of digital arteries and nerves

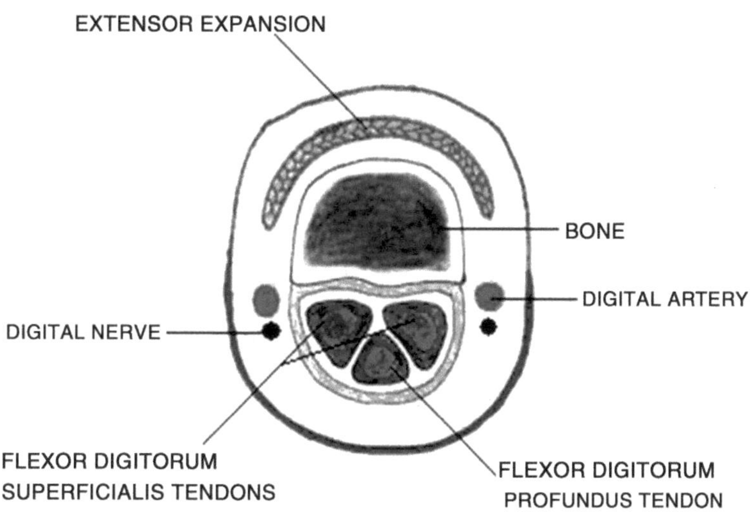

Fig. 10.4 Cross section of a finger

cartilage is to absorb shock and provide an extremely smooth surface to facilitate motion. There is articular cartilage essentially everywhere that two bony surfaces move against one another, or articulate.

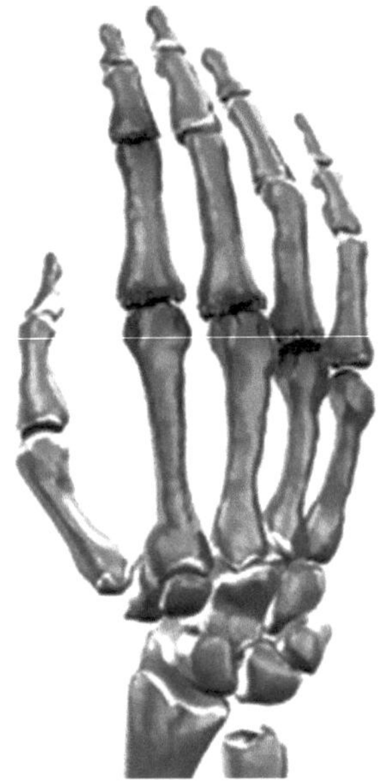

Fig. 10.5 Arthritic finger joint with inflammation

Degenerative arthritis is a condition in which a joint wears out, or degenerates, usually slowly over a period of many years. The term **arthritis** means joint inflammation causing pain, redness, heat, and swelling. The term **degenerative arthritis** means inflammation of a joint due to wear and tear. Some doctors believe that degenerative arthritis is not a true arthritis and that the term can cause confusion. The term *arthritis* is used to describe true inflammatory conditions of a joint, such as gout, infection, and rheumatoid arthritis. Injury to a joint, such as a bad sprain or fracture, can cause damage to the articular cartilage. An injury to any of the joints of the fingers, even if it does not injure the articular cartilage directly, can alter the working of the joint. This is true for a fracture that involves the joint when the bone fragments do not quite "line up" correctly and heal differently from the way they were before the fracture occurred. When an injury changes the way the joint lines up and move, force can start to press against the surface of the articular cartilage. This is similar to how a rotary machine part that is out of balance wears out faster.

Over time, this imbalance in the joint can lead to damage to the articular surface. Since articular cartilage cannot heal itself very well, the damage adds up. Eventually, the joint can no longer compensate for the increasing damage, and symptoms begin. The damage in the joint starts well before the symptoms of arthritis appear.

10.2.1 Symptoms

Pain is the main problem with arthritis. At first, the pain usually only causes problems when one begins an activity. Once the activity progresses, the pain eases.

But after the individual has rested for several minutes, the pain and stiffness increase. When the arthritis condition worsens, pain may be felt even at rest. The sensitive joint may feel enlarged and warm to the touch from inflammation. In rheumatoid arthritis, the fingers often become deformed as the disease progresses. The MCP joints of the fingers may actually begin to point sideways (toward the little finger). This is called ulnar drift. Ulnar drift can cause weakness and pain, making it difficult to use the hands for daily activities.

10.2.2 Nonsurgical Treatment

Treatment usually requires mild anti-inflammatory medications, such as aspirin or ibuprofen. Reducing the activity, or changing from occupations that require heavy repetitive hand and finger motions, may be necessary to help control the symptoms.

An injection of cortisone into the finger joint can give temporary relief. Cortisone is a very powerful anti-inflammatory medication; when injected into the joint itself, it can help relieve the pain. Pain relief is temporary and usually only lasts several weeks to months.

10.3 Finger Joint Replacement

10.3.1 Introduction

We have already discussed that all mobile joints can be replaced by artificial ones. However, complications such as the lack of functional motion, recurrent joint deformity, implant loosening, wear, and component fracture are still common, and revision surgery is often required. An ideal solution for any implant system is still awaited with eager anticipation.

The greatest achievement in joint replacement surgery is probably the total hip arthroplasty (THA) introduced by John Charnley's low-friction arthroplasty based on the articulating of a metallic femoral head and polymeric acetabular cup, in 1969 in the UK. Gunston employed this metal-on-plastic concept to develop a polycentric knee joint design, which is still widely accepted as the gold standard in total knee arthroplasty (TKA). **Charnley's implant reported long-term survival of about 90% at 10 years and 80% at 20 years in follow-up studies.**

There are artificial joints available for the fingers; they are made of silicone with a hinge joint between the two parts. These implants are used by hand surgeons primarily to replace the MCP joints, which are commonly referred to as the knuckles. The implant, or "prosthesis," acts as a spacer to fill the gap created when the arthritic surfaces of the MCP joint are removed. In 1972, Swanson introduced an arthroplasty

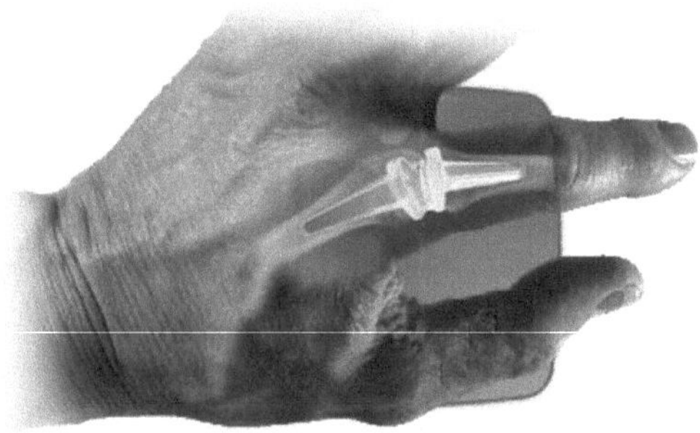

Fig. 10.6 Silastic finger implanted in the middle finger

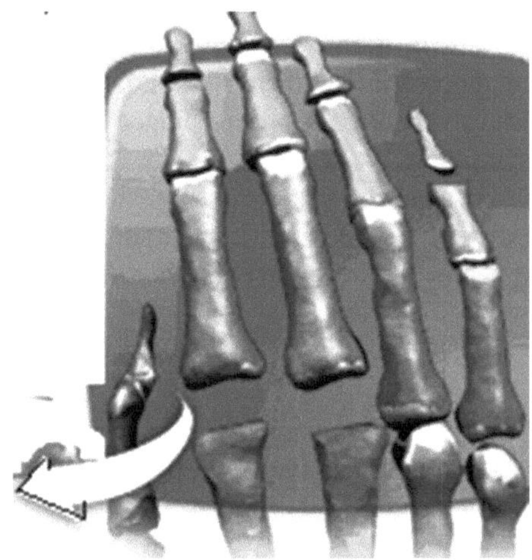

Fig. 10.7 Arthritic joint surfaces (cartilage) are removed

replacing arthritic finger joints with a joint spacer made of silicon rubber. Swanson's Silastic® implant has two components with a hinge knuckle joint, as shown in Fig. 10.6.

To perform a joint replacement of the MCP joint, the surgeon first makes an incision in the back of the hand over the joints or between the first and middle fingers and between the ring and little fingers. Each joint that needs to be replaced is then opened so that the surgeon can see the joint surfaces. The cartilage is removed from both joint surfaces using an oscillating saw to leave two ends of bone flat. Next, a small cutting tool called a burr that has cutting edges is used to create holes in the bones of the finger joint closer to the size of the joint prosthesis (Figs. 10.7 and 10.8).

The artificial finger joint has a stem on each side that is inserted into the canals created in the bones of the finger and the metacarpal joint using the cutting burr (Fig. 10.9).

The surgeon then completes the operation by using the tendons and ligaments around the joint to form a tight sack to hold the implant in place. The skin is sutured together and a splint is applied. Patients will probably be in a splint, brace, or cast

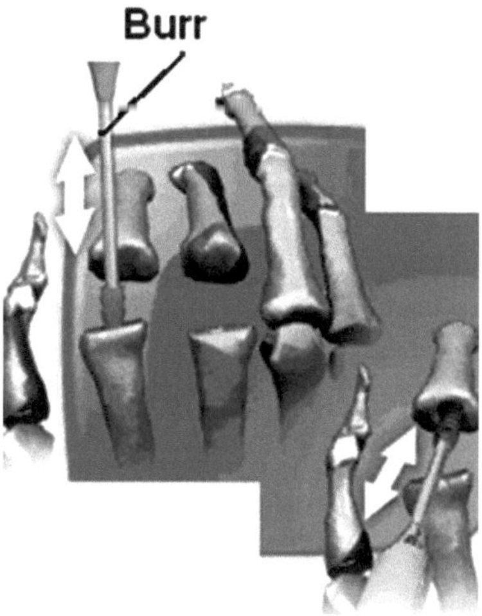

Fig. 10.8 A burr is used to cut holes in the metacarpals to facilitate insertion of the stem

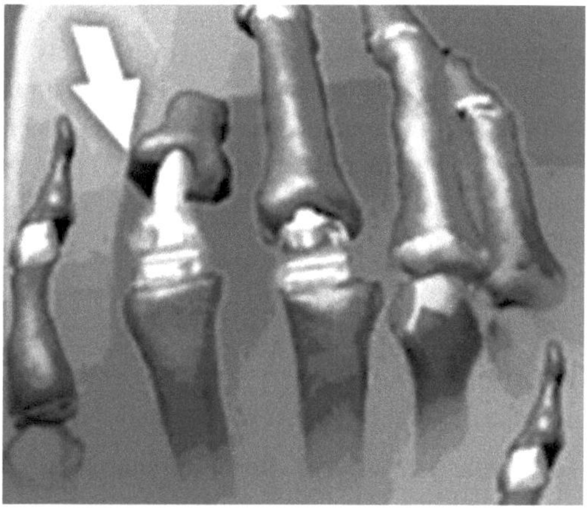

Fig. 10.9 The artificial finger joints are inserted

for 6 weeks. As with any medical treatment, individual results may vary. Only an orthopedic surgeon can determine whether an orthopedic implant is an appropriate course of treatment. There are potential risks, and the recovery takes time. The performance of the new joint depends on weight, activity level, age, and other factors related to the patient. Silastic is an elastomer used in medicine to a great extent for tubing; properties of various sizes of Silastic are provided by Dow Corning (USA), a major producer of such materials.

The values listed in Table 10.1 are not intended for use in preparing specifications. These are indicative and not precise. A review of the literature regarding long-term complications with Swanson silicone finger joint implants was done by Foliart & Associates (Moraga, CA, USA). The review focused on seven broad categories of complications: synovitis, lymphadenopathy, bone change, implant fracture, implant loosening, infection, and implant removal. Outcome data on 15,556 small joint

Table 10.1 Properties of Silastic polymer

			Result		
	CTM[a]-test	Unit	Rx-SO	Rx-6S	Rx-8O
0099	Durometer hardness[b]	Shore A	50	65	78
0137A	Tensile strength at break, die D[c]	MPa	9.6	9.0	7.1
		psi	1,388	1,301	1,030
0137A	Elongation at break, die D[c]	%	815	613	568
0137A	Modulus at 200%, die D[c]	MPa	2.1	2.7	4.0
		psi	299	391	581
0159A	Tear strength, die B[b]	kN/m^2	46	45	42
		psi	263	257	240

[a]CTM (Corporate Test Method): corresponds to American Standard for Testing Materials (ASTM)
[b]Typical properties for the elastomers
[c]Typical properties of the extruding tubing

implants were identified in 70 pertinent articles. The prevalence rates of complications associated with finger implants were all very low. Implant fracture and bone changes occurred in 2% and 4% (respectively) of finger implants. Each of the other complications developed in 1% or fewer of implants, including particulate synovitis (0.06% of implants). Particle-wear lymphadenopathy was reported in 0.08% of implants. The literature review identified no reports of immunologic reactions, connective tissue disease, or other systemic effects associated with the use of Swanson finger implants.

Problems

1. Silastic (i.e., silicone rubber implants) is used in finger joint replacement and has worked fine. This material is not easily available in India. Can you think of an alternative indigenously available material that may be used? Could we use metals in finger joint arthroplasty? Justify your opinion. Could you suggest any other polymer? A standard finger joint with the operative procedure is shown in Fig. 10.10 for your reference.
2. Doctoral scholar Sandip Bag (Jadavpur University, 2007) developed UHMWPE–alumina ceramic material and coated it with hydroxyapatite and also Bioglass. How can these materials be used for finger joint replacement? Design the finger joint prosthesis for appropriate function. What type of problem(s) may arise during function? Figure 10.11 shows the metal–polyethelene, cement PIP joint implant for arthritic patients.

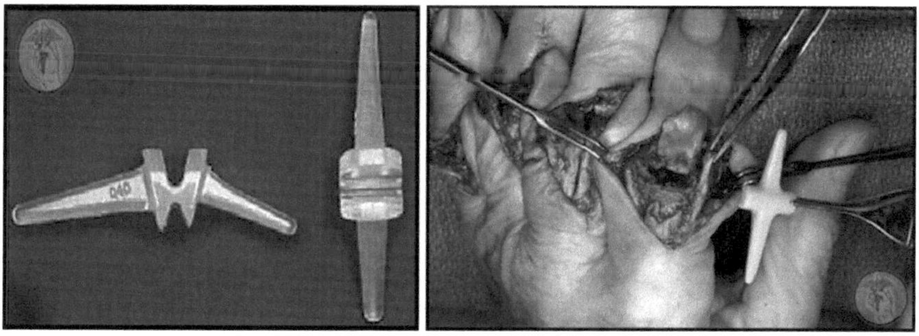

Fig. 10.10 Silicone rubber implant and surgical procedure for such an implant

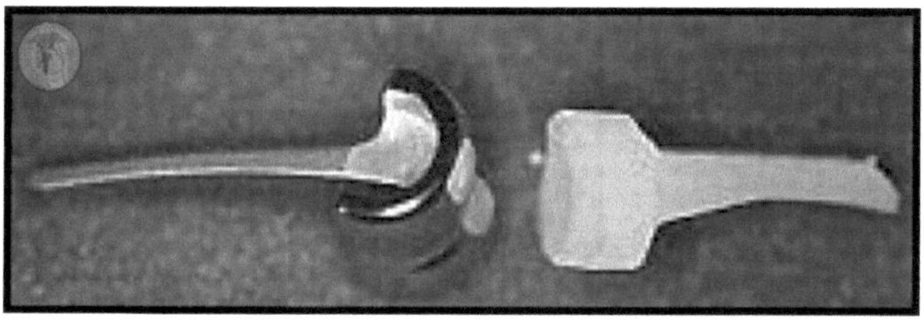

Fig. 10.11 Polyethelene–cement PIP joint implant for arthritic patients

References

1. (1998) The NeuFlex MCP finger joint implant: designed for comfort and performance. DePuy White Paper, Publication 0601-15-050
2. Harvey Kliman (1998) Internal Dupont Report—Flex Fatigue Testing of the NeuFlex MCP Implant. On File at DePuy
3. Allieu Y, Dimeglio A, Pech J (1974) Les arthroplasties des métacarpo-phalangiennes avec implants de Swanson dans la main rhumatismale. Evaluation critique des résultats. Ann Chir 28(10):873–882 (This is in French)
4. Griffiths RW, Nicolle FV (1975) Three years' experience of metacarpophalangeal joint replacement in the rheumatoid hand. Hand 7(3):275–283
5. Hagert CG, Eiken O, Ohlsson NM, Aschan W, Movin A (1975) Metacarpophalangeal joint implants. I. Roentgenographic study on the Silastic finger joint implant, Swanson design. Scand J Plast Reconstr Surg 9(2):147–157
6. Mannerfelt L, Andersson K (1975) Silastic arthroplasty of the metacarpophalangeal joints in rheumatoid arthritis. J Bone Joint Surg Am 57(4):484–489
7. Murray J (2003) New-generation implant arthroplasties of the finger joints. Am Acad Orthop Surg 11:295–301

Chapter 11
The Hip Joint and Its Artificial Replacement

11.1 Introduction

The **hip joint** is located lateral to the gluteal region (i.e., the buttock), inferior to the iliac crest, and overlying the greater trochanter of the thigh bone. In adults, three of the bones of the pelvis have fused into the hip bone, which forms part of the hip region.

The **hip joint**, medically referred to as the **acetabulofemoral joint**, is the joint between the femur and acetabulum of the pelvis. Its primary function is to support the weight of the body in both static (e.g., standing) and dynamic (e.g., walking or running) postures.

11.2 Anatomy of the Hip Region

As Fig. 11.1 shows, the bones of the hip region are the hip bone (or innominate bone) and the femur (or thigh bone). Prominent palpable bony structures of the hip bone include the iliac crest, the anterior superior (ASIS) and posterior superior iliac spines (PSIS), the posterior inferior iliac spine (PIIS), the five or so tubercles and the lower lateral borders of the sacrum, and the ischial tuberosity ("sitting bone") [3].

Proximally, the femur is largely covered by muscles, and, as a consequence, the greater trochanter is often the only palpable bony structure. Distally on the femur, some more palpable bony structures are the condyles [4].

11.2.1 Articulation

The hip joint is a synovial joint formed by the articulation of the rounded head of the femur and the cuplike acetabulum of the pelvis. It forms the primary connection between the bones of the lower limb and the axial skeleton of the trunk and pelvis.

S. Pal, *Design of Artificial Human Joints & Organs*,
DOI 10.1007/978-1-4614-6255-2_11, © Springer Science+Business Media New York 2014

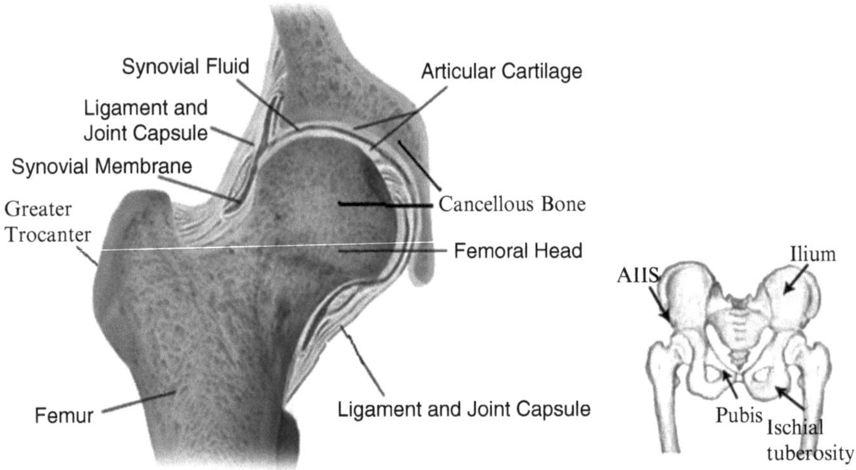

Fig. 11.1 Simplified anatomy of the hip joint with main parts

Both joint surfaces are covered with a strong but lubricated layer called the articular hyaline cartilage. The cuplike acetabulum forms at the union of three pelvic bones—the ilium, pubis, and ischium [5]. The Y-shaped growth plate that separates them, the triradiate cartilage, is fused definitively at ages 14–16 [6]. It is a special type of spheroidal or ball-and-socket joint where the roughly spherical femoral head is largely contained within the acetabulum and has an **average radius of curvature of 25 mm** [7], especially among the Western populace. This will be nearly 20–23 mm in Asia. The acetabulum grasps almost half the femoral ball, a grip augmented by a ring-shaped fibrocartilaginous lip, the acetabular labrum, which extends the joint beyond the equator [5]. The head of the femur is attached to the shaft by a thin neck region that is often prone to fracture in the elderly, which is mainly due to the degenerative effects of osteoporosis (Fig. 11.2).

The **transverse angle** of the acetabular inlet can be determined by measuring the angle between a line passing from the superior to the inferior acetabular rim and the horizontal plane. This angle normally measures 51° at birth and 40° in adults; it affects the acetabular lateral coverage of the femoral head and several other parameters. The **sagittal angle** of the acetabular inlet measures 7° at birth and increases to 17° in adults [8].

11.2.2 Femoral Neck Angle

The angle between the longitudinal axes of the femoral shaft and neck, called the **caput-collum-diaphyseal angle** or CCD angle, normally measures approximately 150° in newborns and **126° in adults (coxa norma)** [9]. An abnormally small angle

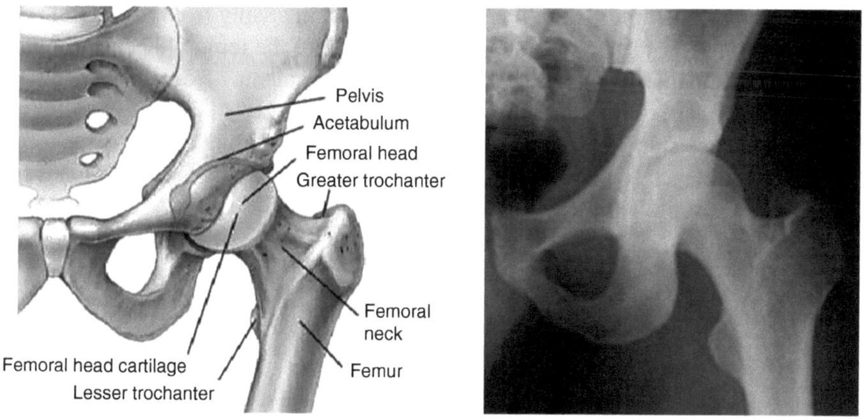

Fig. 11.2 Pictorial and radiographic views of a healthy human hip joint

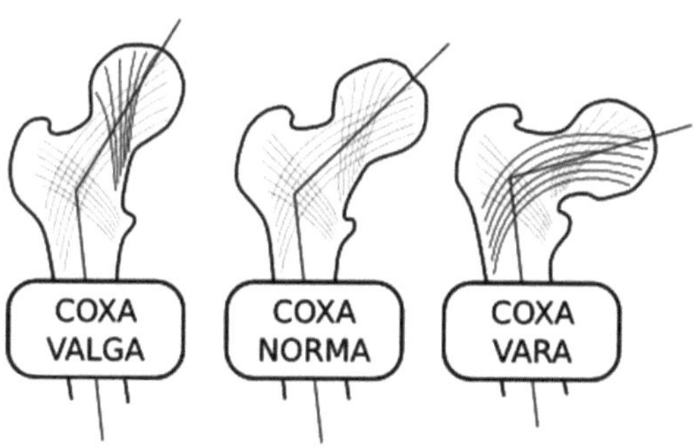

Fig. 11.3 Various neck shaft angles

is known as **coxa vara** and an abnormally large angle as **coxa valga**. Because changes in the shape of the femur naturally affect the knee, *coxa valga* is often combined with *genu varum* (bow-leggedness), while *coxa vara* leads to *genu valgum* (knock knees) (Fig. 11.3) [10].

Changes in trabecular patterns due to an altered CCD angle, coxa valga leads to more compression trabeculae, coxa vara to more tension trabeculae [9]. Changes in CCD angle are the result of changes in the stress patterns applied to the hip joint. Such changes, caused, for example, by a dislocation, change the trabecular patterns inside the bones. Two continuous trabecular systems emerging on the auricular surface of the sacroiliac joint meander and crisscross each other down through the hip bones, the femoral head, neck, and shaft.

In the hip bone, one system arises on the upper part of the auricular surface to converge onto the posterior surface of the greater sciatic notch, from where its trabeculae are reflected to the inferior part of the acetabulum. The other system emerges on the lower part of the auricular surface, converges at the level of the superior gluteal line, and is reflected laterally onto the upper part of the acetabulum.

In the femur, the first system lines up with a system arising from the lateral part of the femoral shaft to stretch to the inferior portion of the femoral neck and head. The other system lines up with a system in the femur stretching from the medial part of the femoral shaft to the superior part of the femoral head [11].

On the lateral side of the hip joint, the fascia lata is strengthened to form the iliotibial tract, which functions as a tension band and reduces the bending loads on the proximal part of the femur [9].

11.2.3 Capsule

The capsule attaches to the hip bone outside the acetabular lip, which thus projects into the capsular space. On the femoral side, the distance between the head's cartilaginous rim and the capsular attachment at the base of the neck is constant, which leaves a wider extracapsular part of the neck at the back than at the front [12, 13]. The strong but loose fibrous capsule of the hip joint permits the hip joint to have the second-largest range of movement (second only to the shoulder) and yet support the weight of the body, arms, and head. The capsule has two sets of fibers: longitudinal and circular. The circular fibers form a collar around the femoral neck called the zona orbicularis. The longitudinal retinacular fibers travel along the neck and carry blood vessels.

11.2.4 Ligaments

The hip joint is reinforced by five ligaments, four of which are extracapsular and one intracapsular. The **extracapsular** ligaments are the iliofemoral, ischiofemoral, and pubofemoral ligaments attached to the bones of the pelvis (the ilium, ischium, and pubis, respectively). All three strengthen the capsule and prevent an excessive range of movement in the joint. Of these, the Y-shaped and twisted iliofemoral ligament is the strongest ligament in the human body [13]. In the upright position, it prevents the trunk from falling backward without the need for muscular activity. In the sitting position, it becomes relaxed, thus permitting the pelvis to tilt backward into its sitting position. The ischiofemoral ligament prevents medial rotation, while the pubofemoral ligament restricts abduction in the hip joint [14]. The zona orbicularis, which lies like a collar around the narrowest part of the femoral neck, is covered by the other ligaments, which partly radiate into it. The zona orbicularis acts like a buttonhole on the femoral head and assists in maintaining the contact in the joint [13].

The **intracapsular** ligament, the ligamentum teres, is attached to a depression in the acetabulum (the acetabular notch) and a depression on the femoral head (the fovea of the head). It is only stretched when the hip is dislocated and may then prevent further displacement [13]. It is not that important as a ligament but can often be vitally important as a conduit of a small artery to the head of the femur. This arterial branch is not present in everyone but can become the only blood supply to the bone in the head of the femur when the neck of the femur is fractured or disrupted by injury in childhood.

11.2.5 Blood and Nerve Supply

The hip joint is supplied with blood from the medial circumflex femoral and lateral circumflex femoral arteries, which are both usually branches of the deep artery of the thigh (profunda femoris), but there are numerous variations, and one or both may also arise directly from the femoral artery. There is also a small contribution from a small artery in the ligament of the head of the femur that is a branch of the posterior division of the obturator artery, which becomes important to avoid avascular necrosis of the head of the femur when the blood supply from the medial and lateral circumflex arteries is disrupted (e.g., through fracture of the neck of the femur along their course).

The hip has two anatomically important anastomoses, the cruciate and trochanteric anastomoses, the latter of which provides most of the blood to the head of the femur. These anastomoses exist between the femoral artery or profunda femoris and the gluteal vessels.

11.2.6 Muscles and Movements

The hip muscles act on three mutually perpendicular main axes, all of which pass through the center of the femoral head, resulting in three degrees of freedom and three pairs of principal directions: flexion and extension around a transverse axis (left–right); lateral rotation and medial rotation around a longitudinal axis (along the thigh); and abduction and adduction around a sagittal axis (forward–backward) and a combination of these movements (i.e., circumduction, a compound movement in which the leg describes the surface of an irregular cone) [14]. It should be noted that some of the hip muscles also act on either the vertebral joints or the knee joint; that with their extensive areas of origin and/or insertion, different part of individual muscles participate in very different movements; and that the range of movement varies with the position of the hip joint. Additionally, the inferior and superior gemelli may be termed *triceps coxae* together with the obturator internus, and their function simply is to assist the latter muscle.

The movement of the hip joint is thus performed by a series of muscles that are presented here in order of importance with the range of motion from the neutral zero-degree position indicated:

- **Lateral** or **external rotation** (30° with the hip extended, 50° with the hip flexed): gluteus maximus; quadratus femoris; obturator internus; dorsal fibers of gluteus medius and minimus; iliopsoas (including psoas major from the vertebral column); obturator externus; adductor magnus, longus, brevis, and minimus; piriformis; and sartorius.
- **Medial** or **internal rotation** (40°): anterior fibers of gluteus medius and minimus; tensor fascia latae; the part of adductor magnus inserted into the adductor tubercle; and, with the leg abducted, also the pectineus.
- **Extension** or **retroversion** (20°): gluteus maximus (if put out of action, active standing from a sitting position is not possible, but standing and walking on a flat surface are); dorsal fibers of gluteus medius and minimus; adductor magnus; and piriformis. Additionally, the following thigh muscles extend the hip: semimembranosus, semitendinosus, and long head of biceps femoris.
- **Flexion** or **anteversion** (140°): iliopsoas (with psoas major from vertebral column); tensor fascia latae, pectineus, adductor longus, adductor brevis, and gracilis. Thigh muscles acting as hip flexors: rectus femoris and sartorius.
- **Abduction** (50° with hip extended, 80° with hip flexed): gluteus medius; tensor fascia latae; gluteus maximus with its attachment at the fascia lata; gluteus minimus; piriformis; and obturator internus.
- **Adduction** (30° with hip extended, 20° with hip flexed): adductor magnus with adductor minimus; adductor longus, adductor brevis, gluteus maximus with its attachment at the gluteal tuberosity; gracilis (extends to the tibia); pectineus, quadratus femoris; and obturator externus. Of the thigh muscles, the semitendinosus is especially involved in hip adduction.

11.3 Hip Joint Replacement

In total hip replacement, the head of the **femur** is removed along with the surface layer of the socket in the **pelvis** (the two large bones that rest on the lower limbs and support the spinal column) (Figs. 11.4 and 11.5).

A total hip replacement (THR)—also called a hip **arthroplasty**—is a surgical procedure that reforms the hip joint. The head of the femur, which is situated within the pelvis socket, is replaced with a metal ball and stem. This stem fits into the shaft of the femur. The socket is replaced with a polymer (UHMWPE) or a metal-backed and polymer cup.

For nearly a century, doctors have been putting various materials into diseased and painful hip joints to relieve pain. Until the 1960s, outcomes had been unreliable. At that time, the metal ball and plastic socket for the replacement of the hip joint were introduced. Today, the artificial components used in THR are stronger and better to the designer's eye, and more than 50 different designs are available.

There are many different shapes, sizes, and designs of artificial components of the hip joint. For the most part, these are composed of chrome–cobalt alloy, titanium

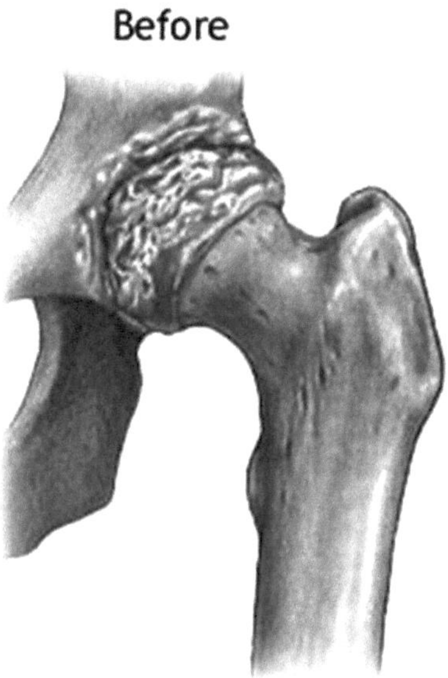

Fig 11.4 Diseased joint; the cartilage is degenerated

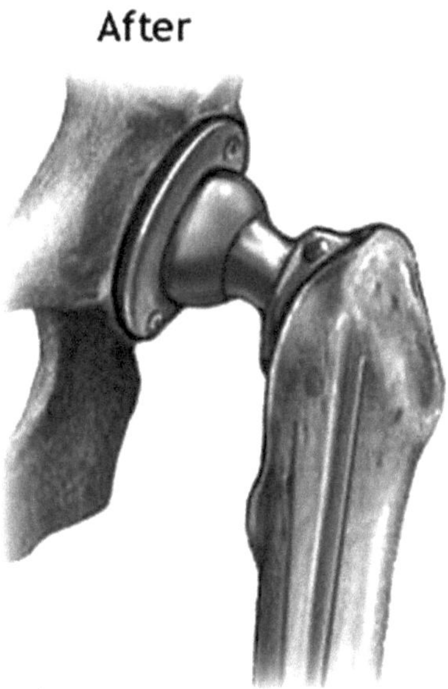

Fig. 11.5 Total hip-replaced joint

alloy, or ceramic materials. Some surgeons are also using custom-made components to improve the fit in the femur (Fig. 11.6).

Total hip replacement (THR), or total joint arthroplasty, is a reconstructive orthopedic surgery generally conducted to relieve arthritis pain or fix severe physical joint damage as part of hip fracture treatments due to trauma or avascular necrosis.

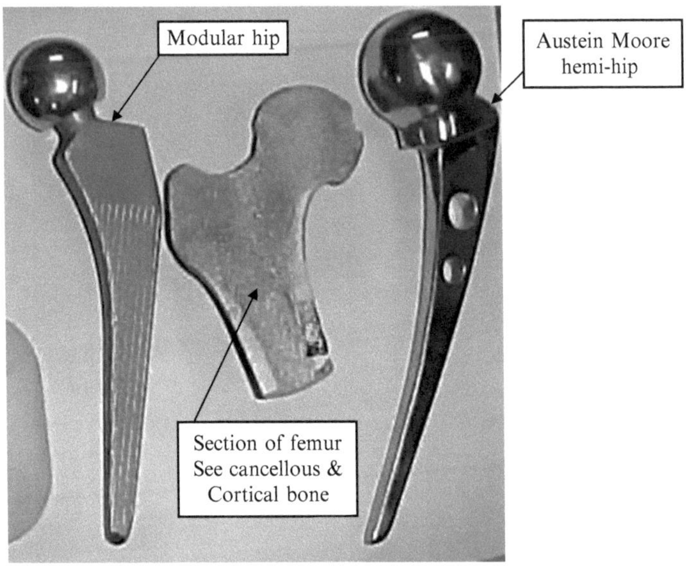

Fig. 11.6 Two examples of hip prostheses and a cutaway section of a femur head. Notice the cancellous bone in the head area

In this surgical procedure, the hip joint is replaced by a prosthetic implant. Replacing the hip joint consists of replacing both the acetabulum and the femoral head. A hip replacement is currently the most successful and reliable orthopedic operation, with 97% of patients reporting improved outcome, especially in the Western world.

Total hip replacement is most commonly used to treat joint failure caused by osteoarthritis. Other indications include rheumatoid arthritis, avascular necrosis, traumatic arthritis, protrusio acetabuli, certain hip fractures, benign and malignant bone tumors, arthritis associated with Paget's disease, ankylosing spondylitis, and juvenile rheumatoid arthritis. The aims of the procedure are pain relief and improvement in hip function. Hip replacement is usually considered only after other therapies, such as physical therapy and pain medications, have failed.

11.3.1 History

The earliest recorded attempts at hip replacement (Gluck 1891), which were carried out in Germany, used ivory to replace the femoral head (the ball on the femur).

In 1940, at Johns Hopkins Hospital in the US, Austin T. Moore (1899–1963), an American surgeon, reported and performed the first metallic hip replacement surgery. The original prosthesis he designed was a proximal femoral replacement, with a large fixed head, made of the cobalt–chrome alloy Vitallium. It was about a foot in length and it bolted to the resected end of the femoral shaft (hemiathroplasty). This was unlike later (and current) hip replacement prostheses, which are inserted within the medullary canal of the femur. A later version of Dr. Moore's prosthesis, the so-called Austin Moore, introduced in 1952, is still in use today.

In 1960, a Burmese orthopedic surgeon, Dr. San Baw (1922–1984), introduced the use of ivory hip prostheses to replace the fractured hip bone of an 83-year-old Burmese nun, Daw Punya. Dr. Baw was the chief of orthopedic surgery at Mandalay General Hospital in Burma. He used over 300 ivory hip replacements from the 1960s to the 1980s. He presented a paper entitled, "Ivory hip replacements for un united fractures of the neck of femur" at the conference of the British Orthopaedic Association held in London in September 1969. An 88% success rate was discerned in that Dr. Baw's patients, ranging in ages from 24 to 87, were able to walk, squat, ride a bicycle, and play football a few weeks after their fractured hip bones were replaced with ivory prostheses. Ivory may have been used because it was cheaper than metal at that time in Burma and also was thought to have good biomechanical properties, including biological bonding of ivory with the surrounding human tissues. A synopsis of Dr. Baw's paper was presented at the British Orthopaedic Association's 1969 conference and was published in the *Journal of Bone and Joint Surgery* (British edition) in February 1970. With modern hip replacement surgery, patients can walk immediately after the operation.

11.4 Modern Developments

The modern artificial joint owes much to the work of John Charnley at Wrightington Hospital; his work in the field of tribology resulted in a design that almost completely replaced the other designs by the 1970s. Charnley's design consisted of three parts (see Fig. 11.7):

1. A metal (originally stainless steel) femoral component
2. A teflon acetabular component, the wear debris of which resulted in a condition called osteolysis, and so it was replaced by ultra-high-molecular-weight polyethylene (UHMWPE) in 1962, both of which were fixed to the bone using
3. PMMA (acrylic) bone cement and/or screws, which consist of two components.

The replacement joint, which was known as the low-friction arthroplasty, was lubricated with synovial fluid. The small femoral head [7/8″ (22.2 mm)] was chosen for its decreased wear rate; however, this has relatively poor stability (the larger the head of a replacement, the less likely it is to dislocate, but the more wear debris produced due to the increased surface area attributed to the higher surface velocity). For over 2 decades, the Charnley low-friction arthroplasty and subsequent similar designs were the most used systems in the world, far surpassing the other available options (such as the McKee and ring). Recently, the use of a polished tapered cemented hip replacement (e.g., the Exeter) and uncemented hip replacements have become more popular. Cemented stems are commonly used in older patients due to their lower cost, including the Austin Moore proximal femoral replacement for Medicaid patients, while more modern and longer-lasting "cementless" stems, often

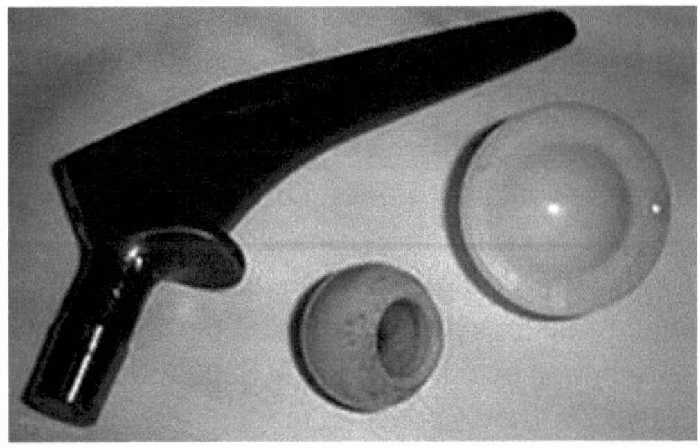

Fig. 11.7 A titanium modular hip prosthesis, with a ceramic head and polyethylene acetabular cup

coated with hydroxyapatite ceramic, are used in "younger" and more physically active patients. Once an uncommon operation, hip replacement is now common, even among active athletes, including the racecar drivers Bobby Labonte and Dale Jarrett.

11.5 Costs

The cost of a product in this world of inequality is an important issue, and the designer should always try to make the product cost-effective with respect to the socioeconomic status of the country.

In 2008, hip replacements in India cost in the range of INR10,000 (US$400) to 1 lakh ($2,000), depending on the prosthesis type in a general hospital; in the U.S., the cost was about $45,000. The Austin Moore hemi-hip used without cement is the cheapest.

In 2008, one source reported that hip replacements cost US$7,000–9,000 in India at an internationally accredited hospital, whereas in a county in Florida (USA), it cost $41,597–$56,258. Most likely, the total costs reported included the hospital stay and considered cementless devices.

Surgery costs vary from country to country, with the U.S. typically being among the highest-cost markets, and Thailand, Cuba, Argentina, and India being among the lowest. In 2012, the cost has increased not less than 75% to 100% Sarah Kliff (Washington post) in February, 2013 indicated it cost between $10,000 and $125,000.

11.6 Aseptic Loosening of Hip Prosthesis

In the long term, many problems arise from osteolysis from acrylic bone cement debris and/or wear debris. An inflammatory process causes bone resorption and subsequent loosening or fracture, often requiring revision surgery. Very hard ceramic-bearing surfaces are being used in the hope that they will have less wear and less osteolysis, with better long-term results. Large metal heads are also used for similar reasons, as these also have excellent wear characteristics and benefit from a different mode of lubrication. However, large fixed metal heads, such as the

Austin Moore devices, can result in protrusio acetabuli. A greater head-to-neck ratio also contributes to stability. These new prostheses do not always have the long-term track record of established metal-on-poly bearings.

Postoperative sciatic nerve palsy is another possible complication. A few patients who have had a hip replacement suffer chronic pain after the surgery despite normal imaging.

11.7 Techniques of Surgery

There are several different incisions, defined by their relation to the gluteus medius. The approaches are posterior (Moore), lateral (Hardinge or Liverpool) [5], antero-lateral (Watson–Jones) [6], anterior (Smith–Petersen) [7], and greater trochanter osteotomy. There is no compelling evidence in the literature for any particular approach, but the consensus of professional opinion favors either a modified anteri-olateral (Hardinge) or posterior approach.

- **The posterior (Moore) approach** accesses the joint through the back, taking the piriformis muscle and the short external rotators off the femur. This approach gives excellent access to the acetabulum and preserves the hip abductors. Critics cite a higher dislocation rate, although repair of the capsule and the short external rotators negates this risk.
- **The lateral approach** is also commonly used for hip replacement. The approach requires elevation of the hip abductors (gluteus medius and gluteus minimus) in order to access the joint. The abductors may be lifted up by osteotomy of the greater trochanter and reapplying it afterward using wires (as per Charnley) or may be divided at their tendinous portion, or through the functional tendon (as per Hardinge) and repaired using sutures.
- **The anterolateral approach** develops the interval between the tensor fasciae latae and the gluteus medius.
- **The anterior approach** utilizes an interval between the sartorius muscle and tensor fascia latae.

The double-incision surgery and minimally invasive surgery seek to reduce soft tissue damage through reducing the size of the incision. However, component positioning accuracy is impaired, and surgeons using these approaches are advised to use computer guidance systems.

11.8 Alternatives to Hip Replacement

The first-line approach as an alternative to hip replacement is conservative management, which involves a multimodal approach of medication, activity modification, and physical therapy. Conservative management can prevent or delay the need for hip replacement.

Hip resurfacing is an alternative to hip replacement surgery. It is a bone-conserving procedure that places a metal cap on the femoral head instead of amputating it. There is no long stem placed down the femur, so it is more like a natural hip and allows patients a full return to all activities, including marathons and triathlons. Some patients have even completed Ironman and Ultraman competitions following hip resurfacing surgery although patients must have good bone quality to qualify for it. It has been used in Europe for over 17 years; the first device, the BHR or Birmingham hip resurfacing device, was approved by the FDA on May 9, 2006.

The potential advantages of hip resurfacing compared to THR include less bone removal (bone preservation), a reduced chance of hip dislocation due to a relatively larger femoral head size (giving the patient an anatomically correct femoral head size), and easier revision surgery for any subsequent revision to a THR device because a surgeon will have more original bone stock available. The potential disadvantages of hip resurfacing are femoral neck fractures (rate of 0–4%), aseptic loosening, and metal wear. Due to the retention of the patient's complete femoral neck, other advantages exist: Surgeon-induced discrepancies in leg length (as could happen with THR) are minimized. Also, the toe-in or toe-out faults that could occur interoperatively with THR do not exist,r because the femoral neck, which determines the foot direction, is left undisturbed with hip resurfacing.

Current alternatives also include viscosupplementation, or the injection of artificial lubricants into the joint [10]. Some believe the future of osteoarthritis treatment is bioengineering, targeting the growth and/or repair of the damaged, arthritic joint. Centeno et al. [10] have reported on the partial regeneration of an arthritic human hip joint using mesenchymal stem cells in one patient. It has yet to be shown that this result will apply to a larger group of patients and result in significant benefits.

The FDA has stated that this procedure is being practiced without conforming to regulations, but Centeno's group claims it is exempt from FDA regulation. It has not been shown in controlled clinical trials to be effective, and it costs over $7,000. A set of pictorial views depicting the hip replacement surgery is shown here as Figs. 11.8, 11.9, 11.10 and 11.11.

Diameters of the hip ball for both types are 35–55 mm, in intervals of 1 mm.

11.8.1 Apollo Hip System: Simple, Straight-Stem Design for the Low-Demand Patient

The Apollo hip system's simple cobalt–chrome straight-stem design and all-polyethylene acetabular component are the perfect low-demand option. The instrumentation is simple, precise, and quick. A one-for-one ream, broach, and cement technique is easy to remember (Fig. 11.12).

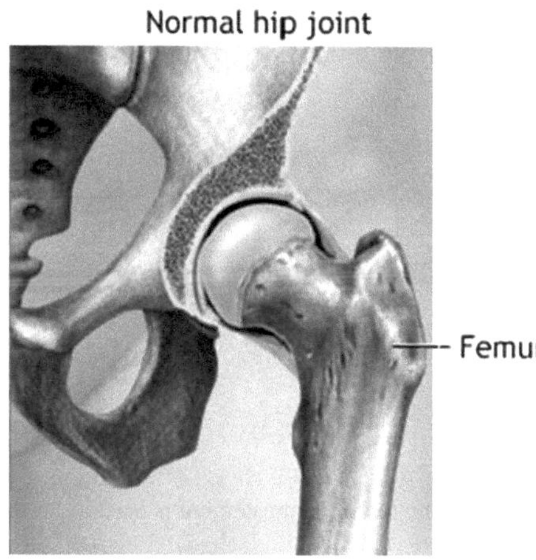

Fig. 11.8 A cutaway view of a normal hip joint and an overview of the hip region (insert)

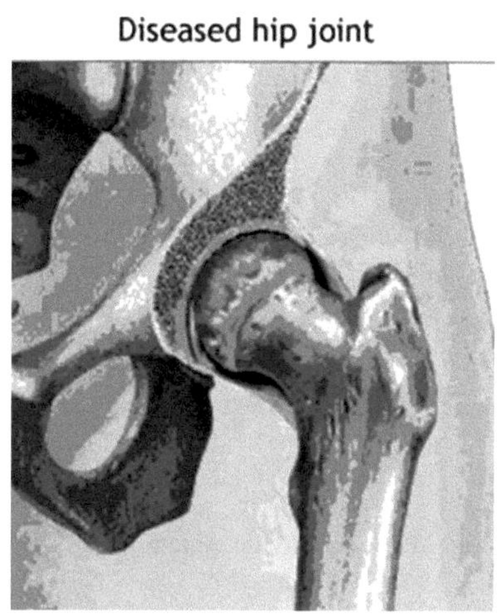

Fig. 11.9 A cutaway view of a degenerated cartilage of the hip head

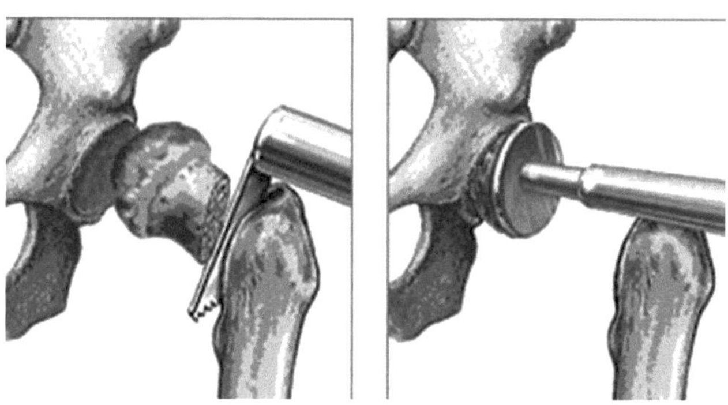

Fig. 11.10 The head of the femur is removed using an oscillating saw, and the pelvis socket is being milled using a milling tool to remove the diseased cartilage

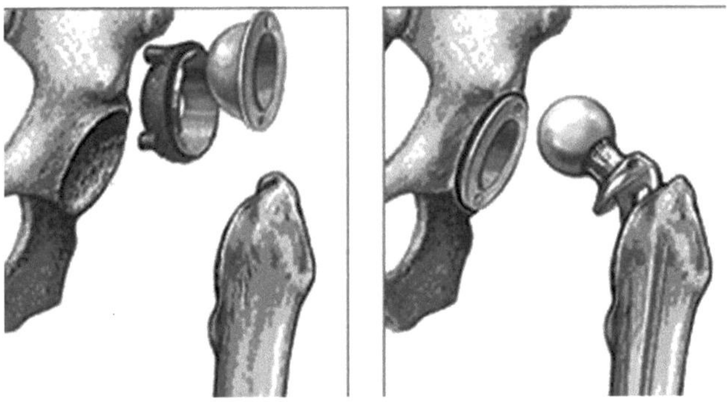

Fig. 11.11 A UHMWPE socket is placed in the enlarged pelvis cup and a metal ball and stem are placed in the femur. Bone cement is used to fix the stem

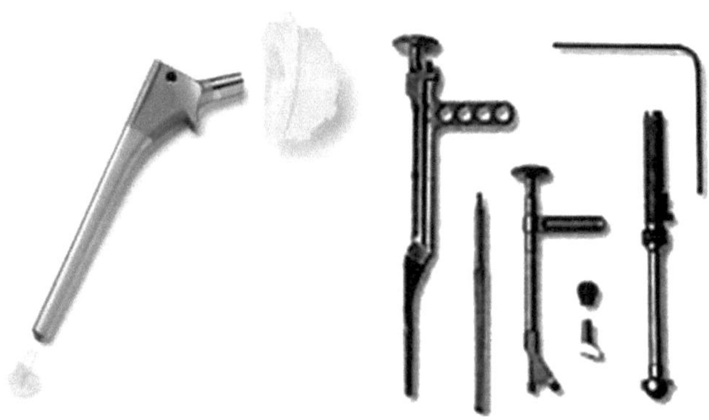

Fig. 11.12 Apollo hip system

11.8.2 Dr. K. H. Sancheti's Hip Prosthesis

Dr. Sancheti, an orthopedic surgeon par excellence from Pune, has established a hospital Research Centre and Manufacturing unit of artificial hip and knee joints. He has expertise in surgery and understands the manufacturing aspect as well. Some illustrations of his work are given in Fig. 11.13 with his kind permission. He personally demonstrated the surgical operative technique in different hospitals under the banner of West Bengal Orthopaedic Society, of which the author is a member. These are indigenous efforts based on Western studies.

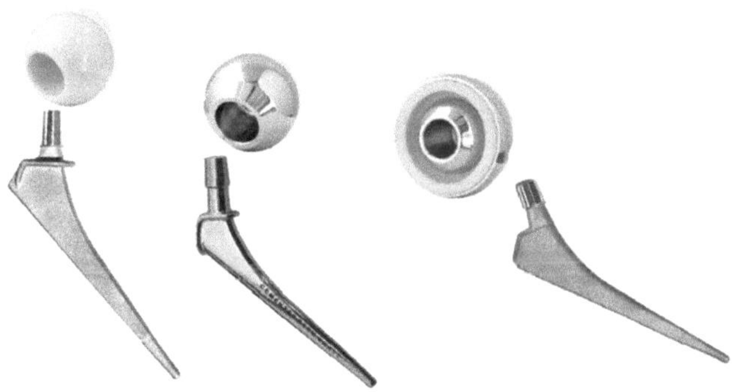

Fig. 11.13 Dr. Sancheti's design and manufactured hip prosthesis based on his work in Pune, India. Notice the ceramic ball, metallic ball, and modular UHMWPE acetabular part and ball

11.9 Design Modification Using Hollow Section

11.9.1 Stress Shielding

With cementless stem implantation in total hip arthroplasty, bone resorption of the surrounding proximal femur is regularly seen due to the effect of a stiff prosthesis stem implanted in bone, which is more elastic: The phenomenon is known as stress shielding [11]. Stress shielding of the femur is known to be a principal factor in aseptic loosening of hip replacements. This was addressed by Gross et al. [12]. Using finite-element (FE) modeling, the stresses in the proximal femur using different shapes of a hollow stem were compared with those produced using comparable sizes of a solid stem with different values of elastic modulus. Given present constraints in the design of hip prostheses, their theoretical study has shown that significant reductions in stress shielding at the proximal end can be achieved by the use of a hollow stem. Their design approach offers a more practical means of varying the rigidity of the stem that does the alternative of using a material, such as a composite, with varying elastic modulus, investigated by Kuiper et al. from Holland [12]. Although it is now a commonly held view that alternative methods to shear-loading transfer systems for femoral components of hip prostheses must be found, until new types of hip prostheses become available commercially, small improvements to the existing technology may be worthwhile. The hollow stem offers the potential of an incremental improvement to current designs of the prosthetic stem. However, the observation made by Rohlmann et al. [13], that optimal hip implant fixation is undetermined, remains true today, and long-term fixation relies on a range of design factors, not only on the rigidity and shape of the stem of the implant.

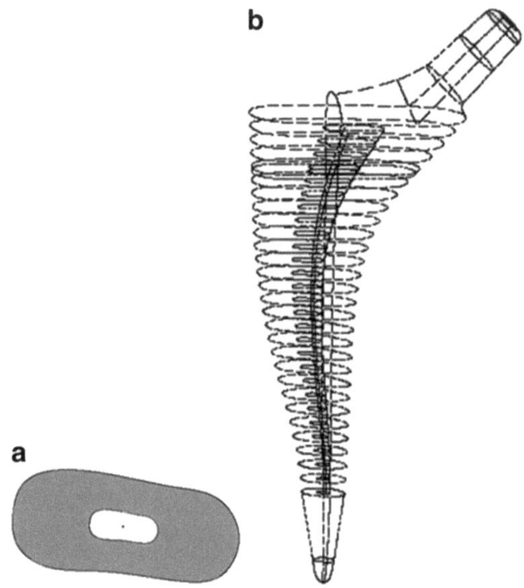

Fig. 11.14 (**a**) A typical hollow section of the designed custom hip; (**b**) outline of the designed custom hollow hip stem

11.9.2 Hollow Design

As discussed in the previous section, it was found that the hollow nature of the implant not only reduces the weight of the implant, but also increases the overall elasticity of the implant. This results in a drastic reduction in stress shielding in the intertrochanteric region and reduces bone remodeling at the distal tip after implantation of a cementless stem in the femoral canal. This contributes to the longer survival life of the implant and thereby reduces the number of revision surgeries required due to implant loosening.

11.9.2.1 Design Methodology

The customized hip implant was designed from the CT-scan data of a particular patient. This lead to a close matching of the outer boundary of the implant with the medullary canal of the femur. The intramedullary boundary created in MIMICS® from CT-scan data was exported to ANSYS® for creation of the outer boundary of the implant. However, there was no precursor for creation of the inner boundary of the hollow prosthesis. With the limitations of the iterations performed in this study, it was decided to scale the outer boundary cross sections of the designed hip implant by a certain percentage to get the inner boundary of the hollow implant.

A 30% reduction in sectional areas obtained from the solid custom implant was used to design the hollow implant. A regular thick hollow implant was designed to get an evenly distributed stress pattern along the entire stem. With the help of the

ANSYS® preprocessor, the center of mass of each of the sections was calculated. Keeping this center as the center of scaling for each section, the outer boundary layers were scaled down by 30%. Figure 11.14a shows a typical section of the hollow custom hip stem. The center of mass of the section is indicated with a dot at the central white portion of the section. This gives a narrow stem-like structure at the central line of the stem. This narrow stem was then subtracted from the solid custom stem to get a hollow stem. Figure 11.14b shows the outlines of the hollow custom hip implant. A portion of about 20 mm was left out on the distal end of the stem, as this portion is relatively narrow in the solid custom stem.

Now there is specialized software available, MIMICS, which utilizes the CT data of a patient's bone(s) and joint and then reconstructs the joint with its inner and outer sections; a customized prosthesis can be designed and subsequently produced using CAD-ACM technology. This author and his team had undertaken customized hip design based on a DST, Govt. of India-funded project, earlier.

References & Reading List

1. Gomez PF, Morcuende JA (2005) Early attempts at hip arthroplasty—1700s to 1950s. Iowa Orthop J 25:25–9, PMID 16089067. http://www.pubmedcentral.nih.gov/articlerender.fcgi?artid=1888777#R13
2. Daniel Stiles: Ivory carving in Myanmar. http://www.asianart.com/articles/ivory/index.html. Retrieved 2007-11-26.
3. That must be Bob. I hear his new hip squeaking. New York Times, May 11, 2008
4. Pai VS (1997) A comparison of three lateral approaches in primary total hip replacement. Int Orthop 21(6):393–398, http://link.springer.de/link/service/journals/00264/bibs/8021006/80210393.htm
5. Jones W (2007) Anterolateral approach to hip joint. Wheeless' textbook of orthopaedics. http://www.wheelessonline.com/ortho/anterolateral_approach_to_hip_joint_watson_jones. Retrieved 2007-11-26.
6. Petersen S. Anterior approach to the hip—Wheeless' textbook of orthopaedics. http://www.wheelessonline.com/ortho/anterior_approach_to_the_hip_smith_peterson. Retrieved 2007-11-26.
7. Cibulka MT, White DM, Woehrle J, Harris-Hayes M, Enseki K, Fagerson TL, Slover J, Godges JJ (2009) Hip pain and mobility deficits—hip osteoarthritis: clinical practice guidelines linked to the international classification of functioning, disability, and health from the Orthopaedic Section of the American Physical Therapy Association. J Orthop Sports Phys Ther 39(4):A1–A25. doi:10.2519/jospt.2009.0301
8. http://www.fda.gov/cdrh/mda/docs/p040033.html
9. van den Bekerom MP, Lamme B, Sermon A, Mulier M (2008) What is the evidence for viscosupplementation in the treatment of patients with hip osteoarthritis? Systematic review of the literature. Arch Orthop Trauma Surg 128(8):815–23
10. Centeno CJ, Kisiday J, Freeman M, Schultz JR (2006) Partial regeneration of the human hip via autologous bone marrow nucleated cell transfer: a case study. Pain Physician 9(3):253–6
11. Maistrelli GL, Fornasier V, Binnington A, McKenzie K, Sessa V, Harrington I (1991) Effect of stem modulus in a total hip arthroplasty model. J Bone Joint Surg Br 73(1):43–46
12. Kuiper JH (1993) Numerical optimization of artificial hip joint designs. Ph.D. Thesis, University of Nijmegen, Nijmegen, The Netherlands

13. Rohlmann A, Mossner U, Bergmann G, Hees G, Kolbel R (1987) Effects of stem design and material properties on stresses in hip endoprostheses. J Biomed Engineer 9:77–83
14. Christopher J, Centeno MD, Faulkner SBA (2011) Review of the literature, current research, and regulatory landscape. J Am Phys Surg 16(2) Summer

Problems

1. Make a list of the varieties of designs of hip prostheses, and develop sketches of the designs. Describe the specific motivation behind each design.
2. Give the reasoning for cementless hemihip arthoplasty. You will observe two holes in the Austin Moore stem. What is the rationale for having those two holes? The author has observed that breakage of such a hip occurred through such holes. Some surgeons argued new bones were formed through the holes for better anchorage. Do some research, getting X-rays of patients after several years of use of the hemihip.
3. Hip joint replacement is the most successful and popular surgery worldwide. In terms of cost, this surgery is the cheapest in India. India has some world-class hospitals. Do a comparative study, and determine how health tourism could be improved to boost the Indian economy.
4. What modifications will be desired in the hip for the Eastern populace when compared to Westerners? Remember, in Asia people squat and sit cross-legged (hint: ante-version angle of the hip).
5. How will you design and manufacture a customized (i.e., personalized) hip prosthesis?

Hint: This can be done using computerized tomographic data of the hip in question using image processing software like MIMICS in conjunction with ANSYS. Then it has to be given to a computer-aided manufacturing center that has a CNC workstation. This can also be done using LENS (laser-engineered net shaping) technology.

Readers may look into the Ph.D. (BESU) thesis of Ujjal Bhanu Ghosh, entitled, “Personalized Human Joint Prosthesis Design, Analysis and Manufacture Using CT-Data, FEM and NC-Technology,” which was completed it in January 2011 under a DST, Govt. of India-funded project of the author conducted at Jadavpur University.

Chapter 12
The Knee Joint and Its Artificial Replacement

12.1 Anatomy

The knee is the largest synovial joint in the body. It is composed of three bones and three joints although two of the three joints share a common cavity. The bones of the knee consist of the femur (thigh bone), the tibia (shin bone), the patella (kneecap), and, to a lesser degree, the fibula. The knee joint is made up of the tibiofemoral joint, which itself is comprised of a medial compartment and a lateral compartment. The true knee joint also includes the patellofemoral joint. Another important component of the knee joint complex, although not part of the true knee joint, is the superior tibiofibular joint. Figure 12.1 shows the details of the knee joint and its associated tendons and ligaments. The tibial plateau and cartilage are also visible.

12.1.1 Cruciate Ligaments

The cruciate ligaments consist of a highly organized collagen matrix that accounts for approximately three fourths of their dry weight. The majority of the collagen is type 1 (same as in bone 90%), and the remainder is type 3 (10%). Water constitutes 60% of the net weight under physiological conditions.

The cruciate ligaments are named based on their attachments on the tibia and their relationships to the intercondylar prominence of the proximal tibia. They are essential to the function of the knee joint. The cruciate ligaments act to stabilize the knee joint and prevent anteroposterior displacement of the tibia and the femur. They also contain numerous sensory endings, implying an important role in proprioceptive function. These ligaments are intraarticular, but because they are covered by synovium, they are considered extrasynovial. They receive their blood supply from branches of the middle genicular artery and both inferior genicular arteries.

S. Pal, *Design of Artificial Human Joints & Organs*,

DOI 10.1007/978-1-4614-6255-2_12,

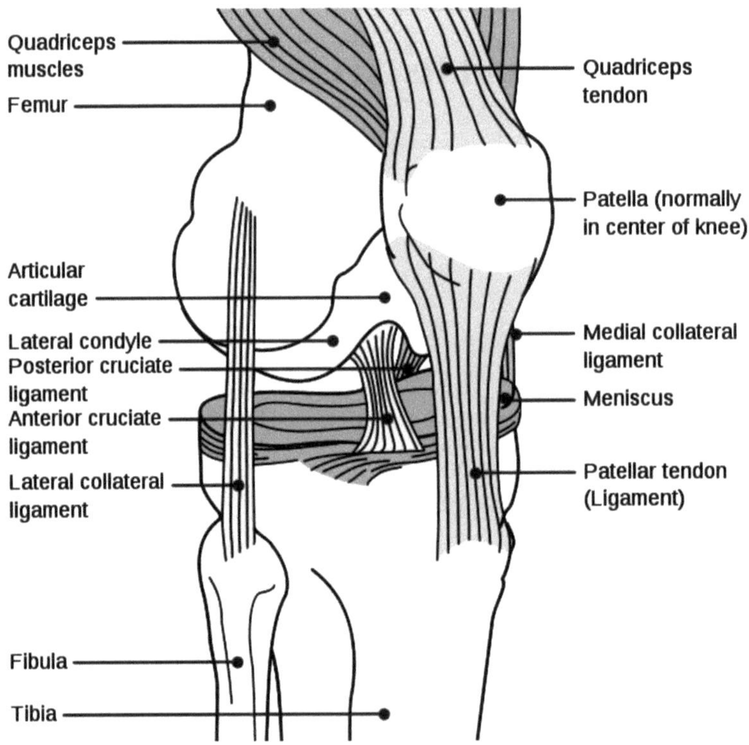

Fig. 12.1 Anatomy of the knee joint

12.1.2 Anterior Cruciate Ligament

The anterior cruciate ligament (ACL) originates from the medial surface of the lateral femoral condyle posteriorly in the intercondylar notch in the form of a segment of a circle. The ligament courses anteriorly, distally, and medially toward the tibia. Over the length of its course, the fibers of the ligament undergo a slight external rotation. The average length of the ligament is 38 mm and the average width 11 mm (Western data). The tibial attachment is a wide depressed area anterior and lateral to the medial tibial tubercle in the intercondylar fossa. The tibial attachment is more robust than the femoral attachment. There is a well-marked slip to the anterior horn of the lateral meniscus. Morphologically it is felt that the ACL is composed of two bands, the anteromedial band and the posterolateral band.

In extension, the posterolateral band of the ACL is taut and the anteromedial band is lax, whereas in flexion, the anteromedial band is tight and the posterolateral band is relatively relaxed.

The ACL is a prime static that stabilizes against anterior translation of the tibia on the femur, accounting for up to 86% of the total force resisting anterior draw. The ACL also plays a lesser role in resisting internal and external rotation.

The maximum tensile strength of the ACL is approximately 1,725 ± 270 N. This is less than the peak forces that occur in vigorous athletic activity. Stability is enhanced by dynamic stabilizers such as muscles that exert a force across the knee joint. The ACL plays an important proprioceptive function because of the variety of receptors in the anterior cruciate ligament. In people with ACL-deficient knees, a significantly higher threshold for detecting passive motion of the involved knee has been reported.

On the MRI scan, the normal ACL has a relatively low visibility. With a tear, the ACL on MRI appears bright with lots of fiber definition. The articular cavity of the knee joint is the largest joint space of the body. The cavity includes a space between and around the tibial and femoral condyles but also extends upward behind the patella to include the patellofemoral articulation and further into the suprapatellar bursa, which lies between the tendon of the quadriceps femoris muscle and the femur.

The synovial membrane lines the articular capsule and reflects onto the bone as far as the edges of the articular cartilage. It also lines the suprapatellar bursa and may also line any other bursas that communicate with the knee joint. The synovial membrane also covers the cruciate ligament, except where the PCL is attached to the back of the capsule. The anterior cruciate ligament is therefore an intraarticular structure that is within the synovial cavity of the knee joint, whereas the posterior cruciate ligament (PCL) is an intraarticular structure that is outside the synovial cavity of the knee joint.

Many folds and recesses within the knee are potential sites for the collection of wear debris, loose bodies, and bacterial contamination. All these recesses need to be assessed arthroscopically in these conditions.

12.1.3 Action of Muscles

The movements of the knee are flexion, extension, and rotation. Flexion is performed by the hamstrings and biceps femoris and, to a lesser extent, by the gastrocnemius and popliteus. Flexion is limited by the soft tissues at the back of the knee. Extension is performed by the quadriceps; because of the shape of the articulation and ligament attachments, the femur rotates medially on the tibia in terminal extension, the screw home mechanism that locks the joint. This movement is purely passive, as are other rotatory movements occurring during flexion/extension, and is due to articular geometry and static stabilizers.

The exception is lateral rotation of the femur, which precedes flexion by unlocking the joint. This movement is performed by the popliteus muscle. The sartorius, gracilis, and hamstrings are weak rotators of the knee but probably do not act as such primarily.

12.1.4 Knee Joint Stabilization

The stability of the knee joint is dependent upon static and dynamic conditions. The static stabilizer includes passive structures such as the knee joint capsule and various ligaments and other associated structures, such as the menisci, the coronary ligaments, and the meniscopatella and patellofemoral ligaments. The ligaments, which all act as static stabilizers, include the medial collateral ligament, the lateral collateral ligament, the ACL, PCL, the oblique popliteal ligament, and the arcuate ligament. The iliotibial band is also considered a static stabilizer.

The bony geometry also contributes to the static stability of the knee. The contribution is variable but can be made worse by certain anatomic variants such as a flat lateral femoral trochlea, which will predispose to lateral instability of the patella.

The dynamic stabilizers of the knee are all the muscles and their aponeuroses, including (1) quadriceps femoris and extensor retinaculum, (2) pesanserinus, (3) popliteus, (4) biceps femoris, and (5) semi-membranosis. The structures on the medial, anteromedial, and posteromedial sides of the knee are medial compartment structures and stabilizers, and the structures on the respective lateral side are lateral compartment stabilizers.

Knee Motion. The knee joint is a modified hinge joint (ginglymus). The active movements of the knee joint are described as **flexion, extension, medial rotation, and lateral rotation.**

The flexion and extension at this joint differ from those of a true hinge as the axis about which the movement occurs is not fixed, but translates upward and forward during extension and backward and downward during flexion. The knee joint possesses limited inherent stability from the bony architecture. The lack of conformity between bony surfaces allows six degrees of freedom of motion about the knee, including translation in three planes (mediolateral, anteroposterior, proximodistal) and rotation in three planes (flexion/extension, internal/external, varus/valgus).

With the foot fixed on the ground, the last 30° of extension is associated with medial rotation of the femur. Compared with the medial femoral condyle, the articular surface of the smaller lateral femoral condyle is rounder and flattens more rapidly anteriorly. Consequently, it approaches a more fully congruent relationship with its opposed tibial meniscal surface, some 30° before full extension has been obtained. To achieve full extension, the lagging medial compartment must medially rotate about a fixed vertical axis while moving backward in an arc.

There is a progressive increase in passive mechanism that resists further extension. In full-extension parts of both cruciate ligaments, the collateral ligaments, the posterior capsular and oblique posterior ligament complex, and the skin and fascia are all taught. There is also passive or active tension in the hamstrings, gastrocnemius muscles, and the ITB. In addition, the anterior parts of the menisci compress between the femoral condyles and the tibia.

At the beginning of flexion, the knee "unlocks" with an external rotation of the femur on the tibia. This is partly related to the opposite interplay of the meniscal articular and ligamentous structures involved but is also brought about by the contraction of the popliteus muscles. It pulls downward and posteriorly on its attachment to the lateral condyle of the femur, helping greater roll back in this compartment, which occurs with flexion. Through its meniscal attachment, the popliteus pulls on the posterior horn of the lateral meniscus. In this way, while rolling back, the posterior motion of the menisci occurs in both compartments, and the greater motions laterally can be facilitated. The menisci, which are squeezed between the joint surface in extension, are moved posteriorly with the femur in flexion, the lateral more so than the medial. With terminal extension achieved and the knee locked by the femur rotating internally on the tibia, this is called the screw home mechanism.

Flexion is checked by the quadriceps mechanism, by the anterior parts of the capsule and the PCL, and by the compression of the soft tissue structures in the popliteal fossa.

12.2 Knee Replacement, or Knee Arthroplasty

12.2.1 Introduction

Knee replacement is a surgical procedure to replace the weight-bearing surfaces of the knee joint to relieve the pain and disability of osteoarthritis [1]. It may be performed for other knee diseases such as rheumatoid arthritis and psoriatic arthritis. In patients with severe deformity from advanced rheumatoid arthritis, trauma, or long-standing osteoarthritis, the surgery may be more complicated and carry a higher risk. Osteoporosis does not typically cause knee pain, deformity, or inflammation and is not a reason to perform knee replacement. Figure 12.2 shows the degenerated knee and the same knee after arthroplasty with two components, namely, the femoral part of metal alloy and the bearing tibial component made of UHMWPE (Fig. 12.3).

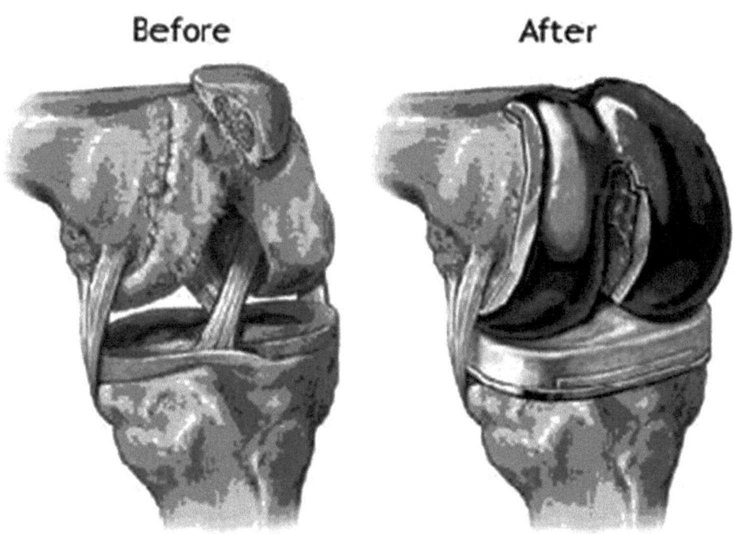

Fig. 12.2 The diseased knee before and after arthroplasty with femoral and tibial parts

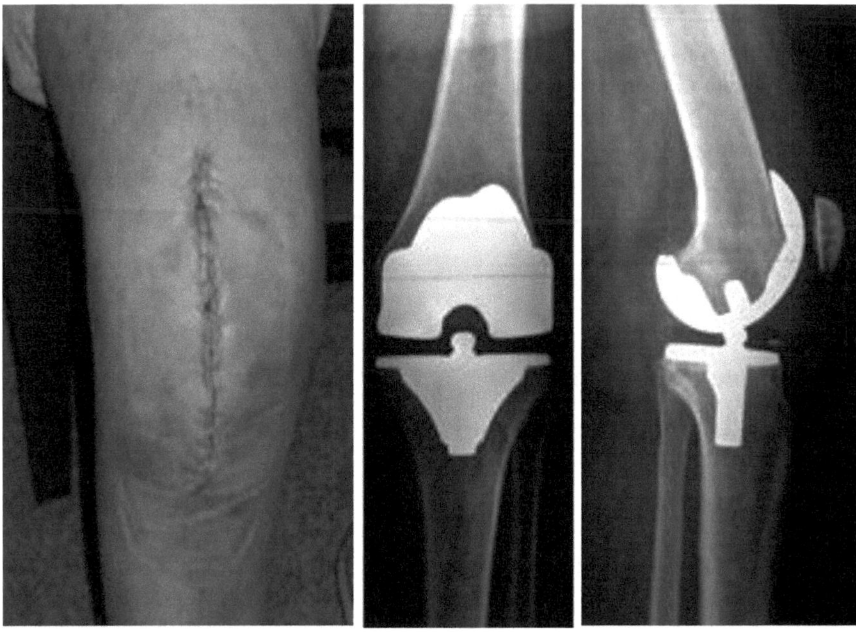

Fig. 12.3 Total knee replacement: the incision for knee replacement surgery and the AP view and lateral view (X-ray)

Other major causes of debilitating pain include meniscus, cartilage, and ligament tears. Debilitating pain from osteoarthritis is much more common in the elderly. Knee replacement surgery can be performed as a partial or total knee replacement [2]. In general, the surgery consists of replacing the diseased or damaged joint surfaces of the knee with metal and polymer components shaped to allow continued motion of the knee. The recovery period is 6 weeks or longer and involves use of a walker and then a cane [3].

12.3 History

Following John Charnley's success with hip replacement in the 1960s, numerous attempts were made to design knee replacements. Gunston and Marmor were pioneers in North America. Marmor's design allowed for unicompartmental operations, but these designs did not last long. In the 1970s, the "geometric" design found favor as well as John Insall's condylar knee design. Hinged knee replacements for salvage date back to Guepar but did not stand up due to excessive wear. The history of knee replacement is the story of continued innovation to try to limit the problems of wear, loosening, and loss of range of motion (Fig. 12.4).

Fig. 12.4 Hinged knee

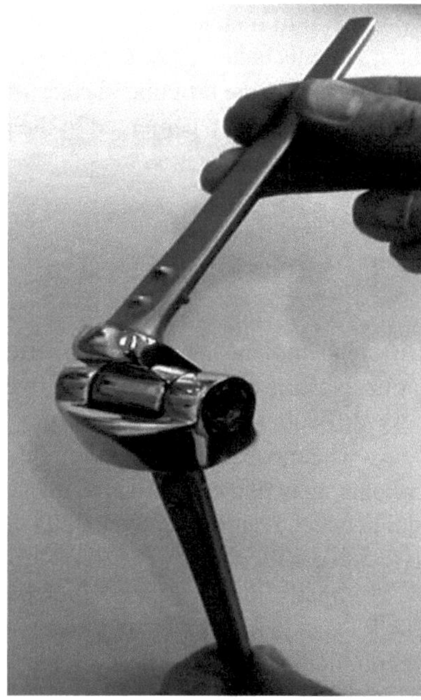

Knee replacement surgery is most commonly performed in people with severe osteoarthritis. It should be considered when conservative treatments did not work. Physical therapy quite often improves function and may delay or prevent the need for knee replacement [4].

12.3.1 Technique

The surgery involves exposure of the front of the knee, with detachment of part of the quadriceps muscle (vastus medialis) from the patella. The patella is displaced to one side of the joint, allowing exposure of the distal end of the femur and the proximal end of the tibia. The ends of these bones are then accurately cut to shape using cutting guides and an oscillating saw oriented to the long axis of the bones. The cartilages and the anterior cruciate ligament are removed; the posterior cruciate ligament may also be removed, but the tibial and fibular collateral ligaments are preserved. Metal components are then impacted onto the bone or fixed using polymethyl methacrylate (PMMA) bone cement. A roundedend implant is used for the femur, mimicking the natural shape of the bone. On the tibia the component is flat, although it often has a smaller tapered stem that goes down inside the tibial bone for further stability. A slightly dished high-density polyethylene part is then inserted onto the tibial component so that the weight is transferred from metal to plastic, not from metal to metal. During the operation, any deformities must be corrected, and the ligaments balanced so that the knee has a good range of movement and is stable. In some cases, the articular surface of the patella is also removed and replaced by a polyethylene button cemented to the posterior surface of the patella. In other cases, the patella is kept as it is.

12.4 Variations in Design

Different implant manufacturers offer slightly different instrumentation and technique. No consensus has emerged over which design is the best. Clinical studies are very difficult to perform, requiring large numbers of cases followed over many years. The most significant variations are between cemented and uncemented components, between operations that spare or sacrifice the posterior cruciate ligament, and between resurfacing the patella or not. Some also study patient satisfaction data associated with pain.

Minimally invasive procedures have been developed in total knee replacement (TKR) that do not cut the quadriceps femoris muscle. There are different definitions of minimally invasive knee surgery, which may include a shorter incision length, retraction of the patella (kneecap) without eversion (rotating out), and specialized instruments. There are few randomized trials, but studies have found less postoperative pain, shorter hospital stays, and shorter recovery. However, no studies have shown long-term benefits [5].

12.4.1 Partial Knee Replacement

Partial knee replacement is a procedure in which orthopedic surgeons replace part of a painful, dysfunctional knee joint that has only either the inner (medial) or outer (lateral) part of the joint damaged. A limited surgical procedure can resurface only the affected area with a metal implant and can preserve the remainder of the otherwise healthy knee.

Unicompartmental arthroplasty (UKA), also called partial knee replacement, is an option for some patients. The knee is generally divided into three "compartments": medial (the inside part of the knee), lateral (the outside), and patellofemoral (the joint between the kneecap and the thighbone). Most patients with arthritis severe enough to consider knee replacement have significant wear in two or more of the above compartments and are best treated with total knee replacement. A minority of patients (the exact percentage is hotly debated but is probably between 10–30%) have wear confined primarily to one compartment, usually the medial, and may be candidates for unicompartmental knee replacement. Advantages of UKA compared to TKA include smaller incision; easier post-op rehabilitation; shorter hospital stay; less blood loss; lower risk of infection, stiffness, and blood clots; and easier revision if necessary. While most recent data suggest that UKA in properly

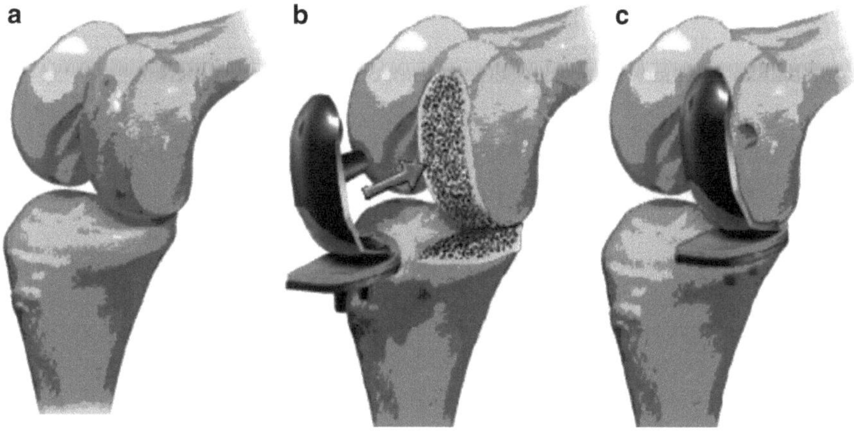

Fig. 12.5 (**a**) Diseased knee; (**b**) partial condylar and tibial cartilage removed; (**c**) the partial components are fixed

selected patients has survival rates comparable to TKA, most surgeons believe that TKA is the more reliable long-term procedure. Persons with infectious or inflammatory arthritis (rheumatoid arthritis, lupus, psoriatic arthritis) or marked deformity are not candidates for this procedure. Partial knee replacement is usually recommended in patients who have severe arthritis of the knee and who have not benefited from conservative treatments such as medication, injections, strengthening exercises, and weight loss. Figure 12.5 illustrates the partial knee replacement.

After the surgery, patoents can expect to start rehab almost immediately. Patients usually stay in the hospital for one or two nights, but putting weight on the knee is usually permitted immediately. Exercise training will depend on the age and activity level of the patient and is customized to the patient's abilities. At a follow-up appointment with the surgeon, the patient will get a better idea of the actual length of time to full recovery, and it depends largely on regaining muscle strength. Good muscle strength before the replacement is desirable for satisfactory performance of the knee.

Several studies have demonstrated the advantages of this treatment option over the more conventional total knee replacement. These include a smaller scar, less pain, a potentially shorter hospital stay, a faster rehab and recovery time, and possibly a greater range of motion when compared to standard total knee procedures. In addition, because a greater part of the knee remains untouched, patients often report that the knee feels quite normal.

12.5 Risks and Complications

The most serious complication is infection of the joint, which occurs in less than 1% of patients. Deep vein thrombosis occurs in up to 15% of patients and is symptomatic in 2–3%. Nerve injuries occur in 1–2% of patients. Persistent pain or stiffness occurs in 8–23% of patients. Prosthesis failure occurs in approximately 2% of patients at 5 years [6]. Periprosthetic fractures are becoming more frequent with the aging patient population and can occur intraoperatively or postoperatively.

12.5.1 Loss of Motion

The knee at times may not recover its normal range of motion (0–135° flexion usually) after total knee replacement. Much of this is dependent on preoperative function. Most patients can achieve a 0–110° range of motion, but stiffness of the joint can occur in some situations. Manipulation of the knee under anesthetic is used to improve postoperative stiffness. There are also many implants that are designed to be "high-flex" knees, offering a greater range of motion.

12.5.2 Instability

In some patients, the kneecap is unstable postsurgery and dislocates to the outer side of the knee. This is painful and usually needs to be treated by surgery to realign the kneecap. However, this is quite rare. In the past, there was a considerable risk of the implant components loosening over time as a result of wear. As medical technology has improved, however, this risk has fallen considerably. One implant manufacturer claims to have reduced this risk of wear by 79% in fixed-bearing knees. Another implant manufacturer claims to have reduced the risk of wear by 94% in mobile-bearing, also known as rotating-platform, knees. Knee replacement implants can last up to 20 years in many patients; whether or not they actually survive that long depends largely in part upon how active the patient is after surgery. **Sometimes, the Gamma irradiation for sterilization causes brittleness of UHMWPE, leading to higher wear. Irradiation at an inert liquid (nitrogen) environment gives** a **better result with respect to wear**.

12.5.3 Infection

The current classification of AAOS divides prosthetic infections into four types [7]:

- Type 1 (positive intraoperative culture)
- Type 2 (early postoperative infection): infection occurring within first month after surgery
- Type 3 (acute hematogenous infection): hematogenous seeding of site of previously well-functioning prosthesis
- Type 4 (late chronic infection): chronic indolent clinical course; infection present for more than a month.

While it is relatively rare, periprosthetic infection remains one of the most challenging complications of joint arthroplasty. A detailed clinical history and physical assessment remain the most reliable tools to recognize a potential periprosthetic infection. In some cases, the classic signs of fever, chills, painful joint, and a draining sinus may be present, and diagnostic studies are simply done to confirm the diagnosis. In reality, though, most patients do not present with those clinical signs, and, in fact, the clinical presentation may overlap with other complications, such as aseptic loosening. In those cases, diagnostic tests can be useful in confirming or excluding infection.

According to a recent review, several new tests can be used in the diagnosis of a periprosthetic infection [8]:

- Conventional radiograph: Rules out other conditions such as loosening and/or osteolysis.
- Radionucleotide imaging: Technetium-99m sulfur imaging combined with indium-111-labeled leukocytes probably offers improved specificity than either test alone. Gallium 67 scans alone have a low sensitivity for infection. FDG-PET imaging has been shown to have variable specificity and sensitivity.
- Serology: Elevated serum C-reactive protein (CRP) and erythrocyte sedimentation rate (ESR) more than three months following arthroplasty are good screening tests [9].
- Cultures: High sensitivity and specificity, but only if done two weeks following antibiotic discontinuation. Gram stains have low specificity and sensitivity. The predictive value of a positive culture increases if the culture is performed in a patient with a high clinical suspicion, rather than a screening test.
- Joint fluid leukocyte counts: A joint fluid white blood cell count of more than 500/μl is suggestive of an infection.

The choice of treatment depends on the type of prosthetic infection [10, 11]:

Positive intraoperative cultures: Antibiotic therapy alone.
Early postoperative infections: debridement, antibiotics, and retention of prosthesis.
Late chronic: delayed exchange arthroplasty. Surgical debridement and parenteral antibiotics alone in this group has limited success, and standard of care involves exchange arthroplasty [12].
Acute hematogenous infections: debridement, antibiotic therapy, retention of prosthesis.

12.6 Developmental Work on a Congruent Tibial Component

Over the years, scientists have been continuously trying to improve upon the problems stemming from the failure of knee joint prostheses. There is a competition among the manufacturers to produce a better design; accordingly, designers are trying to improve their design and market it with a special feature.

Fig. 12.6 Tibial part of PE (courtesy JB-Am 2009, 2105, Sigma DePuy)

In the natural knee joint (as in the artificial total knee prosthesis), the joint surfaces between the tibia and femur are not conforming. Nature has placed cartilage—meniscus—between these two joint surfaces to provide conformity. The meniscus, which takes up and distributes the stresses evenly between the nonconforming surfaces of the tibial and femoral joint condyles, moves a little during knee joint bending and stretching. This small movement helps the meniscus to conform better to the joint surfaces.

1. The surgeons have an identical problem with the lacking congruence between the femoral and tibial components of a total knee. They tried to improve this lacking congruence between the artificial knee joint's surfaces' congruence in three different designs, described here.

   First design. The first tibial components were manufactured of **all-polyethylene (PE) components** and were fixed by cement directly to the bone tissue of the tibia. They are still in use today, although their form may be changed a little. Soon, the surgeons and material scientists became concerned that the component made of polyethylene alone was too soft. It needed support from something harder (Fig. 12.6).

   Second design. Next came the metal-backed tibial component. The polyethylene plate articulating with the femoral component received a metal backing. This hard backing was supposed to prohibit the deformation of the soft polyethylene. The polyethylene inlay was rigidly fixed to the metal backing plate and thus immobile. Soon, this rigid fixation was considered a disadvantage, and development continued to a mobile polyethylene plate (Fig. 12.7).

   Third design. The peak of the development is at present the mobile-bearing tibial component. The polyethylene plate that articulates with the femoral component lost its fixation to the metallic back-up plate and became mobile (Fig. 12.8).
2. **Mobile-bearing total knee prosthesis** has a polyethylene bearing plate that is very conforming with the femoral condyles. The increased conformity is possible because the polyethylene bearing plate moves on the surface of a highly polished metallic tray during extension and flexion of the knee joint. The metallic tray is itself affixed to the tibia with bone cement (Fig. 12.9).

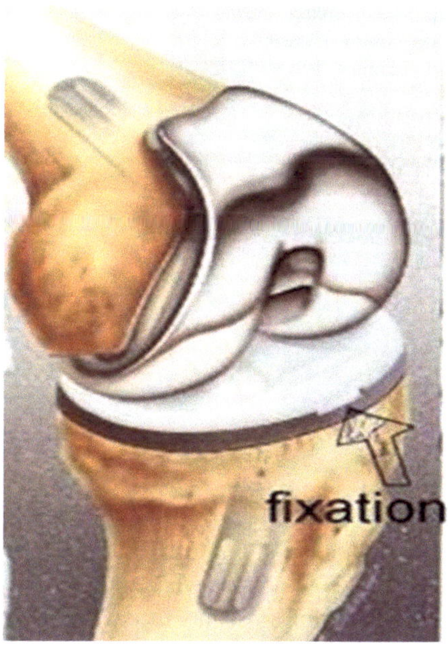

Fig. 12.7 Fixed bearing plate (courtesy Miller-Galante TK, Zimmer)

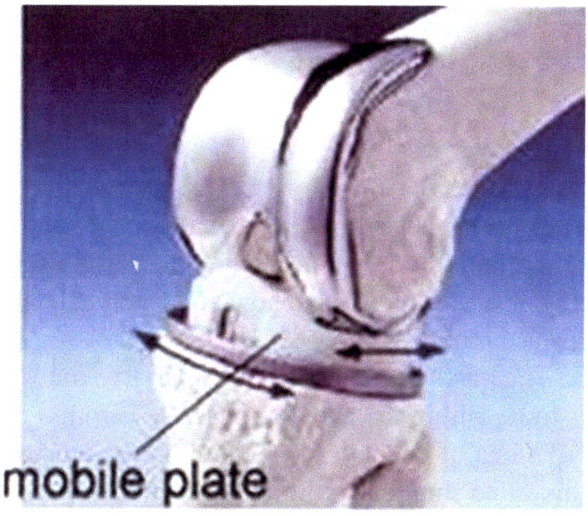

Fig. 12.8 Movable bearing plate (courtesy: Corin Rotaglide TK)

12.6.1 This Design has Two Advantages

First: The polyethylene bearing plate has a large contact area with the femoral component during the whole range of knee joint movement. This large contact area reduces the contact stresses on the polyethylene bearing plate substantially, thereby reducing the wear debris generation and thus improving the service life of total knee prostheses. The contact areas for stable bearing varies between 200–300 mm^2 and between 1,000–1,500 mm^2 for mobile bearing, that is, five times the fixed bearing. The wear debris production will also reduce substantially.

Fig. 12.9 Moving bearing plate (courtesy Mathys Ltd.–Balansys PS)

Second: The bearing plate moves freely and does not restrict the natural movements of the femoral component. The "unrestricted" movements of the plate, the stresses transmitted on the area where the total knee prosthesis is fixed to the tibia, are low and the risk for loosening of knee components is lower as well. Additional advantages are a more natural gait pattern and a larger range of movement achieved with these total knee prostheses.

12.6.2 Disadvantages

The polyethylene mobile plate has two wearing surfaces: One is the surface opposed to the femoral component; the other is the surface opposed to the polished tibial tray. It is as yet uncertain how many polyethylene particles this doubling of wearing surface produces.

The maximum load may be up to five times the body weight. If the bearing area is to be within the above limits, one can find out the bearing stress by the relation $5\,W$/(area of contact), depending on the type of bearing, as mentioned earlier, and this value should be within the allowable value. Similarly, the metal tray should be designed for bearing and bending stress, as these will be the failure criteria.

12.7 Indian Effort

As indicated earlier, Dr. K. H. Sancheti, a surgeon cum engineer in Pune, also has developed and manufactured the total knee prosthesis, shown in Fig. 12.10, which is known as the Indus knee joint. This design is similar to the Miller–Galante TK,

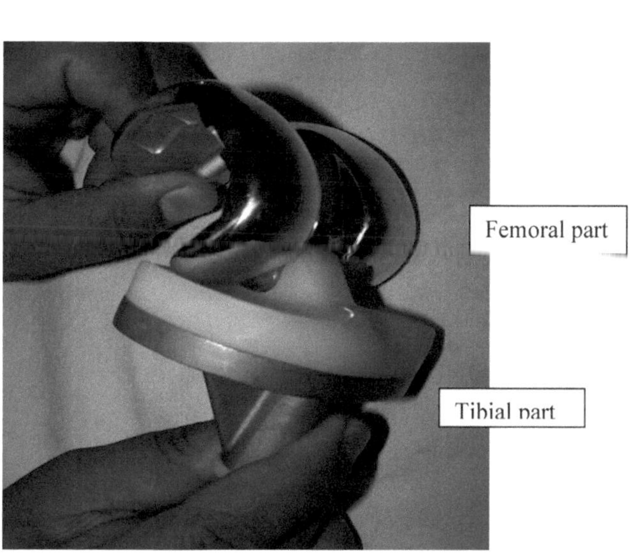

Fig. 12.10 Total knee design by Dr. K. H. Sancheti of Pune, with a UHMWPE liner and with stiffeners of the tibial part to strengthen it

Zimmer. It has a very strong tibial component with a rectangular cross section and additional stiffeners that require more bone space, but it will prevent bending deflection of the tibial part. The UHMWPE liner will reduce the friction, and the femoral condylar part is made of Co-Cr-Mo alloy. It may be made of Ti6Al4V as well. The metallic parts are best produced by casting. It may be sand mould-cast or pressure die-cast.

References

1. Plamer SH, Cross MJ (2004-01-30). "Total knee arthroplasty." eMedicine. http://www.emedicine.com/orthoped/topic347.htm. Retrieved 2008-10-28
2. Total knee replacement – Ortho info – AAOs. http://orthoinfo.aaos.org/topic.cfm?topic=A00389
3. Leopold SS (2009) Minimally invasive total knee arthroplasty for osteoarthritis. N Engl J Med 360(17):1749–1758
4. Deyle GD, Henderson NE, Matekel RL, Ryder MG, Garber MB, Allison SC (2000) Effectiveness of manual physical therapy and exercise in osteoarthritis of the knee. Ann Intern Med 132(3):173–181
5. Leopold SS (2009) Minimally invasive total knee arthroplasty for osteoarthritis. N Engl J Med 360(17):1749–1758
6. Leopold SS (2009) Minimally invasive total knee arthroplasty for osteoarthritis. N Engl J Med 360(17):1749–1758
7. Leone JM, Hanssen AD (2006) Management of infection at the site of a total knee arthroplasty. Instr Course Lect 55:449–461
8. Bauer TW, Parvizi J, Kobayashi N, Krebs V (2006) Diagnosis of periprosthetic infection. J Bone Joint Surg Am 88(4):869–882. doi:10.2106/JBJS.E.01149. PMID 16595481

9. Spangehl MJ, Masri BA, O'Connell JX, Duncan CP (1999) Prospective analysis of preoperative and intraoperative investigations for the diagnosis of infection at the sites of 202 revision total hip arthroplasties. J Bone Joint Surg Am 81(5):672–683
10. Corless CE, Guiver M, Borrow R, Edwards-Jones V, Kaczmarski EB, Fox AJ (2000) Contamination and sensitivity issues with a real-time universal 16S rRNA PCR. J Clin Microbiol 38(5):1747–1752
11. Segawa H, Tsukayama DT, Kyle RF, Becker DA, Gustilo RB (1999) Infection after total knee arthroplasty. A retrospective study of the treatment of 81 infections. J Bone Joint Surg Am 81(10):1434–1445
12. Chiu FY, Chen CM (2007) Surgical debridement and parenteral antibiotics in infected revision total knee arthroplasty. Clin Orthop Relat Res 461:130–135

Problems

1. Discuss the design procedure of a partial knee-bearing joint. How do you propose to find the dimensions of the part, and what method of insertion and fixation would you propose?
2. Why is a softer liner (UHMWPE) used in an artificial knee joint? How does the Gamma sterilization method affect the service life of the implant? How can this problem be solved?
3. What is a cruciate-sparing design of a knee prosthesis? How it is related to knee stability? What is the life expectancy of the TKR? What is the approximate cost of the TKR?
4. Make a comparative study of the various designs of total knee replacement prostheses, indicating the kinematics and service life as well as the wear debris production. How can wear be reduced?

Chapter 13
The Ankle Joint and Its Artificial Replacement

13.1 Introduction

The ankle joint acts like a hinge. But it's much more than a simple hinge joint. The ankle is actually made up of several important bones and their associated assembly. The exceptional assemblage of the ankle makes it a very stable joint. This joint has to be stable in order to withstand 1.5 times our body weight when we walk slowly and up to eight times our body weight when we run at a high speed or jump.

Normal ankle function is needed to walk with a smooth and nearly effortless gait. The muscles, tendons, and ligaments that support the ankle joint work together to propel the body. Conditions that disturb the normal way the ankle works can make it difficult to do daily activities without pain.

13.1.1 Structure

The important structures of the ankle can be divided into several categories. These include bones and joints, ligaments and tendons, muscles, nerves, and blood vessels. The top of the foot is called the dorsal surface. The sole of the foot is the plantar surface.

13.2 Bones and Joints

The ankle joint is formed by the connection of three bones. The ankle bone is called the talus. The top of the talus fits inside a socket that is formed by the lower end of the tibia (shinbone) and the fibula (the small bone of the lower leg). The bottom of the talus sits on the heel bone, called the calcaneus (Fig. 13.1).

S. Pal, *Design of Artificial Human Joints & Organs*,
DOI 10.1007/978-1-4614-6255-2_13,

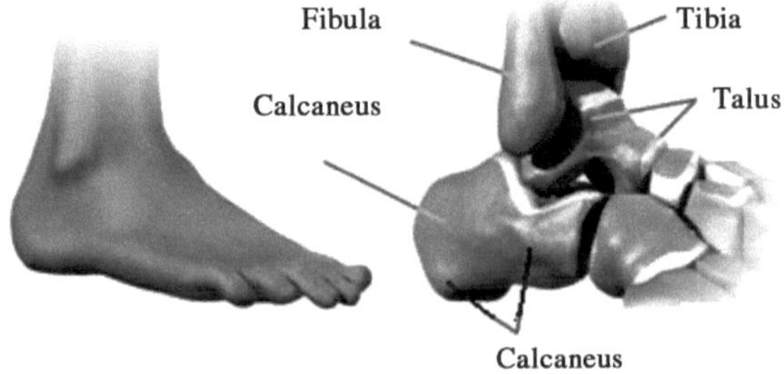

Fig. 13.1 The important bones of the ankle joint

The talus works like a hinge inside the socket to allow the foot to move up (**dorsiflexion**) and down (**plantarflexion**). Carpenters use a similar construction, called **a mortise and tenon**, to create stable structures. They routinely use it to make strong and sturdy items, such as furniture and buildings. Inside the joint, the bones are covered with articular cartilage that allows the bones to move smoothly against one another in the joints of the body. The cartilage lining is about 6 mm thick in most joints that carry body weight, such as the ankle, hip, or knee. It is soft enough to allow for shock absorption but tough enough to last a lifetime, as long as it is not injured or diseased.

13.3 Ligaments and Tendons

Ligaments are the soft tissues that attach bones to bones. Ligaments are very similar to tendons. The difference is that tendons attach muscles to bones. Both of these structures are made up of small fibers of a material called collagen. The collagen fibers are bundled together to form a ropelike structure. Ligaments and tendons come in many different sizes and, like rope, are made up of many smaller fibers. The thickness of the ligament or tendon determines its strength.

Ligaments on both sides of the ankle joint help hold the bones together. Three ligaments make up the lateral ligament complex on the side of the ankle farthest from the other ankle. ("Lateral" means farther away from the center of the body.) These include the anterior talofibular ligament (ATFL), the calcaneofibular ligament (CFL), and the posterior talofibular ligament (PTFL). A thick ligament, called the deltoid ligament, supports the medial ankle (the side closest to the other ankle) (Fig. 13.2).

Ligaments also support the lower end of the leg where it forms a hinge for the ankle. This series of ligaments supports the ankle syndesmosis, the part of the ankle where the bottom end of the fibula meets the tibia. Three main ligaments support this area. The ligament crossing just above the front of the ankle and connecting the tibia to the fibula is called the anterior inferior tibiofibular ligament (AITFL). The posterior fibular ligaments attach across the back of the tibia and fibula. These ligaments include the posterior inferior tibiofibular ligament (PITFL) and the transverse ligament. The interosseous ligament lies between the tibia and fibula

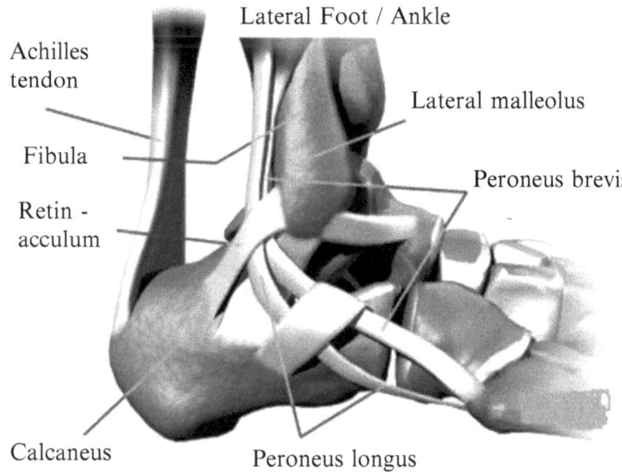

Fig. 13.2 The ankle joint and the important tendons and ligaments

("interosseous" means between bones). The interosseus ligament is a long sheet of connective tissue that connects the entire length of the tibia and fibula, from the knee to the ankle.

The ligaments that surround the ankle joint help form part of the joint capsule. A joint capsule is a watertight sac that forms around all joints. It is made up of the ligaments around the joint and the soft tissues between the ligaments that fill in the gaps and form the sac.

The ankle joint is also supported by nearby tendons. The large Achilles tendon is the most important tendon for walking, running, and jumping. It attaches the calf muscles to the calcaneus (heel bone) and allows us to stand on our toes. The posterior tibial tendon attaches one of the smaller muscles of the calf to the underside of the foot. This tendon helps support the arch and allows us to turn the foot inward. The anterior tibial tendon allows us to raise the foot. Two tendons run behind the outer bump of the ankle (the lateral malleolus). These two tendons, called the peroneals, help turn the foot down and out.

13.4 Muscles

Most of the motion of the ankle is caused by the stronger muscles in the lower leg, whose tendons pass by the ankle and connect in the foot. Contraction of the muscles in the leg is the main way that we move our ankle when we walk, run, and jump.

The key ankle muscles have been discussed earlier in the section on ligaments and tendons. These muscles and their actions are also listed here:

The **peroneals (peroneus longus and peroneus brevis)** on the outside edge of the ankle and foot bend the ankle down and out.

The **calf muscles (gastrocnemius and soleus)** connect to the calcaneus by the Achilles tendon. When the calf muscles tighten, they bend the ankle down.

The **posterior tibialis muscle** supports the arch and helps turn the foot inward. The **anterior tibialis** pulls the ankle upward.

13.5 Nerves

The nerve supply of the ankle is from nerves that pass by the ankle on their way into the foot. The tibial nerve runs behind the medial malleolus. Another nerve crosses in front of the ankle on its way to top of the foot. There is also a nerve that passes along the outer edge of the ankle. The nerves on the front and outer edge of the ankle control the muscles in this area, and they give sensation to the top and outside edge of the foot.

13.6 Blood Vessels

The ankle gets blood from nearby arteries that pass by the ankle on their way to the foot. The dorsalis pedis runs in front of the ankle to the top of the foot. (The pulse can be felt where this artery runs in the middle of the top of the foot.) Another large artery, called the posterior tibial artery, runs behind the medial malleolus. It sends smaller blood vessels to the inside edge of the ankle joint. Other, less important arteries entering the foot from other directions also supply blood to the ankle.

The anatomy of the ankle is clearly very complex. When every part works together, the ankle functions correctly. When one part becomes damaged, it can affect every other part of the ankle and foot, leading to problems with standing, walking, and movements.

13.7 Artificial Replacement of the Ankle Joint

Each artificial ankle prosthesis is made of two parts:

- The tibial component is the part of the artificial joint that replaces the socket portion of the ankle (the upper portion).
- The talus component replaces the outer part of the talus.

The tibial component is usually made up of two parts: a flat metal piece (stainless steel or Ti-6Al-4V alloy) called a metal tray that is attached directly to the tibia bone inside a slot prepared surgically, and a plastic cup (usually UHMWPE) that fits onto the metal piece, forming a socket for the artificial ankle joint. The talus component is made of the same metal alloy as the tibial component and fits into the socket of the tibial component, as shown in Fig. 13.3.

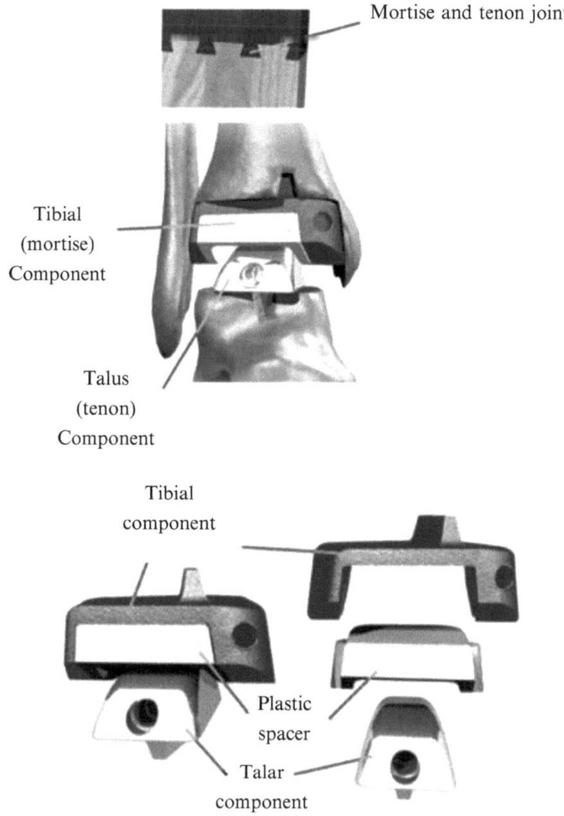

Fig. 13.3 The ankle joint, an ideal mortise and tenon joint, an artificial joint fitted into bones, and an exploded view of the ankle joint. The metallic parts are of stainless steel or Ti-6Al-4V alloy and plastic HDPE or UHMWPE

The surgeon uses a bone cement (PMMA) to attach the metal components to the bone. This is called a **cemented prosthesis**, as usual.

Some surgeons prefer to put the new joint in without using cement. This is called an **uncemented prosthesis**. The surface of this type of prosthesis, which comes directly in contact with bone, bears a sand-blasted or porometal surface that allow osteocytes to grow into the pores and attach the prosthesis to the bone. The contact surface with bone may be treated with a plasma spray of metal or hydroxyapatite to improve osseointegration.

13.8 The Operative Procedure

During the surgery, the patient will be placed under either general anesthesia or a spinal type of anesthesia. The surgeon makes an **incision** through the skin on the front of the ankle; the nerves, blood vessels, and tendons are protected and moved to the side. An incision is then made into the joint capsule that encloses the ankle joint. The surgeon opens the joint to prepare the surfaces to be replaced.

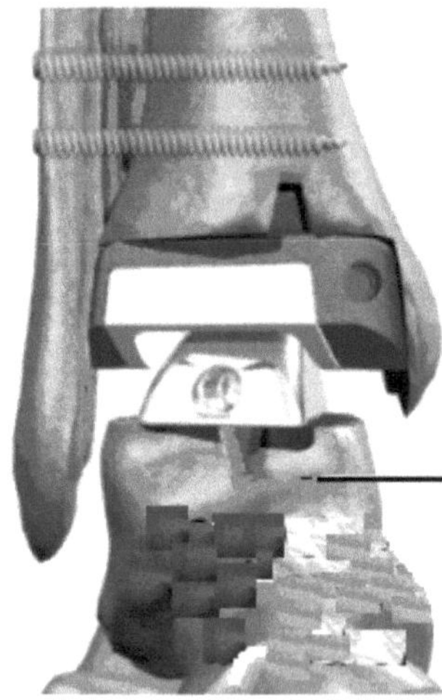

Fig. 13.4 Artificial ankle joint components are installed

To fit the metal socket in place, the ends of the ankle bones are sawed. The **tibia and fibula** are shaped first. Next, the top of the **talus is shaped** so the metal talus component can be inserted. Finally, a trial piece of the **artificial ankle joint** is **put in place**, and the ankle is tested to make sure the pieces fit properly. To make sure that the ankle socket or the tibial component fits tightly, **two screws are placed** through the fibula and the tibia just above the artificial ankle joint.

Bone is grafted between the fibula and tibia (where the bone was removed) to create a fusion between them. When the surgeon feels that everything is satisfactory, the joint capsule is sewn, and the skin is stitched together. A large bandage and splint are placed on the lower leg to protect the new ankle joint as the bony structure heals.

This surgery is not performed as often as knee or hip replacements. Still, when necessary, this operation can reduce the pain from arthritis of the ankle. Recent advances in the design of the artificial ankle and changes in the way the operation is performed have made artificial ankle replacement a growing alternative to ankle fusion for the treatment of ankle arthritis. The tibial component is usually made up of two parts: a flat metal piece called a metal tray that is attached directly to the tibia bone, and a UHMWPE cup that fits onto the metal piece, forming a socket for the artificial ankle joint. The talus component is made of stainless steel or titanium alloy and fits into the socket of the tibial component. Figure 13.4 gives an assembly view of the total ankle joint replacement.

13.9 Complications

As with all major surgical procedures, complications may occur. We highlight some of the most common problems. Some of the most common complications following artificial ankle replacement are infection, loosening, and nerve injury.

13.9.1 Infection

Infection can be a very serious complication following an artificial joint surgery. The chance of getting an infection following artificial ankle replacement is probably 2–4%.

13.9.2 Loosening

The major reason that artificial joints eventually fail continues to be a process of loosening of the metal or cement from the bone. Great advances have been made in extending the life of an artificial joint, but most will eventually loosen and require a revision.

13.9.3 Nerve Injury

Since the operation is performed so close to these important structures, it is possible to injure either the nerves or the blood vessels during surgery, but it will not lead to permanent injury.

Aside from that, wear and strength failure of the cement–metal complex are not uncommon.

Study List

1. Saltzman CL (1998) Total ankle arthroplasty: state of the art. J Bone Joint Surg Am 80(10):1410–1420; (2000) J Orthop Sports Phys Ther 30(2):56–67
2. Haddad SL, Coetzee JC, Estok R, Fahrbach K, Banel D, Nalysnyk L (2007) Intermediate and long-term outcomes of total ankle arthroplasty and ankle arthrodesis. A systematic review of the literature. J Bone Joint Surg Am 89(9):1899–1905
3. Kitaoka HB, Patzer GL (1996) Clinical results of the Mayo total ankle arthroplasty. J Bone Joint Sur Am 78A(11):1658–1664
4. Lachiewicz PF (1994) Total ankle arthroplasty. Indications, techniques, and results. Orthop Rev 23(4):315–320
5. Giannini S, Romagnoli M, O'Connor JJ, Malerba F, Leardini A (2010) Total ankle replacement compatible with ligament function produces mobility, good clinical scores, and low complication rates: an early clinical assessment. Clin Orthop Relat Res 468:2746–2754

Problems

1. What are the ranges of movement of the ankle joint? What are the force variations during normal walking? How can an engineer approach finding them? Search the Internet for a possible solution.
2. Sketch the components of the ankle joint in an isometric view and then develop the top view and front view in half-section and side views.

3. How will you approach the development of an artificial ankle joint of the left foot, including the ankle joint, which was irreversibly damaged in a road accident? Sketch each part and indicate the material to be used. How will you get the dimensions of the joints? What methods are to be applied to manufacture the components? Consult Fig. 13.3 for help.

Chapter 14
The Eye and Its Artificial Replacement

The human eye is a marvelous organ that is complex in function. Due to diseases, trauma, and the natural aging process, vision is impaired and reduced to a great extent. Medical scientists, working in tandem with material scientists, have developed many operative techniques and devices for correcting vision and giving relief to patients. To understand these developments, we need to start with the anatomy of the eye.

14.1 Anatomy of the Eye

When looking into someone's eyes, we can easily see several parts that are in the outer surface and exposed to the atmosphere. Figure 14.1 gives details of the human eye.

- A black-looking aperture, **the pupil**, allows light to enter the eye (it appears dark because of the absorbing pigments in the retina).
- A colored circular muscle, **the iris**, is beautifully pigmented, giving us our eye color (the central aperture of the iris is the pupil). This circular muscle controls the size of the pupil so that more or less light, depending on environment, is allowed to enter the eye. Eye color, or more correctly, iris color, is due to variable amounts of eumelanin (brown/black melanins) and pheomelanin (red/yellow melanins) produced by melanocytes. More eumelanins are in brown-eyed people and more pheomelanins in blue- and green-eyed people. The melanocortin-1 receptor (MC1R) gene is a regulator of eumelanin production and is located on a chromosome. Point mutations in the MC1R gene will affect melanogenesis. The presence of point mutations in the MC1R gene alleles is a common feature in light-skinned and blue- and green-eyed people.
- A transparent external surface, **the cornea**, covers both the pupil and the iris. This is the first and most powerful lens of the optical system of the eye; together with **the crystalline lens**, it allows the production of a sharp image at the retinal photoreceptor level.

S. Pal, *Design of Artificial Human Joints & Organs*,
DOI 10.1007/978-1-4614-6255-2_14, © Springer Science+Business Media New York 2014

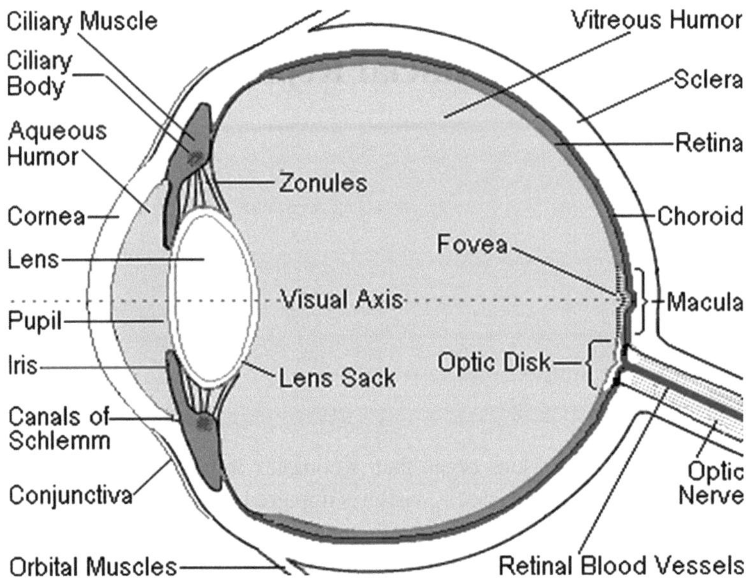

Fig. 14.1 Sagittal cutaway view of the human eye

- The "white of the eye," **the sclera**, forms part of the supporting wall of the eyeball. The sclera is continuous with the cornea. Furthermore, this external covering of the eye is in continuity with the dura of the central nervous system.

When we remove the eye from the orbit, we can see that the eye is a slightly asymmetrical sphere with an approximate sagittal **diameter or length of 24–25 mm and a transverse diameter of 24 mm**. **It has a volume of about 6.5 cc**.

A cross-sectional view of the eye shows the following:

Three Different Layers

1. The external layer, formed by the **sclera and cornea**
2. The intermediate layer, divided into two parts: anterior (**iris and ciliary body**) and posterior (**choroid**)
3. The internal layer, or the sensory part of the eye, **the retina**

   - **Three chambers of fluid**: **anterior chamber** (between cornea and iris), **posterior chamber** (between iris, zonule fibers, and lens), and **vitreous chamber** (between the lens and the retina). The first two chambers are filled with aqueous humor, whereas the vitreous chamber is filled with a more viscous fluid, the vitreous humor.
   - The sagittal section of the eye also reveals **the lens,** which is a transparent body located behind the iris. The lens is suspended by ligaments (called zonule fibers) attached to the anterior portion of the ciliary body. The contraction

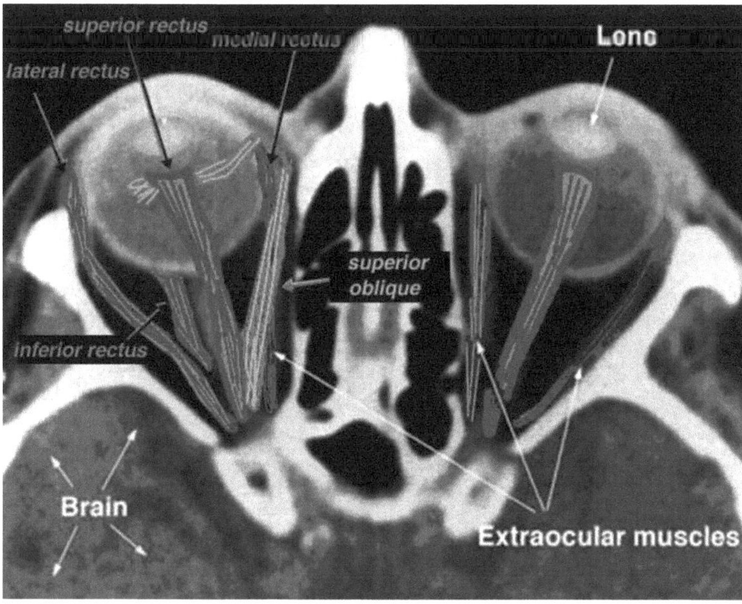

Fig. 14.2 Horizontal transverse CT scan at the plane of the brain and orbits

or relaxation of these ligaments as a consequence of ciliary muscle actions changes the shape of the lens, a process called **accommodation**, which allows us to form a sharp image on the retina.

Light rays are focused through the transparent cornea and lens upon the retina. The central point for image focus (the visual axis) in the human retina is the fovea. Here a maximally focused image initiates resolution of the finest detail and direct transmission of that detail to the brain for the higher operations needed for perception. Slightly more nasally than the visual axis is the optic axis projecting closer to the optic nerve head. The optic axis is the longest sagittal distance between the front or vertex of the cornea and the farthest posterior part of the eyeball. The eye is rotated about the optic axis by the eye muscles. Instead of a fovea, some vertebrate retinas have another specialization of the central retina, known as an area centralis or a visual streak (Fig. 14.2).

14.1.1 Extraocular Muscles

Each eyeball is held in position in the orbital cavity by various ligaments, muscles, and fascial expansions that surround it (Fig. 14.3).

Inserted into the sclera are three pairs of muscles (six muscles altogether). Two pairs are rectus muscles running straight to the bony orbit of the skull orthogonal to

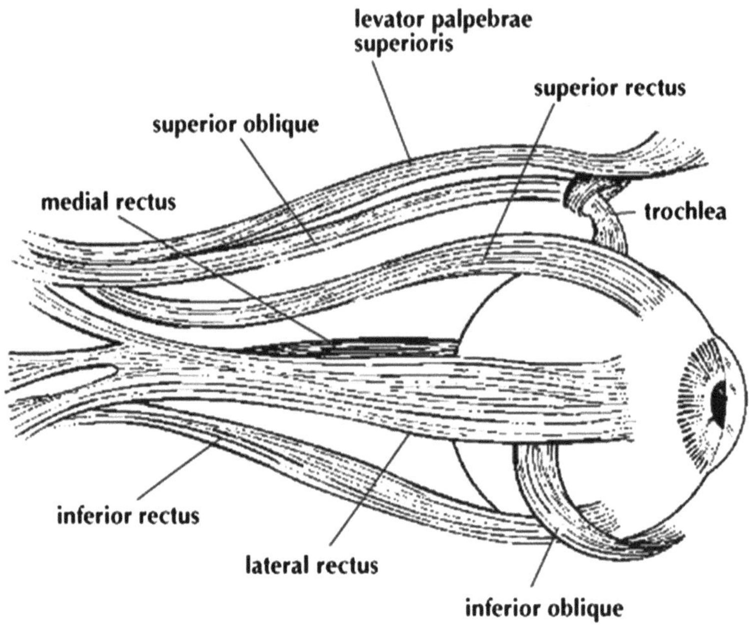

Fig. 14.3 Muscles of the eye

each other (the superior rectus, the inferior rectus, the lateral rectus, and the medial rectus muscles). A further pair of muscles, the oblique muscles (superior oblique and inferior oblique), are angled, as the name implies. These muscles, named **extra-ocular muscles**, rotate the eyeball in the orbits and allow the image to be focused at all times on the fovea of the central retina.

14.2 Development of the Eye

The retina is a part of the central nervous system and an ideal region of the vertebrate brain to study, because like other regions of the central nervous system, **it derives from the neural tube**. The retina is formed during development of the embryo from **optic vesicles** outpouching from two sides of the developing neural tube. The primordial optic vesicles fold back in upon themselves to form the **optic cup**, with the inside of the cup becoming the retina and the outside remaining a single monolayer of epithelium known as the retinal pigment epithelium. Initially, both walls of the optic cup are one cell thick, but the cells of the inner wall divide to form a neuroepithelial layer number of cells thick: **the retina** (Fig. 14.4).

Sensory retinal development begins as early as the optic vesicle stage, with the migration of cell nuclei to the inner surface of the sensory retina. Additional retinal development is characterized by the formation of further layers arising from **cell division and subsequent cell migration**. The retina develops in an inside-to-outside manner: Ganglion cells are formed first, and photoreceptors cells

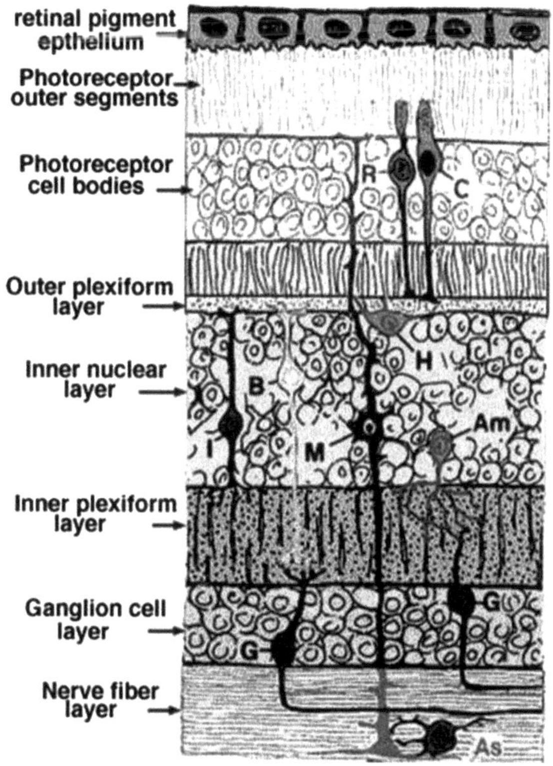

Fig. 14.4 Schema of the layers of the developing retina around five months' gestation (adopted from Odgen 1989)

become fully mature last. Further changes in retinal morphology are accomplished by the simultaneous formation of multiple complex intercellular connections. Thus, by 5 months of gestation, most of the basic neural connections of the retina have been established (Mann 1964).

The functional synapses are made almost exclusively in the two plexiform layers, and the perikarya of the nerve cells are distributed in the three nuclear layers.

Photoreceptor cell maturation begins with the formation of outer segments (OS) containing visual pigment from multiple infoldings of the plasma membrane of each cell. Outer-segment formation proceeds and the eye becomes sensitive to light at about 7 months' gestation. The final portion of the sensory retina to mature is the fovea, where the ganglion cell layer thickening begins during mid-gestation. The outer nuclear layer is also wider here than elsewhere in the retina and consists almost entirely of developing cone cells. The ganglion cell nuclei migrate radially outward in a circle, leaving the fovea free of ganglion cell nuclei. Cell–cell attachments persist, however, and foveal cone cells alter their shape to accommodate the movement of ganglion cells. Foveal development continues with cell rearrangements and alteration in cone shape until about 4 years after birth. Surface membranes cover the eye cup and develop into the lens, iris, and cornea with the three chambers of fluid-filled and vitreous humors. Now let us talk about some common problems associated with the eye. The following terms are common for various impairments.

14.2.1 Disorders of the Eye

- Myopia—nearsightedness
- Hyperopia—farsightedness
- Astigmatism—refractions/not focused

14.2.2 Disorders of the Eye Muscles

- Strabismus—two images received by the brain, possible cause for a nonfunctional eye
- Nystagmus—involuntary movements of the eye that interfere with bringing objects into focus

14.2.3 Disorders of the Cornea, Iris, and Lens

- Glaucoma—fluid
- Aniridla—extremely sensitive to light
- Cataract—a cloudy film over the lens of the eye

14.2.4 Disorders of the Retina

- Diabetic retinopathy—changes in the eye's blood vessels caused by diabetes.
- Macular degeneration—damage to a small area near the center of the retina; difficulty in reading and writing.
- Retinopathy of prematurity (ROP)—excess oxygen.
- Retinitis pigmentosa—detachment of the retina interrupts transmission of visual information to the brain.
- Retinoblastoma—tumor.

14.2.5 Visual Impairments

- **Refractive disorders:** The way the eye focuses light is impaired, as in myopia (nearsightedness), hyperopia (farsightedness), and astigmatism (blurred vision).
- **Muscle disorders:** The ability to control eye movements is impaired, as in strabismus (crossed eyes).

- **Receptive disorders:** The ability to receive and process signals from light is impaired, as in retinal detachment caused by glaucoma or a blow to the eye.
- **Mixed losses:** A combination of conductive and sensorineural impairments.

Related Literature for Study on the Human Eye Anatomy

1. Hendrickson AE, Youdelis C (1984) The morphological development of the human fovea. Ophthalmology 91:603–612
2. Mann I (1964) The development of the human eye. Grune and Stralton, New York
3. Ogden TE (1989) Retina: basic science and inherited retinal disease, vol. 1. CV Mosby, St. Louis, MO
4. Curcio CA, Hendrickson AE (1991) Organization and development of the primate photoreceptor mosaic. Prog Ret Ret 10:89–120

To mitigate some of the above problems related to vision, corrective and assistive implants or explants will be discussed in the subsequent sections of this chapter.

14.3 Contact Lens

A **contact lens** (also known simply as a **contact**) is a corrective, cosmetic, or therapeutic lens usually placed on the cornea of the eye. Leonardo da Vinci is credited with describing and sketching the first ideas for contact lenses in 1508, but it was more than 300 years later before contact lenses were actually fabricated and worn on the eye. Modern soft contact lenses were invented by the Czech chemist Otto Wichterle and his assistant Drahoslav Lím, who also invented the first gel used for their production (Figs. 14.5 and 14.6).

Contact lenses usually serve the same corrective purpose as glasses, but are lightweight and virtually invisible—many commercial lenses are tinted a faint blue to

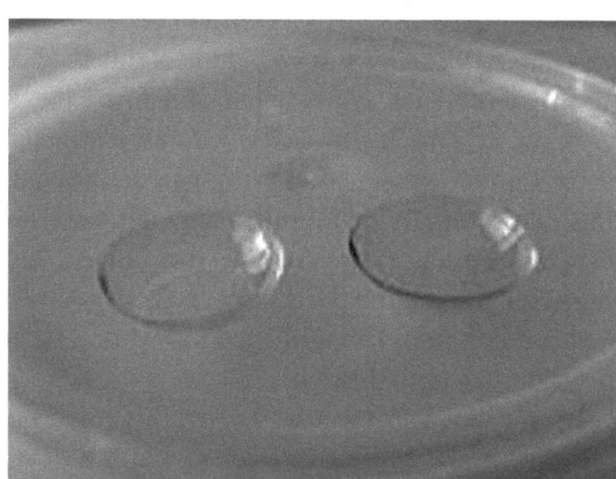

Fig. 14.5 A pair of contact lenses, positioned with the concave side facing upward

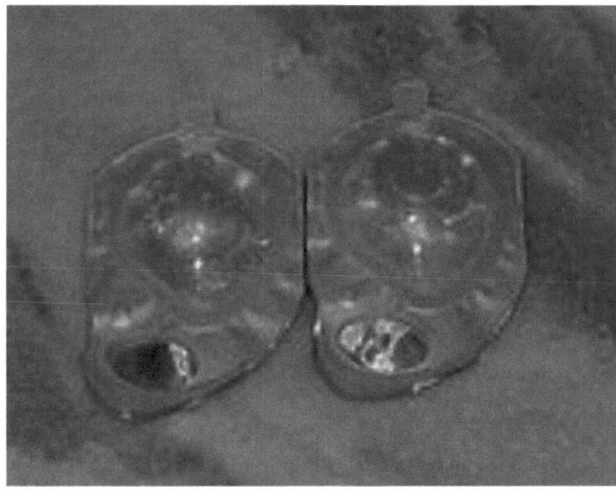

Fig. 14.6 One-day disposable blue-color contact lens in packaging

make them more visible when immersed in cleaning and storage solutions. Some cosmetic lenses are deliberately colored to alter the appearance of the eye. Some lenses now have a thin surface treatment, which is a UV coating; this helps to reduce UV damage to the eye's natural lens.

It has been estimated that 125 million people use contact lenses worldwide (2%) [1], including 28–38 million in the United States [1] and 13 million in Japan [2]. The types of lenses used and prescribed vary markedly between countries, with rigid lenses accounting for over 20% of currently prescribed lenses in Japan, the Netherlands, and Germany, but less than 5% in Scandinavia [1].

People choose to wear contact lenses for many reasons, often due to their appearance and practicality [2]. When compared with eyeglasses, contact lenses are less affected by wet weather, do not steam up, and provide a wider field of vision. They are more suitable for a number of sporting activities.

Leonardo da Vinci is credited with introducing the idea of contact lenses in his 1508 **Codex of the eye, Manual D.** In 1801, while conducting experiments concerning the mechanisms of accommodation, scientist Thomas Young constructed a liquid-filled "eyecup," which could be considered a predecessor to the contact lens. It was not until 1887 that a German glassblower, F. E. Muller, produced the first eye covering to be seen through and tolerated [3]. In 1887, the German physiologist Adolf Eugen Fick constructed and fitted the first successful contact lens. While working in Zürich, he described fabricating afocal scleral contact shells, which rested on the less sensitive rim of tissue around the cornea, and experimentally fitting them: initially on rabbits, then on himself, and lastly on a small group of volunteers. These lenses were made from heavy blown glass and were **18–21 mm in diameter**. Fick filled the empty space between the cornea/callosity and glass with a dextrose solution. He published his work, "Contactbrille," in the journal *Archiv für Augenheilkunde* in March 1888.

Fick's lens was large and unwieldy and could only be worn for a few hours at a time. August Müller in Kiel, Germany, corrected his own severe myopia with a more convenient glass-blown scleral contact lens of his own manufacture in 1888 [4].

In 1936, optometrist William Feinbloom introduced plastic lenses made of transparent PMMA (Perspex), making them lighter and more convenient [5]. These lenses were a combination of glass and plastic.

In 1949, the first "corneal" lenses were developed [6–9]; they were much smaller than the original scleral lenses, as they sat only on the cornea rather than across the entire visible ocular surface, and could be worn up to 16 h per day. PMMA corneal lenses became the first contact lenses to have mass appeal through the 1960s, as lens designs became more sophisticated with improving manufacturing (lathe) technology.

One important disadvantage of PMMA lenses is that no oxygen is transmitted through the lens to the conjunctiva and cornea, which can cause a number of adverse clinical effects. By the end of the 1970s, and through the 1980s and 1990s, a range of **oxygen-permeable** but rigid materials were developed to overcome this problem. Collectively, these **polymers** are referred to as "rigid gas-permeable" or "RGP" materials or lenses. Although all the above lens types—sclerals, PMMA lenses, and RGPs—could be correctly referred to as being "hard" or "rigid," the term "hard" is now used to refer to the original PMMA.

The principal breakthrough in soft lenses was made by the Czech chemists Otto Wichterle and Drahoslav Lim, who published their work "Hydrophilic gels for biological use" in *Nature* in 1959 [10]. This led to the launch of the first soft (hydrogel) lenses in some countries in the 1960s and the first approval of the "Soflens" material by the U.S. Food and Drug Administration (FDA) in 1971. These lenses were soon prescribed more often than rigid lenses, mainly due to the immediate comfort of soft lenses; by comparison, rigid lenses require a period of adaptation before full comfort is achieved. The polymers from which soft lenses are manufactured improved over the next 25 years, primarily in terms of increasing the oxygen permeability by varying the ingredients making up the polymers.

In 1999, an important development was the launch of the first silicone hydrogels on the market. These new materials encapsulated the benefits of silicone—which has extremely high oxygen permeability—with the comfort and clinical performance of the conventional hydrogels that had been used for the previous 30 years. These lenses were initially advocated primarily for extended (overnight) wear, although more recently, daily (no overnight) wear silicone hydrogels have been launched.

In a slightly modified molecule, a polar group is added without changing the structure of the silicone hydrogel. This is referred to as the Tanaka monomer because it was invented and patented by Kyoichi Tanaka of Menicon Co. of Japan in 1979. Second-generation silicone hydrogels, such as Galyfilcon A (Acuvue Advance, Vistakon) and Senofilcon A (Acuvue Oasys, Vistakon), use the Tanaka monomer. Vistakon improved the Tanaka monomer even further and added other molecules, which serves as an internal wetting agent [11].

Comfilcon A (Biofinity, CooperVision) was the first third-generation polymer. The patent claims that the material uses two siloxy macromers of different sizes that, when used in combination, produce a very high oxygen permeability (for a given water content). Enfilcon A (Avaira, CooperVision) is another third-generation material that's naturally wettable. The Enfilcon A material is 46% water [11].

Contact lenses are classified in many different ways, depending on their uses [12, 13].

14.3.1 Corrective Contact Lenses

A corrective contact lens is designed to improve vision. In many people, there is a mismatch between the **refractive power of the eye and the length of the eye, leading to a refraction error**. A contact lens neutralizes this mismatch and allows for correct focusing of light onto the retina. Conditions correctable with contact lenses include myopia (near- or shortsightedness), hypermetropia (far- or longsightedness), astigmatism, and presbyopia. Contact wearers must usually take their contact lenses out every night or every few days, depending on the brand and style of the contact. Recently, there has been renewed interest in orthokeratology, the correction of myopia by deliberate overnight flattening of the cornea, leaving the eye without contact lens or eyeglasses correction during the day.

For those with certain color deficiencies, a red-tinted "X-Chrom" contact lens may be used. Although the lens does not restore normal color vision, it allows some colorblind individuals to distinguish colors better [14, 15]. ChromaGen lenses have been used and have been shown to have some limitations with vision at night although otherwise producing significant improvements in color vision [16]. An earlier study showed very significant improvements in color vision and patient satisfaction [17]. Later work that used these ChromaGen lenses with dyslexics in a randomized, double-blind, placebo-controlled trial showed highly significant improvements in reading ability over reading without the lenses [18].

14.3.2 Cosmetic Contact Lenses

A cosmetic contact lens is designed to change the appearance of the eye. These lenses may also correct the vision, but some blurring or obstruction of vision may occur as a result of the color or design. In the United States, the FDA frequently calls noncorrective cosmetic contact lenses **decorative contact lenses**. These types of lenses tend to cause mild irritation on insertion, but after the eyes become accustomed, they tend to cause no long-term damage. Though it is advised that these lenses not be worn too much, research has shown them to have no direct link to any forms of eye degradation.

Scleral lenses cover the white part of the eye (i.e., sclera) and are used in many theatrical lenses. Due to their size, these lenses are difficult to insert and do not move very well within the eye. Similar lenses have more direct medical applications. For example, some lenses can give the iris an enlarged appearance, or mask defects such as the absence of (aniridia) or damage to (dyscoria) the iris.

Although many brands of contact lenses are lightly tinted to make them easier to handle, cosmetic lenses worn to change the color of the eye are far less common, accounting for only 3% of contact lens fits in 2004 [19].

14.3.3 Therapeutic Contact Lenses

Soft lenses are often used in the treatment and management of nonrefractive disorders of the eye. A bandage contact lens protects an injured or diseased cornea from the constant rubbing of blinking eyelids, thereby allowing it to heal [20]. They are used in the treatment of conditions including **bullous keratopathy, dry eyes, corneal ulcers and erosion, keratitis, corneal edema, descemetocele, corneal ectasis, Mooren's ulcer**, **anterior corneal dystrophy, and neurotrophic keratoconjunctivitis** [21]. Contact lenses that deliver drugs to the eye have also been developed [22].

14.4 Classification of Contact Lenses

14.4.1 By Construction Material

Contact lenses, other than the cosmetic variety, become almost invisible once inserted in the eye. The first contact lenses were made of glass, which caused eye irritation and were not wearable for extended periods of time. But when William Feinbloom introduced lenses made from transparent **polymethyl methacrylate (PMMA or Perspex/Plexiglas**), contacts became much more convenient. These PMMA lenses are commonly referred to as "hard" lenses (this term is not used for other types of contact lens).

Rigid lenses offer a number of unique properties. In effect, the lens is able to replace the natural shape of the **cornea** with a new refracting surface. This means that a regular (spherical) rigid contact lens can provide a good level of vision in people who have astigmatism or distorted corneal shapes, as with keratoconus.

The principal breakthrough in soft lenses made by Otto Wichterle led to the launch of the first soft (hydrogel) lenses in some countries in the 1960s and the approval of the "Soflens" material (polymacon) by the U.S. FDA in 1971. Soft lenses are immediately comfortable, while rigid lenses require a period of adaptation before full comfort is achieved. The polymers from which soft lenses are manufactured improved over the next 25 years, primarily in terms of increasing the oxygen permeability by varying the ingredients in making the polymers.

In 1999, **silicone hydrogels** became available. Silicone hydrogels have both the extremely high **oxygen permeability of silicone** and the comfort and clinical performance of the conventional hydrogels. These lenses were initially advocated primarily for extended (overnight) wear, although more recently daily-wear (no overnight) silicone hydrogels have been approved [23] and launched. While it provides the oxygen permeability, the silicone also makes the lens surface highly hydrophobic and less "wettable." This frequently results in discomfort and dryness during lens wear. In order to compensate for the hydrophobicity, hydrogels are added (hence the name "silicone hydrogels") to make the lenses more hydrophilic. However, the lens surface may still remain hydrophobic. Hence, some of the lenses undergo surface modification processes by plasma treatments, which alter the hydrophobic nature of the lens surface. Other lens types incorporate internal rewetting agents to make the lens surface hydrophilic.

14.4.2 By Time of Wear

A **daily wear** contact lens is designed to be removed prior to sleeping. An *extended-wear* (EW) contact lens is designed for continuous overnight wear, typically for six or more consecutive nights. Newer materials, such as silicone hydrogels, allow for even longer-wear periods of up to 30 consecutive nights; these longer-wear lenses are often referred to as **continuous wear (CW).** Generally, extended-wear lenses are discarded after the specified length of time. Extended- and continuous-wear contact lenses can be worn for such long periods of time because of their high **oxygen permeability** (typically 5–6 times greater than conventional soft lenses), which allows the eye to remain healthy. Extended-lens wearers may have an increased risk for corneal infections and corneal ulcers.

14.4.3 By Frequency of Replacement

The various soft contact lenses available are often categorized by their replacement schedule. The shortest replacement schedule is single-use (daily disposable) lenses, which are disposed of each night. Shorter replacement cycle lenses are commonly thinner and lighter, due to lower requirements for durability against wear and tear, and may be the most comfortable in their respective class and generation.

14.4.4 By Design

A **spherical** contact lens is one in which both the inner and outer optical surfaces are portions of a sphere. A **toric lens** is one in which either or both of the optical surfaces have the effect of a cylindrical lens, usually in combination with the effect of a spherical lens. Myopic (nearsighted) and **hypermetropic (farsighted)** people who also have astigmatism and who have been told they are not suitable for regular contact lenses may be able to use toric lenses. Toric lenses are made from the same materials as regular contact lenses but have a few extra characteristics:

- They correct for both spherical and cylindrical aberration.
- They may have a specific "top" and "bottom," as they are not symmetrical around their center and must not be rotated. Lenses must be designed to maintain their

orientation regardless of eye movement. Often lenses are thicker at the bottom, and this thicker zone is pushed down by the upper eyelid during blinking to allow the lens to rotate into the correct position (with this thicker zone at the 6 o'clock position on the eye). Toric lenses are usually marked with tiny striations to assist their fitting.

- They are usually more expensive to produce than **nontoric** lenses; therefore, they are usually meant for extended wear. The first disposable toric lenses were introduced in 2000 by Vistakon.

Rigid gas-permeable (RGP) bifocal contact lenses most commonly have a small lens on the bottom for the near correction. When the eyes are lowered to read, this lens comes into the optical path. RGPs must translate (move vertically) to work properly, and thus the gaze of the eye can change from the near to the distant sections, much like bifocal eyeglasses.

Multifocal soft contact lenses are more complex to manufacture and require more skill to fit. All soft bifocal contact lenses are considered "simultaneous vision" because both far- and near-vision corrections are presented simultaneously to the retina, regardless of the position of the eye.

14.4.5 Implantation

Intraocular lenses, also known as **implantable contact lenses**, are special small corrective lenses surgically implanted in the eye's posterior chamber behind the iris and in front of the lens to correct higher degrees of myopia and hyperopia.

14.5 Manufacturing of Contact Lenses

Most contact lenses are mass-produced. The common methods are discussed below.

- **Spin-cast lenses**—A spin-cast lens is a soft contact lens manufactured by whirling liquid silicone in a revolving mold at high speed [24].
- **Lathe-turned**—A lathe-turned contact lens is cut and polished on a **CNC lathe** [24]. The lens starts out as a cylindrical disk held in the jaws of the lathe. The lathe is equipped with an industrial-grade **diamond** as the cutting tool. The CNC lathe rotates at nearly 6,000 RPM (revolutions per minute) as the cutter removes the desired amount of material from the rod to form the lens. The **concave** (inner) surface of the lens is then polished with some fine **abrasive** paste, oil, and a small polyester cotton ball turned at high speeds. In order to hold the delicate lens in the reverse manner, wax is used as an adhesive. The **convex** (outer) surface of the lens is thus cut and polished by the same process.

- **Molded**—Moulding is used to manufacture some brands of soft contact lenses. Rotating moulds are used and the molten material is added and shaped by centrifugal forces. Injection moulding and computer control are also used to create nearly perfect lenses [25].

Although many companies make contact lenses, there are four major manufacturers:

- Acuvue/Vistakon (Johnson & Johnson)
- Ciba Vision (Novartis)
- Bausch & Lomb
- CooperVision.

14.5.1 Hydrogel Materials

The following materials are used to make contact lenses:

- Asmofilcon A, Balafilcon A, Comfilcon A, Enfilcon A, Galyfilcon A, Hilafilcon A, Hilafilcon B, Hioxifilcon, Lotrafilcon B, Methafilcon A, Omafilcon A, Phemfilcon A, Polymacon, Senofilcon, Tetrafilcon A, and Vifilcon A.

14.5.2 Contact Lens Prescriptions

The parameters specified in a contact lenses prescription may include the following:

- Material [e.g., oxygen permeability/transmissibility (Dk/L, Dk/t), water content, modulus]
- Base curve radius (BC, BCR)
- Diameter (D, OAD)
- Power in dioptres—spherical, cylindrical, and/or reading addition)
- Cylinder axis
- Center thickness (CT)
- Brand.

14.6 Complications

Complications due to contact lens use affect roughly 5% of contact lens wearers each year [26]. Excessive wear of contact lenses, particularly overnight wear, is associated with most of the safety concerns. Problems associated with contact lens wear may affect the eyelid, the conjunctiva, the various layers of the cornea [27], and even the tear film that covers the outer surface of the eye [26].

Studies conducted on side effects from long-term wearing of contact lenses (i.e., over 5 years), such as that by Liu et al. in 2000 [28], conclude that "long-term contact lens wear appears to decrease the entire corneal thickness and increase the corneal curvature and surface irregularity."

The long-term wear of rigid contact lens is associated with a decreased corneal keratocyte density [29] and an increased number of epithelial Langerhans cells [30].

14.6.1 Eyelid

- Ptosis

14.6.2 Conjunctiva

- Giant papillary conjunctivitis
- Superior limbic keratoconjunctivitis

14.6.3 Cornea

- **Epithelium**
  - Corneal abrasion, corneal erosion, corneal ulcer, hypoxia
- **Stroma**
  - Infection and keratitis
    - Bacteria protozoa: *Acanthamoeba*
    - Fungal: *Fusarium*
    - Contact lens acute red eye (CLARE)
    - Keratoconus
- **Corneal endothelium**

14.7 Usage

Before touching the contact lens or your eyes, it is important to thoroughly wash and rinse your hands with a soap that does not contain moisturizers or allergens such as fragrances. The soap should not be antibacterial due to risk of improper hand washing and the possibility of destroying the natural bacteria found on the eye. These bacteria keep pathogenic bacteria from colonizing the cornea. The technique for removing or inserting a contact lens varies slightly depending upon whether the lens is soft or rigid.

Fig. 14.7 Lens cover to store contact lens

In all cases, the insertion and removal of lenses requires some training and practice on the part of the user, in part to overcome the instinctual hesitation against actually touching the eyeball with one's fingertip.

14.7.1 Care

While daily disposable lenses require no cleaning, other types require regular cleaning and disinfecting in order to retain clear vision and prevent discomfort and infections by various microorganisms, including bacteria, fungi, and *Acanthamoeba*, that form a biofilm on the lens surface. Figure 14.7 shows a lens cover to store contact lenses. There are a number of solutions that can be used to perform the above tasks.

The introduction of silicone hydrogel soft contact lens materials in 1999 made the selection of the proper disinfecting solution more important. One study has noted several incompatibilities between these new lens materials and some solutions resulting in corneal staining [31].

14.8 Current Research

A large segment of current contact lens research is directed toward the treatment and prevention of conditions resulting from contact lens contamination and colonization by foreign organisms. It is generally accepted by clinicians that the most significant complication of contact lens wear is microbial **keratitis** and that the most predominant microbial pathogen is ***Pseudomonas aeruginosa*** [32]. Other organisms are also major causative factors in bacterial keratitis associated with contact lens wear, although their prevalence varies across different locations. These include both the *Staphylococcus* species (*aureus* and *epidermidis*) and the *Streptococcus* species, among others [33, 34]. Microbial keratitis is a serious focal point of current research due to its potentially devastating effect on the eye, including severe vision loss [35].

One specific research topic of interest is how microbes such as *Pseudomonas aeruginosa* invade the eye and cause infection. Although the pathogenesis of microbial keratitis is not well understood, many different factors have been investigated. One group of researchers showed that corneal hypoxia exacerbated *Pseudomonas* binding to the corneal epithelium, internalization of the microbes, and induction of the inflammatory response [36]. One way to alleviate hypoxia is to increase the amount of oxygen transmitted to the cornea. Although silicone hydrogel lenses almost eliminate hypoxia in patients due to their very high levels of oxygen transmissibility [37], they also seem to provide a more efficient platform for bacterial contamination and corneal infiltration than other conventional hydrogel soft contact lenses. A recent study showed that *Pseudomonas aeruginosa* and *Staphylococcus epidermis* adhere much more strongly to silicone hydrogel contact lenses than conventional hydrogel contact lenses and that adhesion of *Pseudomonas aeruginosa* was 20 times stronger than adhesion of *Staphylococcus epidermidis* [38]. This might help to explain one reason why *Pseudomonas* infections are the most predominant.

Another important area of contact lens research deals with patient compliance. Compliance is a major issue surrounding the use of contact lenses because patient noncompliance often leads to contamination of the lens, the storage case, or both [39–41]. The introduction of multipurpose solutions and daily disposable lenses has helped to alleviate some of the problems observed from inadequate cleaning, but new methods of combating microbial contamination are currently being developed. A silver-impregnated lens case has been developed that helps to eradicate any potentially contaminating microbes that come in contact with the lens case [42]. Additionally, a number of antimicrobial agents are being developed that have been embedded into contact lenses themselves. Contact lenses with covalently attached selenium molecules have been shown to reduce bacterial colonization without adversely affecting the cornea of a rabbit eye [43], and octylglucoside used as a contact lens surfactant significantly decreases bacterial adhesion [44]. These compounds are of particular interest to contact lens manufacturers and prescribing optometrists because they do not require any patient compliance to effectively attenuate the effects of bacterial colonization.

14.9 Intraocular Lens

An **intraocular lens** (IOL) is an implantable lens of the eye, usually replacing the existing crystalline lens because it has been clouded over by a cataract, or as a form of refractive surgery to change the eye's optical power. It usually consists of a small

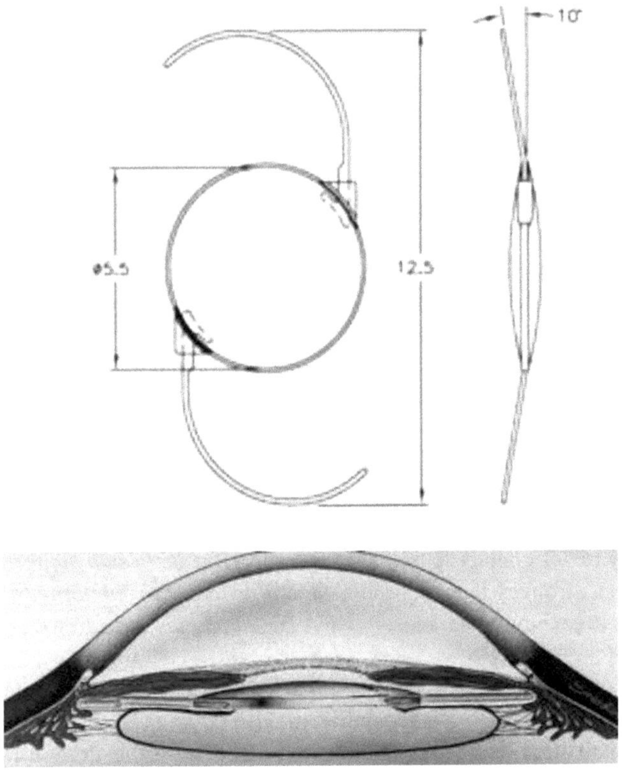

Fig. 14.8 A simple IOL and its fixation in ciliary sulcus

plastic lens with plastic side struts, called haptics, to hold the lens in place within the capsular bag inside the eye (Fig. 14.8). IOLs were traditionally made of an inflexible material (PMMA), although this has largely been superseded by the use of flexible materials. Most IOLs fitted today are fixed monofocal lenses matched to distance vision. However, other types are available, such as multifocal IOLs, which provide the patient with multiple-focused vision at far and reading distance, and adaptive IOLs, which provide the patient with limited visual accommodation.

Inserting an intraocular lens for the treatment of cataracts is the most commonly performed **eye surgical** procedure. The procedure can be done under local anesthesia with the patient awake throughout the operation. The use of a flexible IOL enables the lens to be rolled for insertion into the capsule through a very small incision, thus avoiding the need for stitches, and this procedure usually takes less than 30 min in the hands of an experienced ophthalmologist. The recovery period is about 2–3 weeks. After surgery, patients should avoid strenuous exercise or anything else that significantly increases blood pressure. They should also visit their ophthalmologist regularly for several months so as to monitor the implants.

IOL implantation carries several risks associated with eye surgeries, such as infection, loosening of the lens, lens rotation, inflammation, and nighttime halos. Though IOLs enable many patients to have reduced dependence on glasses, most patients still rely on glasses for certain activities, such as reading.

14.9.1 History

Sir Harold Ridley, a British ophthalmologist, was the first to successfully implant an intraocular lens on November 29, 1949, at St. Thomas' Hospital in London. That first intraocular lens was manufactured by the Rayner Company of England from Perspex polymethyl methacrylate (PMMA) made by Imperial Chemical Industries (ICI). It is said the idea of implanting an intraocular lens came to Dr. Ridley after an intern asked him why he was not replacing the lens he had removed during cataract surgery. Plastic materials were used later, when Dr. Ridley noticed that they were inert, after seeing RAF (Royal Air Force) pilots from World War II with pieces of shattered canopies in their eyes. The intraocular lens did not find widespread acceptance in cataract surgery until the 1970s, when further developments in lens design and surgical techniques had come about. Currently, more than a million IOLs are implanted annually in the United States. It is becoming popular in other countries as well.

14.9.2 Materials Used for Intraocular Lenses

Polymethyl methacrylate (PMMA) was the first material used successfully in intraocular lenses. Advances in technology have brought about the use of silicone and acrylic, both of which are soft, foldable, inert materials. This allows the lens to be folded and inserted into the eye through a smaller incision. PMMA and acrylic lenses can also be used with small incisions and are a better choice in people who have a history of uveitis, have diabetic retinopathy requiring vitrectomy with replacement by silicone oil, or are at high risk of retinal detachment. Acrylic is not always an ideal choice due to its added expense. New FDA-approved multifocal intraocular lens implants allow most postoperative cataract patients the advantage of glass-free vision. These new multifocal lenses can cost the patient upward of $2,800 per eye. The latest advances include IOLs with square-edge design, non-glare-edge design, and yellow dye added to the IOL. In the U.S., a new category of intraocular lenses was introduced in 2003 of multifocal and accommodating lenses, with approval by the FDA.

Multifocal IOLs provide for simultaneous viewing of both distance vision and near vision. Some patients report glare and halos at nighttime with these lenses.

Accommodating IOLs allow for both distance vision and midrange near vision. These IOLs are typically not as strong for closer vision as the multifocal IOLs. To incorporate the strengths of each type of IOL, eye surgeons are increasing using a multifocal IOL in one eye to emphasize close reading vision and an accommodating IOL in the other eye for further midrange vision. This is called "mix and match." Distance vision is not compromised with this approach, while near vision is optimized.

Additionally, blue light–filtering IOLs filter the UV and high-energy blue light present in natural and artificial light, both of which can cause vision problems. Toric IOLs (1998) correct astigmatic vision.

14.9.3 Intraocular Lenses for Correcting Refractive Errors

Intraocular lenses have been used since 1999 for correcting larger errors in myopic (nearsighted), hyperopic (farsighted), and astigmatic eyes. **This type of IOL is also called** a **PIOL (phakic intraocular lens), and the crystalline lens is not removed**.

More commonly, aphakic IOLs (that is, not PIOLs) are implanted via "clear lens extraction and replacement" (CLEAR) surgery. During CLEAR, the crystalline lens is extracted and an IOL replaces it in a process that is very similar to cataract surgery: Both involve lens replacement and local anesthesia; both last approximately 30 min; and both require making a small incision in the eye for lens insertion. People recover from CLEAR surgery 1–7 days after the operation. During this time, they should avoid strenuous exercise or anything else that significantly raises blood pressure. They should also visit their ophthalmologist regularly for several months to monitor the IOL implants. CLEAR has a 90% success rate (risks include wound leakage, infection, inflammation, and astigmatism). CLEAR can only be performed on patients ages 40 and older. This is to ensure that eye growth, which disrupts IOL lenses, will not occur postsurgery.

Once implanted, IOL lenses have three major benefits. First, they are an alternative to LASIK, a form of eye surgery that does not work for people with serious vision problems. Effective IOL implants also entirely eliminate the need for glasses or contact lenses postsurgery. Cataracts will not return, as the lens has been removed. The disadvantage is that the eye's ability to change focus (accommodate) has generally been reduced or eliminated, depending on the kind of lens implanted.

Most PIOLs have not yet been approved by the FDA, but many are under investigation. Some of the risks the FDA has found so far during a three-year study of the Artisan lens, produced by Ophtec Inc. (USA) include

- A yearly loss of 1.8% of the endothelial cells
- A 0.6% risk of retinal detachment
- A 0.6% risk of cataract (other studies have shown a risk of 0.5–1.0%)
- A 0.4% risk of corneal swelling.

Other risks include

- A 0.03–0.05% risk of eye infection, which can lead to blindness in the worst case. This risk exists in all eye surgery procedures and is not unique for IOLs.

- Glaucoma.
- Astigmatism.
- Remaining near or farsightedness.
- Rotation of the lens inside the eye 1 or 2 days after surgery.

One of the causes of these risks is that the lens can rotate inside the eye, because the PIOL is too short, or because the sulcus (a furrow in the surface of an organ) has a slightly oval shape (the height is slightly smaller than the width).

14.9.4 Types of PIOLs

Phakic IOLS (PIOLs) can be either spheric or toric—the latter is used for astigmatic eyes. The difference is that toric PIOLs have to be inserted in a specific angle, or the astigmatism will not be fully corrected, or it can even get worse.

According to the placement site in the eyes, phakic IOLs can be divided into the following types:

- Angle-supported PIOLs: These IOLs are placed in the anterior chamber. They are notorious for their negative impact on the corneal endothelial lining, which is vital for maintaining a healthy dry cornea.
- Iris-supported PIOLs: This type is increasing in popularity. The IOL is attached by claws to the mid-peripheral iris by a technique called enclavation. It is believed to have a lesser effect on the corneal endothelium.
- Sulcus-supported PIOLs: These IOLS are placed in the posterior chamber in front of the natural crystalline lens. They have special vaulting so as not to be in contact with the normal lens. The main complications with this type are their tendency to cause cataracts and/or pigment dispersion.

14.9.5 Accommodating IOLs

One of the major disadvantages of conventional IOLs is that they are primarily focused for distance vision. Though patients who undergo a standard IOL implantation no longer experience clouding from cataracts, they are unable to accommodate, or change focus, from near to far, far to near, and distances in between. Accommodating IOLs interact with ciliary muscles and zonules, using hinges at both ends to “latch on” and move forward and backward inside the eye using the same mechanism for normal accommodation. These IOLs have a 4.5-mm square-edged optic and a long hinged-plate design with polyimide loops at the end of the haptics. The hinges are made of an advanced **silicone called BioSil** that was thoroughly tested to make sure it was capable of unlimited flexing in the eye [45]. There are many advantages to accommodating IOLs. For instance, light comes from and is focused on a single focal point, reducing halos, glares, and other visual aberrations.

Accommodating IOLs provide excellent vision at all distances (far, intermediate, and near), project no unwanted retinal images, and produce no loss of contrast sensitivity or central system adaptation. Accommodating IOLs have the potential to eliminate or reduce the dependence on glasses postcataract surgery. For some, accommodating IOLs may be a better alternative to refractive lens exchange (RLE) and monovision [46].

The FDA approved Eyeonics Inc.'s accommodating IOL, Crystalens AT-45, in November 2003. Bausch and Lomb acquired Crystalens in 2008 and introduced a newer model called Crystalens HD in 2008. Crystalens is the only FDA-approved accommodating IOL currently on the market and is approved in the United States and Europe [47]. Omni Lens Pvt. Ltd. is an Indian manufacturer. There are several others as well.

14.9.6 Criticisms

- The main concern with accommodating IOLs is that there are no long-term, large-scale studies involving their use in patients. Such clinical studies using objective measurement techniques must be done to fully support the claim that accommodating IOLs can restore accommodative vision to the presbyopic eye.
- Though it is rare, potential complications include capsular bag contraction and posterior capsule opacification.
- It is more difficult to implant an accommodating IOL (due to the attachment of hinges), and recovery time may be longer than with a standard IOL.
- Patients should expect that their accommodative abilities will not be restored to perfect or near-perfect function. Though vision is significantly improved, the degree of improvement will not be the same for all, and some will still need glasses after surgery.
- Accommodating IOLs are expensive. Insurance companies do not cover these technologically advanced IOLs, because long-term efficacy is still being determined.

14.9.7 Candidates

Generally, patients over 50 with cataract problems and no serious eye diseases are good candidates for the procedure. The patient must have functional ciliary muscles or zonules for haptic positioning. In addition, the pupils must dilate adequately, as the IOL will induce glares in low-light environments if the pupils dilate too large. Accommodating IOLs are beneficial not only for patients with cataracts, but also for those who wish to reduce their dependency on glasses and contacts due **to myopia, hyperopia, and presbyopia**.

Postoperative care is similar to that of normal IOLs. However, patients must include ophthalmologic exercises such as puzzles and word games as a part of their daily regimen in order to tone up their ciliary muscles and attain the maximum benefit from the accommodating lenses [11]. These exercises should be done consistently for 3–6 months and the patient's performance monitored by their eye care professional.

Other promising multifocal/accommodating IOLs currently in clinical trials include Accommodative 1CU (Human Optics, Erlangen, Germany), Smartlens (Medennium, Irvine, CA), and dual optic accommodating lenses such as Sarfarazi (Bausch and Lomb, Rochester, NY) and Synchrony (Visiogen Inc., Irvine, CA).

14.10 LASIK Eye Surgery

14.10.1 What Is LASIK?

LASIK is used to correct refractive errors. LASIK eye surgery is a treatment that reshapes the cornea in order to produce clear vision. The LASIK refractive eye surgery procedure can treat myopia (nearsightedness), hyperopia (farsightedness), and astigmatism and reduce patients' dependency on contact lenses and glasses. Because refractive errors impede the focusing ability of the eye, patients who suffer from these conditions experience blurred vision. LASIK eye surgery provides the patients with an effective treatment option, where a predetermined amount of corneal tissue is removed. This reshaping of the cornea improves the eye's focusing power and enhances patients' visual acuity.

14.10.2 The LASIK Procedure

The traditional LASIK vision correction procedure begins with the creation of a hinged corneal flap. After the eye is anesthetized with topical eye drops, the LASIK eye surgeon creates the flap on the outer surface of the eye to expose the underlying cornea. The flap can be created with a microkeratome (a handheld device) or with the IntraLase® laser, which allows the surgeon to create the flap using a computer-guided laser. The cornea is then reshaped using an excimer laser and the flap is replaced. After LASIK eye surgery, many patients experience immediate improved vision, although it can take up to 6 months for vision to stabilize. Patients can expect a short LASIK surgery recovery period, but there is usually little to no discomfort following the procedure. Follow-up exams ensure proper healing. LASIK eye surgery is a safe and effective treatment that has continued to gain popularity among patients. A small number of patients—1–5%—experience LASIK eye surgery complications. LASIK risks include sensitivity to light, haloed or glared vision, irregular astigmatism, dry eyes, loss of visual clarity, and sensitivity. Fortunately, advances in LASIK technology have dramatically reduced complications, and many problems can be easily corrected with additional treatment.

The development of the **excimer laser made LASIK eye surgery possible**. Excimer lasers are incredibly precise, only penetrating a microscopic amount of tissue at a time for remarkable accuracy. New-generation microkeratomes provide increased accuracy and safety for LASIK eye surgeons during the creation of the corneal flap. A microkeratome, or handheld blade, allows the surgeon access to the cornea so irregularities can be corrected.

14.10.3 Phacoemulsification

Phacoemulsification cataract surgery is a procedure in which an ultrasonic device is used to break up and then remove a cloudy lens, or cataract, from the eye to improve vision. The insertion of an intraocular lens (IOL) usually immediately follows phacoemulsification. Phacoemulsification, or phaco, as surgeons refer to it, is used to restore vision in patients whose vision has become cloudy from cataracts. In the first stages of a cataract, people may notice only a slight cloudiness as it affects only a small part of the lens, the part of the eye that focuses light on the retina. As the cataract grows, it blocks more light and vision becomes cloudier. As vision worsens, the surgeon will recommend cataract surgery, usually phaco, to restore clear vision. With advancements in cataract surgery such as the IOL, patients can sometimes experience dramatic vision improvement.

14.10.4 Demographics

As people age, cataracts are likely to form. The National Eye Institute (NEI) reported in a 2002 study that more than half of all U.S. residents ages 65 and older have a cataract. People who smoke are at a higher risk for cataracts. Increased exposure to sunlight without eye protection may also be a cause.

Cataracts also can occur anytime because of injury, exposure to toxins, or diseases such as diabetes. Congenital cataracts are caused by genetic defects or developmental problems, or exposure to some contagious diseases during pregnancy.

However, the most common form of cataract is age-related. According to the NEI, cataracts are more common in women than in men, and Caucasians have cataracts more frequently than other races, especially as people age. **People who live close to the equator also are at higher risk for cataracts because of increased sunlight exposure**.

More than 1.5 million cataract surgeries are performed in the United States each year. The NEI reports that the federal government, through **Medicare**, spends more than $3.4 billion each year treating cataracts. Cataract surgery is one of the most common surgeries performed, and also one of the safest and most effective. **Phaco** is currently the most popular version of cataract surgery everywhere. Such data are not available for India.

Phacoemulsification is a variation of **extracapsular cataract extraction**, a procedure in which the lens and the front portion of the capsule are removed. Formerly the most popular cataract surgery, the older method of extracapsular extraction involves a longer incision, about 0.4 in. (10 mm), or almost half of the eye. Recovery from the larger-incision extracapsular extraction also requires almost a week-long hospital stay after surgery and limited physical activity for weeks or even months.

Charles Kelman developed the phacoemulsification technique in the late 1960s. His goal was to remove the cataract with a smaller incision, less pain, and a shorter recovery time. He discovered that the cataract could be broken up, or emulsified, into small pieces using an ultrasound tip. Over the past decades, surgeons have continuously refined phaco to make it even safer and more successful. Innovations in technology such as the foldable IOL also have helped improve outcomes by allowing surgeons to make smaller incisions.

As of 2003, surgeons were beginning to favor the temporal location for the incision because it has proved to be safer. The incision site also varies depending on the size and denseness of the cataract. Once the incision is made, a viscoelastic fluid is injected to reduce shock to the intraocular tissues. The surgeon then makes a microscopic circular incision in the membrane that surrounds the cataract; this part of the procedure is called capsulorhexis. A water stream then frees the cataract from the cortex. The surgeon inserts a small titanium needle, or phaco tip, into the cornea. The ultrasound waves from the phaco tip emulsify the cataract so that it can be removed by suction. The surgeon first focuses on the cataract's central nucleus, which is denser.

While the cataract is being emulsified, the machine simultaneously aspirates the cataract through a small hole in the tip of the phaco probe. The surgeon then removes the cortex of the lens but leaves the posterior capsule, which is used to support the intraocular lens.

The folded IOL is inserted by an injector through a small incision. After the IOL is inserted into the capsular bag, the viscoelastic fluid is removed. No sutures are usually required after the surgery. Some surgeons may recommend that patients wear an eye shield immediately after the surgery. The entire procedure takes about 20 min. The phaco procedure itself takes only minutes. Most surgeons prefer a certain technique for the procedure, although they might vary due to the cataract's density and size. The variations on the phaco procedure involve mostly what part of the nucleus the surgeon focuses on first and how the cataract is emulsified. Some surgeons prefer a continuous "chop," while others divide the cataract into quadrants for removal. One procedure, called the "phaco flip," involves the surgeon inverting and then rotating the lens for removal. Advances in technology also may allow for even smaller incisions; some speculate as small as 0.05 in. (1.4 mm).

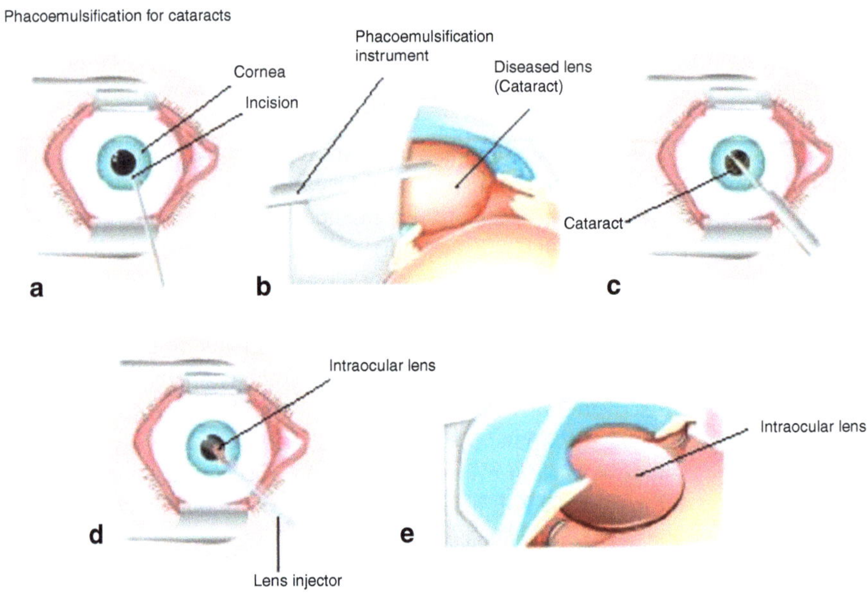

Fig. 14.9 In a phacoemulsification procedure, (**a**) an incision is first made in the cornea, the outer covering of the eye. (**b**) A phacoemulsification instrument uses ultrasonic waves to break up the cataract. (**c**) Pieces of the cataract are then suctioned out. To repair the patient's vision, a folded intraocular lens is (**d**) pushed through the same incision and (**e**) opened in place (illustration adopted from GGS Inc.)

14.10.5 Diagnosis/Preparation

As cataracts develop and worsen, patients may notice these common symptoms:

- Gradual (and painless) onset of blurry vision
- Poor central vision
- Frequent changes in prescription for corrective lenses
- Increased glare from lights
- Near-vision improvement to the point where reading glasses may no longer be needed
- Poor vision in sunlight (Fig. 14.9).

Cataracts grow faster in younger people or diabetics, so doctors will recommend surgery more quickly in those cases. Surgery may also be recommended sooner if the patient suffers from other eye diseases, such as age-related macular degeneration, and if the cataract interferes with complete eye examination.

When symptoms worsen to the point that everyday activities become problematic, surgery becomes necessary. A complete ocular exam will determine the severity of the cataract and what type of surgery the patient will receive. For some denser cataracts, the older method of extracapsular extraction is preferred.

The diagnostic exam should include measurement of visual acuity under both low and high illumination, microscopic examination of eye structures and pupil dilation, assessment of visual fields, and measurement of intraocular pressure (IOP). If cataracts are detected in both eyes, each must be treated separately. The overall patient health must also be considered and how it will affect the surgery's outcome. Surgeons may recommend a complete **physical** before surgery.

An A-scan measurement, which determines the length of the eyeball, will be performed. This helps determine the refractive power of the IOL. Other presurgical testing such as a **chest X-ray**, blood test, or **urine analysis** may be requested if other medical problems are an issue.

The surgeon may also request patients begin using antibiotic drops before the surgery to limit the chance of infection.

14.10.6 Aftercare

Patients are advised to wear an eye shield while sleeping and refrain from rubbing the eye for at least 2 weeks. During that time, the doctor will give the patient special tinted sunglasses or request that he or she wear current prescription eyeglasses to prevent possible eye trauma from accidental rubbing or bumping. Unlike other types of cataract extraction, patients can resume normal activity almost immediately after phaco.

Subsequent exams are usually at 1 week, 3 weeks, and 6–8 weeks following surgery. This can change, however, depending on any complications or any unusual postoperative symptoms. While IOLs can remove the need for myopic correction, patients will probably need new lenses for close work.

14.10.7 Risks

Complications are unlikely but can occur. Patients may experience spontaneous bleeding from the wound and recurrent inflammation after surgery. Flashing and floaters are indications of a problem. Some can easily be treated, while others, such as floaters, may be a sign of a retinal detachment.

Patients may also be concerned that their IOL might become displaced, but newer designs of IOLs also have limited reports of intraocular lens dislocation. Other possible complications are the onset of glaucoma and, in very rare cases, blindness.

Most patients have restored visual acuity after surgery, and some will have the best vision of their lives after the insertion of IOLs. Some patients will no longer require the use of eyeglasses or contact lenses after cataract surgery. Patients will also have better color and depth perception and be able to resume normal activities they may have stopped because of impaired vision from the cataract, such as driving, reading, or sports.

14.10.8 Alternatives

Some older methods of cataract surgery may have to be used if the cataract is too large to remove with a small incision; these methods include the following:

- Extracapsular cataract extraction. While phaco is considered a type of extracapsular extraction, the older version of this technique requires a much larger incision and does not use the phaco machine. It is similar in that the lens and the front portion of the capsule are removed and the back part of the capsule remains. The surgeon might consider this technique if the patient has corneal disease or if the pupil becomes too small during the first stages of surgery.
- Intracapsular cataract extraction. This also requires a larger incision than phaco. It differs in that the lens and the entire capsule are removed. While it is the easiest cataract surgery for the surgeon technically, this method carries an increased risk for the patient, with an increased potential for detachment of the retina and swelling after surgery. Recovery is long, and most patients will have to use large "cataract glasses" to see.

14.11 Bionic Eye

The bionic vision system consists of a camera, attached to a pair of glasses, that transmits high-frequency radio signals to a microchip implanted in the retina. Electrodes on the implanted chip convert these signals into electrical impulses to stimulate cells in the retina that connect to the optic nerve. These impulses are then passed down along the optic nerve to the vision processing centers of the brain, where they are interpreted as an image. To benefit from this technology, patients need to have a functional visual pathway from the retina to the brain along the optic nerve, as well as some intact retinal cells. As such, the two medical conditions that this technology aims to address are retinitis pigmentosa and age-related macular degeneration. This technology was developed in Australia. A company called Second Sight has received FDA approval to begin U.S. trials of a retinal implant system that gives blind people a limited degree of vision.

Related Literature for Study

1. Buettner H (ed) (2002) Mayo Clinic on vision and eye health. Mayo Clinic Health Information, Rochester, MN
2. Cassel GH, Billig MD, Randall HG (1998) The eye book: a complete guide to eye disorders and health. Johns Hopkins University Press, Baltimore, MD
3. Jaffe NS, Jaffe MS, Jaffe GF (1997) Cataract surgery and its complications, 6th edn. Mosby, St. Louis, MO
4. Massengill RK (1987) Supersight: the lens implant miracle. Health Institute Press, Boston, MA

5. Slade SG, Baker RN, Brockman DK (2000) The complete book of laser eye surgery. Sourcebooks, Inc., Naperville, IL
6. Spaeth GL (ed) (1982) Ophthalmic surgery principles and practice. WB Saunders Company, Philadelphia, PA

Study List

1. Barr J (2005) 2004 Annual Report. Contact Lens Spectrum. January 2005
2. Sokol JL, Mier MG, Bloom S, Asbell PA (1990) A study of patient compliance in a contact lens-wearing population. CLAO J 16(3):209–213
3. Pearson RM, Efron N (1989) Hundredth anniversary of August Müller's inaugural dissertation on contact lenses. Surv Ophthalmol 34(2):133–141
4. Mandell RB (1988) Contact lens practice, 4th edn. Charles C. Thomas, Springfield, IL
5. U.S. Patent No. 2,510,438, filed February 28, 1948
6. The corneal lens. The Optician, September 2, 1949, pp. 141–144.
7. Corneal contact lenses. The Optician. September 9, 1949, p. 185.
8. New contact lens fits pupil only. The New York Times, February 11, 1952, p. 27.
9. Wichterle O, Lim D (1960) Hydrophilic gels for biological use. Nature 185:117–118
10. Looking at silicone hydrogels across generations. Optometric Management. http://www.optometric.com/article.aspx?article=101727. Accessed April 5, 2009.
11. Hartenbaum NP, Stack CM (1997) Color vision deficiency and the X-Chrom lens. Occup Health Saf 66(9):36–40
12. Swarbrick HA, Nguyen P, Nguyen T, Pham P (1981) The X-Chrom lens. On seeing red. Surv Ophthalmol 25(5):312–324, PMID 6971497
13. Swarbrick HA, Nguyen P, Nguyen T, Pham P (2001) The ChromaGen contact lens system: color vision test results and subjective responses. Ophthalmic Physiol Opt 21(3):182–196, PMID 11396392
14. Harris D (1997) Colouring sight: a study of CL fittings with colour enhancing lenses. Optician 8 June 1997
15. Harris DA, MacRow-Hill SJ (1999) Application of ChromaGen haploscopic lenses to patients with dyslexia: a double masked placebo controlled trial. J Am Optom Assoc 70:629–640
16. Morgan PB et al. (2005) International contact lens prescribing in 2004: an analysis of more than 17,000 contact lens fits from 14 countries in 2004 reveals the diversity of contact lens practice worldwide. Cont Lens Spectrum. January 2005
17. EyeMDLink.com
18. "45 Coverage Issues—Supplies—Drugs 11–91 45" (PDF). Centers for Medicare and Medicaid Services. http://new.cms.hhs.gov/manuals/downloads/Pub06_PART_45.pdf. Accessed 01 Mar 2006
19. FDA Premarket Notification for "new silicone hydrogel lens for daily wear." July 2008
20. Cassin B, Solomon S (1990) Dictionary of eye terminology. Triad Publishing Company, Gainesville, FL
21. Stamler J (2004) Contact lens complications. *eMedicine.com.* September 1, 2004
22. Efron N (2007) Contact lens-induced changes in the anterior eye as observed in vivo with the confocal microscope. Prog Retin Eye Res 26(4):398–436. doi:10.1016/j.preteyeres.2007.03.003, http://linkinghub.elsevier.com/retrieve/pii/S1350-9462(07)00025-0
23. Liu Z, Pflugfelder SC (2000) The effects of long-term contact lens wear on corneal thickness, curvature, and surface regularity. Ophthalmology 107(1):105–111
24. Hollingsworth JG, Efron N (2004) Confocal microscopy of the corneas of long-term rigid contact lens wearers. Cont Lens Anterior Eye 27(2):57–64

25. Zhivov A, Stave J, Vollmar B, Guthoff R (2007) In vivo confocal microscopic evaluation of langerhans cell density and distribution in the corneal epithelium of healthy volunteers and contact lens wearers. Cornea 26(1):47–54
26. Andrasko G, Ryen K (2007) A series of evaluations of MPS and silicone hydrogel lens combinations. Rev Cornea Cont Lens, 36–42.
27. Robertson DM, Petroll WM, Jester JV, Cavanagh HD (2007) Current concepts: contact lens related *Pseudomonas keratitis*. Cont Lens Anterior Eye 30:94–107
28. Sharma S, Kunimoto D, Rao TN, Garg P, Rao GN (1999) Trends in antibiotic resistance of corneal pathogens: Part II. An analysis of leading bacterial keratitis isolates. Ind J Opthalmol 47:101–109
29. Verhelst D, Koppen C, Looveren JV, Meheus A, Tassignon M (2005) Clinical, epidemiological and cost aspects of contact lens related infectious keratitis in Belgium: results of a seven-year retrospective study. Bull Soc Belge Ophtalmol 297:7–15
30. Burd EM, Ogawa GSH, Hyndiuk RA (1994) Bacterial keratitis and conjunctivitis. In: Smolin G, Thoft RA (eds) The cornea. Scientific foundations and clinical practice, 3rd edn. Little, Brown & Co, Boston, pp 115–167
31. Zaidi T, Mowrey-McKee M, Pier GB (2004) Hypoxia increases corneal cell expression of CFTR leading to increased *Pseudomonas aeruginosa* binding, internalization, and initiation of inflammation. Invest Ophthalmol Vis Sci 45:4066–4074
32. Sweeney DF, Keay L, Jalbert I (2000) Clinical performance of silicone hydrogel lenses. In: Sweeney DF (ed) Silicone hydrogels: the rebirth of continuous wear contact lenses. Butterworth Heinemann, Woburn, MA
33. Kodjikian L, Casoli-Bergeron E, Malet F, Janin-Manificat H, Freney J, Burillon C, Colin J, Steghens JP (2008) Bacterial adhesion to conventional hydrogel and new silicone hydrogel contact lens materials. Graefes Arch Clin Exp Ophthalmol 246:267–273
34. Yung MS, Boost M, Cho P, Yap M (2007) Microbial contamination of contact lenses and lens care accessories of soft contact lens wearers (university students) in Hong Kong. Ophthalmic Physiol Opt 27(1):11–21
35. Midelfart J, Midelfart A, Bevanger L (1996) Microbial contamination of contact lens cases among medical students. CLAO J 22(1):21–24
36. Gray TB, Cursons RT, Sherwan JF, Rose PR (1995) Acanthamoeba, bacterial, and fungal contamination of contact lens storage cases. Br J Ophthalmol 79:601–605
37. Amos CF, George MD (2006) Clinical and laboratory testing of a silver-impregnated lens case. Cont Lens Anterior Eye 29:247–255
38. Mathews SM, Spallholz JE, Grimson MJ, Dubielzig RR, Gray T, Reid TW (2006) Prevention of bacterial colonization of contact lenses with covalently attached selenium and effects on the rabbit cornea. Cornea 25:806–814
39. Santos L, Rodrigues D, Lira M, Oliveira R, Real Oliveira ME, Vilar EY, Azeredo J (2007) The effect of octylglucoside and sodium cholate in *Staphylococcus epidermidis* and *Pseudomonas aeruginosa* adhesion to soft contact lenses. Optom Vis Sci 84:429–434
40. Slade S (2005) Accommodating IOLs: design, technique, results. Rev Ophthalmol. 20 Mar 2009. <http://www.revophth.com/index.asp?page=1_751.htm>
41. USA Eyes (2008) Crystalens accommodating IOL. Council of Refractive Surgery Quality Assurance. 20 March 2009. http://www.usaeyes.org/lasik/faq/crystalens-2.htm>
42. Liz S (2009) Intraocular lenses (IOLs): new advances including AcrySof ReStor, Tecnis, ReZoom, and Crystalens. All about vision. Access Media Group LLC
43. United States Food and Drug Administration. Center for Devices and Radiological Health (CDRH) (2004) Crystalens Model AT-45 Accommodating IOL P030002. New Device Approval. CDRH Consumer Information. Updated Jan 21 2004. http://web.archive.org/http://fda.gov/cdrh/mda/docs/p030002.html
44. Cummings JS et al (2006) Clinical evaluation of the Crystalens AT-45 accommodating interocular lens results of the US Food and Drug Administration clinical trial. J Cataract Refract Surg 32(5):812–825

45. Pepose JS, Qazi MA, Davies J, Doane JF, Loden JC, Sivalingham V, Mahmoud AM (2007) Visual performance of patients with bilateral vs combination Crystalens, ReZoom, and ReSTOR intraocular lens implants. Am J Ophthamol 144(3):347–357
46. Macsai MS et al (2006) Visual outcomes after accommodating intraocular lens implantation. J Cataract Refract Surg 32(4):628–633
47. Sanders DR, Sanders ML (2007) Visual performance results after Tetraflex accommodating intraocular lens implantation. Ophthalmology 114:1679–1684
48. www.insideview.com/ggs-book-services

Chapter 15
The Lung and Its Transplantation and Artificial Replacement

15.1 Introduction

The human thoracic cavity houses a pair of lungs, the left lung and the right lung. The left lung is slightly smaller (since the heart is placed a bit to the left in the body) and has two lobes, and the right lung is bigger, with three lobes. They are spongy and elastic organs that are broad at the bottom and taper at the top. They consist of air sacs, the alveoli. Many alveoli group together and open into a common space. From this space arise the alveolar ducts, which join together to form bronchioles. The bronchioles connect them to the respiratory tract. The lungs also have blood vessels, the branches of the pulmonary artery and veins (Fig. 15.1).

Each lung is enclosed by two membranes called the outer and the inner pleural membrane. The membranes enclose a space called the pleural cavity that contains a fluid. The lungs are capable of expanding and contracting since they are elastic organs. Lubrication for their regular movement is provided by the fluid in the pleural cavity.

The chest wall is made up of 12 pairs of ribs and the intercostal muscles that are attached to the ribs. A thick membranous structure, called the diaphragm, is present below the lungs and separates the thoracic cavity from the abdominal cavity.

15.2 Respiratory Tract

Apart from the lungs, several associated organs and structures together form the respiratory system. The respiratory system is closely linked with the circulatory system, as the transport of the gases takes place through blood.

S. Pal, *Design of Artificial Human Joints & Organs*,

DOI 10.1007/978-1-4614-6255-2_15,

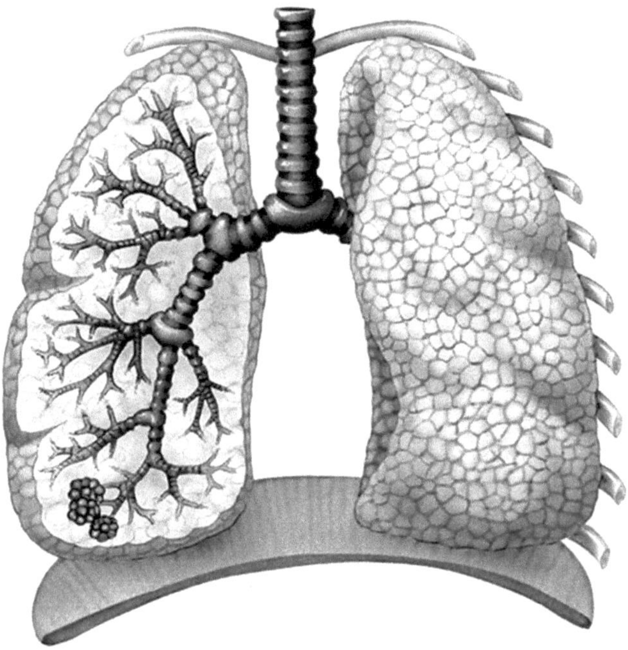

Fig. 15.1 Respiratory organs inside the rib cage

15.2.1 Structure

The respiratory system starts with the **nose,** which covers the **nasal cavity**. The nasal cavity opens to the atmosphere through the openings called the **nostrils**. The nasal cavity is divided into two portions by a cartilagenous septum and is lined by fine hairs that filter the dust particles from the air. The nasal cavity is separated from the mouth by hard and soft palates that form its floor. It opens into the region called the **pharynx** (Fig. 15.2).

The pharynx is a common passage to both food and air. This allows more air whenever required and also allows passage of air in case the nose is blocked. The pharynx continues into the **glottis**.

The glottis is the narrow opening into the larynx. It is guarded by a flap of tissue called the **epiglottis**. Several folds of elastic connective tissue are embedded into the posterior end of the glottis. They are called the **vocal cords**. These extend into the larynx.

The larynx is also called the voice box. The vocal cords stretch across the larynx and vibrate when the air passes through them. This vibration produces various sounds.

The coordinated movement of the lips, cheeks, tongue, and jaws produce specific sounds that result in speech. Speech is an ability that only humans are gifted with,

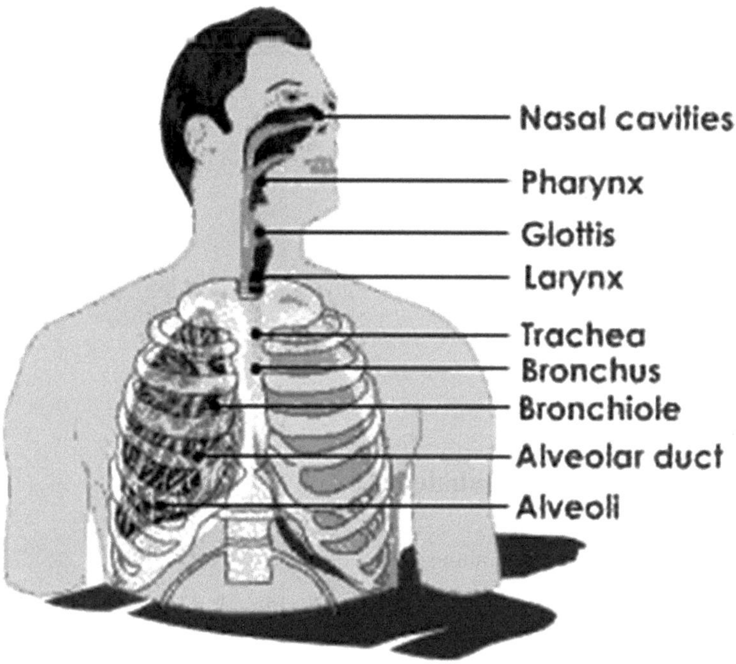

Fig. 15.2 Human respiratory organs

and this is one of the characteristics that have put human beings at the top of the evolutionary pyramid.

The larynx is held open with the help of cartilages. The **"Adam's apple"** is a prominent cartilage of the larynx. The larynx continues as the trachea after the cords.

The **trachea** is also called the **windpipe**. The trachea is held open with the help of C-shaped cartilagenous rings. The open ends of the rings are located toward the esophagus, also known as the foodpipe. The trachea is situated in front of the esophagus. The cartilages keep the larynx and trachea from collapsing even when there is no air in them. The trachea then branch into two main branches called **bronchi**. Each **bronchus** is also supported by the cartilagenous rings. The bronchus then branches into several **bronchioles**. The bronchioles progressively lose the cartilages as they become narrower. The bronchioles end as fine tubules called the **alveolar ducts** (Fig. 15.3).

Each alveolar duct opens into an alveolar sac. An **alveolar sac** is the extended region into which a group of alveoli or air sacs open. Each alveolus is a saclike structure lined by a single layer of epithelial cells. It is bound on the outside by a network of capillaries. All the alveoli on one side are enclosed by the membrane called the pleural membrane and constitute a lung. The pulmonary artery from the heart containing impure blood enters the lungs and branches into minute capillaries that surround the alveoli. These capillaries then join together to form the pulmonary vein, which carries the purified blood back to the heart.

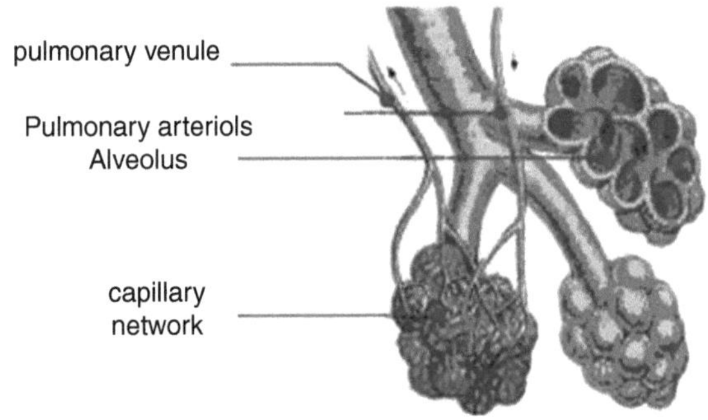

Fig. 15.3 The alveoli and associated capillaries of human lungs

15.3 Path Traced by Inhaled Air

The common composition of atmospheric air that we breathe in is

nitrogen—78 %, oxygen—21 %, carbon dioxide—0.03–0.04 %, hydrogen—traces and noble gases in traces.

Thus, the air naturally contains nearly 500 times more oxygen than carbon dioxide. This oxygen-rich air is taken in by the nostrils. In the nasal cavity, it is filtered by the fine hairs in the nose. The cavity also has a rich supply of blood vessels that keep the air warm. This air then enters the pharynx, then the larynx, and then into the trachea.

The trachea and the bronchi are lined with ciliated epithelial cells and secretory cells (goblet cells). The secretory cells secrete mucus, which moistens the air as it passes through the respiratory tract, and also trap any fine particles of dust or bacteria that have escaped the hairs of the nasal cavity. The cilia beat with an upward motion such that the foreign particles along with the mucus is sent to the base of the buccal cavity, from where it may be either swallowed or coughed out (Fig. 15.4).

The air from the bronchus then enters the bronchioles and then the alveoli. The alveoli form the respiratory surface in humans.

15.4 Gaseous Exchange

The capillaries lining the alveoli have blood that has a low concentration of oxygen. So the oxygen from the air easily diffuses into the blood through the thin barrier of the alveolus wall. Similarly, since the concentration of carbon dioxide is quite high in the blood, the gas easily diffuses out into the alveolar space. From here, the air—which has a comparatively higher concentration of carbon dioxide than the air that entered it—leaves the lungs (Fig. 15.5).

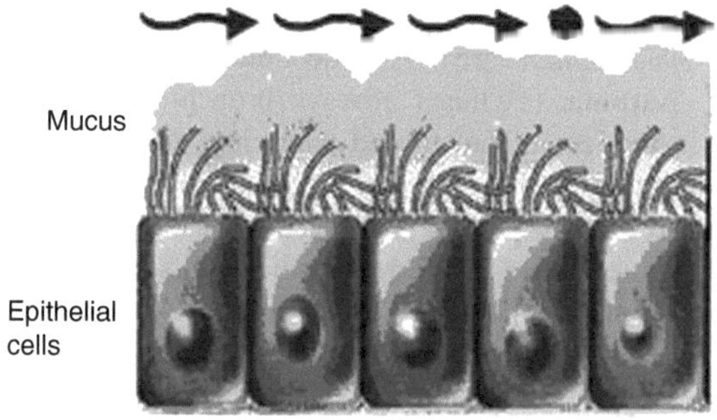

Fig. 15.4 Cilia on the inner lining of windpipe beat to propel a particle outside

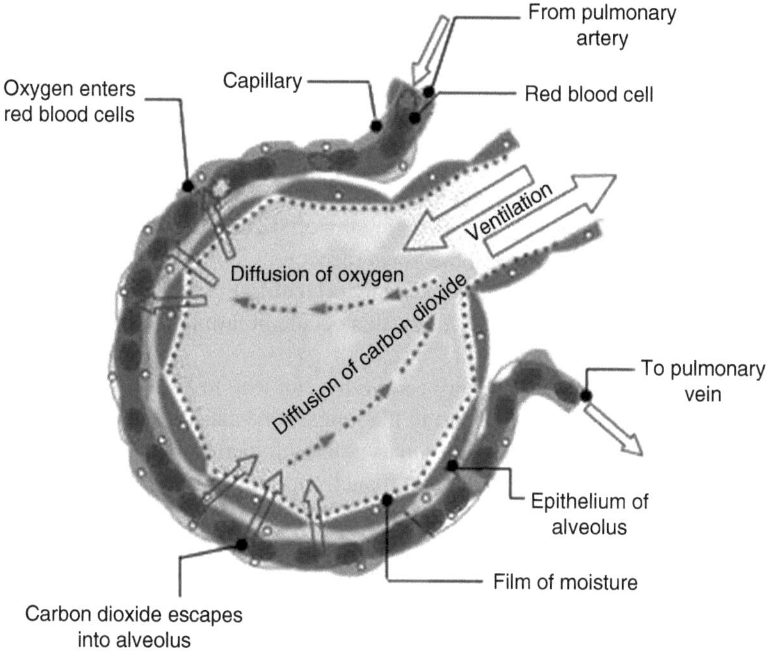

Fig. 15.5 Gaseous exchange in the alveolus of a human

15.5 Common Lung Diseases

The human lung is a very important and strong organ that has to withstand a strong and continuous assault stemming from a wide range of environmental conditions and the lifestyle in our modern world, which cause a multitude of diseases and conditions, as outlined here:

- **Chronic obstructive pulmonary disease (COPD)**: Damage to the lungs results in difficulty blowing air out, causing shortness of breath. Smoking is by far the most common cause of COPD.
- **Emphysema**: A form of COPD usually caused by smoking. The fragile walls between the lungs' air sacs (alveoli) are damaged, trapping air in the lungs and making breathing difficult.

- **Pneumonia**: Infection in one or both lungs. Bacteria, especially *Streptococcus pneumoniae*, are the most common cause.
- **Asthma**: The lungs' airways (bronchi) become inflamed and can spasm, causing shortness of breath and wheezing. Allergies, viral infections, or air pollution often triggers asthma symptoms.
- **Acute bronchitis**: An infection of the lungs' large airways (bronchi), usually caused by a virus. Cough is the main symptom of acute bronchitis.
- **Pulmonary fibrosis**: A form of interstitial lung disease. The interstitium (walls between air sacs) become scarred, making the lungs stiff and causing shortness of breath.
- **Pleurisy**: Inflammation of the lining of the lung (pleura), which often causes pain when breathing in. Autoimmune conditions, infections, or a pulmonary embolism may cause pleurisy.
- **Lung cancer**: Cancer may affect almost any part of the lung. Most lung cancer is caused by smoking.
- **Tuberculosis**: A slowly progressive pneumonia caused by the bacteria *Mycobacterium tuberculosis*. Chronic cough, fever, weight loss, and night sweats are common symptoms of tuberculosis.
- **Acute respiratory distress syndrome (ARDS)**: Severe, sudden injury to the lungs caused by a serious illness. Life support with mechanical ventilation is usually needed to survive until the lungs recover.
- **Hypersensitivity pneumonitis (allergic alveolitis)**: Inhaled dust causes an allergic reaction in the lungs. Usually, this occurs in farmers or others who work with dried, dusty plant material.
- **Pulmonary hypertension**: Many conditions can lead to high blood pressure in the arteries leading from the heart to the lungs. If no cause can be identified, the condition is called idiopathic pulmonary arterial hypertension.
- **Pulmonary embolism**: A blood clot (usually from a vein in the leg) may break off and travel to the heart, which pumps the clot (embolus) into the lungs. Sudden shortness of breath is the most common symptom of a pulmonary embolism.
- **Severe acute respiratory syndrome (SARS)**: A severe pneumonia caused by a specific virus first discovered in Asia in 2002. Worldwide prevention measures seem to have controlled SARS, which has caused deaths in India but no deaths in the U.S.

15.6 Lung Transplantation

Lung transplantation is a surgical procedure to totally or partially replace a patient's diseased lung with a donor's lung. While lung transplants carry certain associated risks, they can also extend life expectancy and enhance the quality of life for end-stage pulmonary patients.

Lung transplantation is the therapeutic measure of last resort for patients with end-stage lung disease who have exhausted all other available treatments without improvement. A variety of conditions may make such surgery necessary. As of

2005, the most common reasons for lung transplantation in the United States were the following [1].

27% from chronic obstructive pulmonary disease (COPD), including emphysema
16% from idiopathic pulmonary fibrosis
14% from cystic fibrosis
12% from idiopathic (formerly known as "primary") pulmonary hypertension
5% from alpha 1-antitrypsin deficiency
2% due to replacing previously transplanted lungs that have failed after a period
12% from other causes.

15.7 Types of Lung Transplants

15.7.1 Lobe

A lobe transplant is a surgery in which part of a living donor's lung is removed and used to replace part of a recipient's diseased lung. This procedure usually involves the donation of lobes from two different people, thus replacing a single lung in the recipient. Donors who have been properly screened should be able to maintain a normal quality of life despite the reduction in lung volume.

15.7.2 Single-Lung Transplant

Many patients can be helped by the transplantation of a single healthy lung. The donated lung typically comes from a donor who has been pronounced brain-dead.

15.7.3 Double-Lung Transplant

Certain patients may require both lungs to be replaced. This is especially the case for people with cystic fibrosis, due to the bacterial colonization commonly found within such patients' lungs; if only one lung were transplanted, bacteria in the native lung could potentially infect the newly transplanted organ.

15.7.4 Heart–Lung Transplant

Some respiratory patients may also have severe cardiac disease that itself would necessitate a heart transplant. These patients can be treated by a surgery in which both lungs and the heart are replaced by organs from a donor or donors. First performed in 1987, this type of transplant typically involves the transplantation of a heart and lungs into a recipient. Prior to operating on the recipient, the transplant surgeon inspects the donor lung(s) for signs of damage or disease. If the lung or lungs are approved, then the recipient is connected to an intervenous line and various monitoring equipment, including pulse oximetry. The patient will be given general anesthesia, and a machine ventilator will breathe for the patient. The donors are usually road accident victims whose organs can be transplanted within 5–6 h of death. There are a large number of patients who are waiting for such transplantation from donors.

15.8 Design of Artificial Lungs

Potkey of VAMC Cleveland, Ohio (2009), recently developed one model of artificial lungs. He indicated more than 35 million Americans are living with chronic lung disease; it is responsible for nearly 350,000 deaths every year in the United States alone [1]. Acute respiratory distress syndrome has a mortality rate of 50% and affects 1.50 lakh Americans each year [2]. Many patients waiting for lung transplants die while on the waiting list. To help combat these problems, artificial lungs have been developed with the goal of replacing or supplementing the respiratory function of the lung. Artificial lungs mimic the function of real lungs, adding oxygen to, and removing carbon dioxide from, the blood. In all cases, however, the performance of artificial lungs is still significantly lower than that of natural lungs. The human lung is a remarkable organ, providing a maximum gas exchange rate for both O_2 and CO_2 of **2–6 l/min** [3]. On the other hand, current artificial lungs are only capable of a maximum gas exchange rate of **0.25–0.40 l/min**, limiting their use to the short-term respiratory support for patients at rest. This insufficiency is due to the smaller surface area, smaller surface-area-to-volume ratio, and greater membrane thickness of artificial lungs compared to the human lung [3].

Recent advances in the micromachining of silicone elastomer (PDMS) have made possible the creation of a new highly efficient artificial lung (Fig. 15.6) with feature sizes similar to or better than those of the human lung. Such a micromachined artificial lung would have an improved gas exchange performance compared to its conventional counterparts, potentially resulting in increased clinical use.

15.9 Design Overview

Silicone has been used as the membrane material in some commercially available artificial lungs due to its biocompatibility, durability, stability, and high permeability to oxygen and carbon dioxide. However, these devices have limited gas-exchange capability mainly due to the membrane's thickness (>50 μm). A significant advantage of silicone membranes is that blood plasma leakage does not occur as it does in microporous hollow fiber oxygenators [3]. In fact, hollow fiber oxygenators are

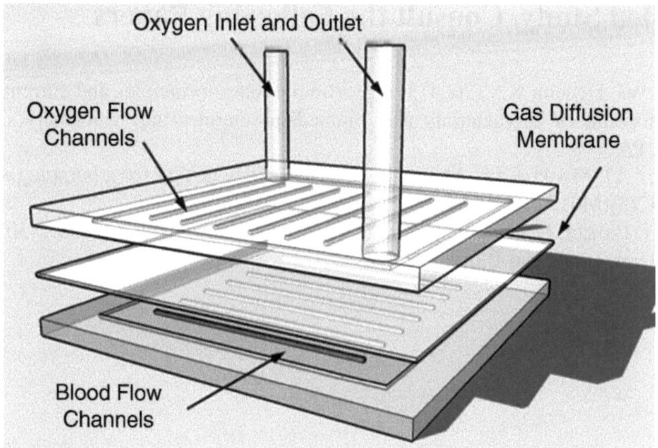

Fig. 15.6 An exploded view of a micromachined artificial lung showing its flow channels and gas diffusion membrane; for clarity, the blood inlet and outlet are not shown

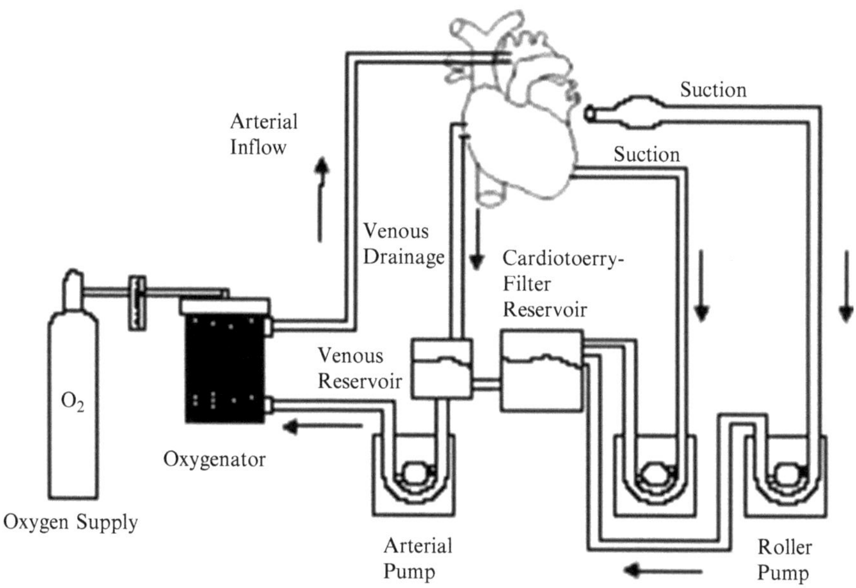

Fig. 15.7 Schema of cardiovascular bypass system

sometimes coated with a thin layer of silicone in order to reduce plasma leakage and increase the device lifetime. Due to these advantages, silicone was utilized as the membrane material in this study (Fig. 15.7).

For Detailed Study, Consult the Following Papers

1. Federspiel WJ, Henchir KA (2004) Lung, artificial: basic principles and current applications. In: Encyclopedia of biomaterials and biomedical engineering. University of Pittsburgh, Pittsburgh, PA
2. Makarewicz AJ, Mockros LF, Mavroudis C (1996) New design for a pumping artificial lung. ASAIO J 42(5):M615–M619
3. Bartlett RH. Project, Development of a total artificial lung, 1 February 2002–30 June 2012 in National Heart, Lung, and Blood Institute

Problems

1. Describe the structure and function of natural lungs. What conditions in a patient create the requirement for an artificial lung?
2. Describe how to develop an artificial lung. Indicate the volume of blood and oxygen and other gases to be handled. Develop a schematic diagram for an artificial lung.

Chapter 16
Design of the Total Artificial Heart

16.1 Introduction

An **artificial heart** is a mechanical device that is implanted into the body to replace the biological heart. The term "artificial heart" has often inaccurately been used to describe ventricular assist devices (VADs), which are pumps that assist the heart but do not replace it. To refresh the reader's memory, a clean view of the normal heart is shown in Fig. 16.1.

An artificial heart is also distinct from a cardiopulmonary bypass machine (CPB), which is an external device used to provide the functions of both the heart and lungs. CPBs are only used for a few hours at a time, most commonly during open heart surgery (Fig. 16.2).

16.2 Artificial Heart

16.2.1 CardioWest Temporary Total Artificial Heart

The CardioWest temporary Total Artificial Heart (TAH) was the first FDA-approved total artificial heart; it is depicted in Fig. 16.3. It received FDA approval on Oct. 15, 2004, following a 10-year pivotal clinical study [1]. 1 Originally designed as a permanent replacement heart, it is currently approved as a bridge to human heart transplant for patients dying because both sides of their hearts are failing (irreversible end-stage biventricular failure) [1]. There have been more than 780 implants of the CardioWest artificial heart, accounting for more than 150 patient-years of life on this device [2].

In the 10-year pivotal clinical study of the CardioWest artificial heart, 79% of patients receiving the artificial heart survived to transplant (*New England Journal of Medicine* 2004; 351: 859–867) [3]. This is the highest bridge-to-transplant rate for any heart device in the world [2].

S. Pal, *Design of Artificial Human Joints & Organs*,

DOI 10.1007/978-1-4614-6255-2_16,

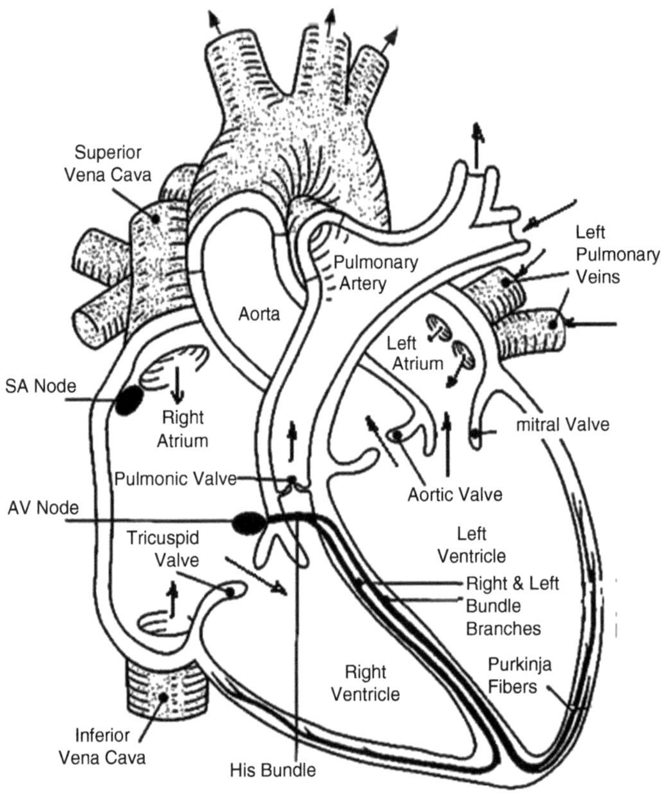

Fig. 16.1 A schematic longitudinal sectional view of the human heart

16.2.2 AbioCor Replacement Heart

The AbioCor Replacement Heart (Abiomed) received FDA approval under a Humanitarian Device Exemption (HDE) on Sept. 5, 2006 [4]. The AbioCor is approved for use in severe biventricular end-stage heart disease patients who are not eligible for heart transplant and have no other viable treatment options [5]. The first implant of the AbioCor as a fully approved FDA device took place on June 24, 2009, at Robert Wood Johnson University Hospital, New Jersey [6]. (See FDA Summary of Safety and Probable Benefit.)

16.3 Origins

A manmade replacement for the heart remains one of the long-sought holy grails of modern medicine. The obvious benefit of a functional artificial heart would be to lower the need for heart transplants, because the demand for organs always greatly exceeds supply.

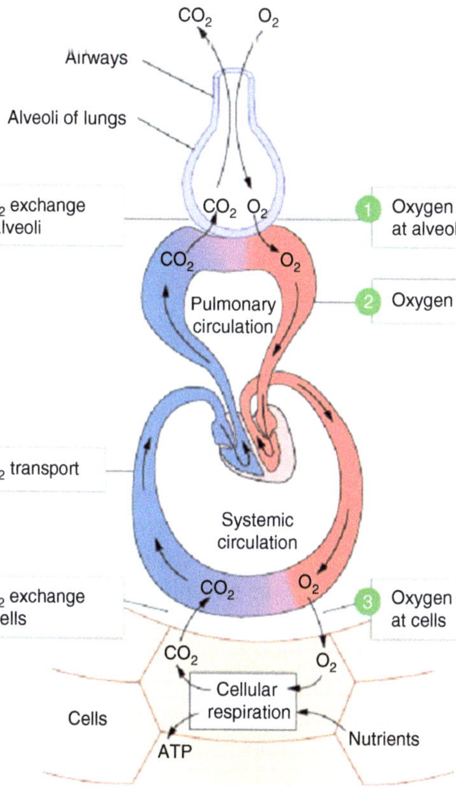

Fig. 16.2 The schematic display of gas exchange in different cellular systems and the circulation system

Fig. 16.3 The CardioWest temporary Total Artificial Heart, which was FDA-approved

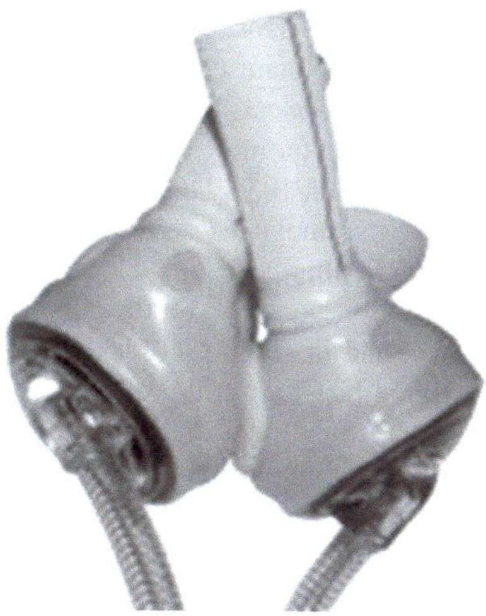

Although the heart is conceptually simple (basically, a muscle that functions as a pump), it embodies subtleties that defy straightforward emulation with synthetic materials and power supplies. Consequences of these issues include severe foreign-body rejection and external batteries that limit patient mobility. These complications limited the lifespan of early human recipients to hours or days.

16.3.1 Early Development

A heart-lung machine was used in 1953 during a successful open heart surgery. Dr. John Heysham Gibbon, the inventor of the machine, performed the operation and developed the heart-lung substitute himself.

On July 3, 1952, 41-year-old Henry Opitek, suffering from shortness of breath, made medical history at Harper University Hospital at Wayne State University in Michigan. The Dodrill-GMR heart machine, considered to be the first operational mechanical heart, was successfully used while performing heart surgery [7, 8].

Dr. Forest Dewey Dodrill used the machine in 1952 to bypass Henry Opitek's left ventricle for 50 min while he opened the patient's left atrium and worked to repair the mitral valve. In Dr. Dodrill's postoperative report, he notes, "To our knowledge, this is the first instance of survival of a patient when a mechanical heart mechanism was used to take over the complete body function of maintaining the blood supply of the body while the heart was open and operated on" [9]. The scientific interest for the development of a solution for heart disease developed in different research groups worldwide.

16.4 Early Designs of Total Artificial Hearts

In 1949, a precursor to the modern artificial heart pump was built by Drs. William Sewell and William Glenn of the Yale School of Medicine using an Erector Set, assorted components from a toystore. The external pump successfully bypassed the heart of a dog for more than an hour [10].

On Dec. 12, 1957, Dr. Willem Kolff, the world's most prolific inventor of artificial organs, implanted an artificial heart into a dog at Cleveland Clinic. The dog lived for 90 min.

In 1958, Domingo Liotta initiated the studies of TAH replacement at Lyon, France, and in 1959–1960 at the National University of Cordoba, Argentina. He presented his work at the meeting of the American Society for Artificial Internal Organs meeting held in Atlantic City in March 1961. At that meeting, Dr. Liotta described the implantation of three types of orthotopic (inside the pericardial sac) TAHs in dogs, each of which used a different source of external energy: an implantable electric motor, an implantable rotating pump with an external electric motor, and a pneumatic pump [11, 12].

In 1964, the National Institutes of Health started the Artificial Heart Program, with the goal of putting a manmade organ into a human by the end of the decade [13].
In 1967, Dr. Kolff left Cleveland Clinic to start the Division of Artificial Organs at the University of Utah and pursue his work on the artificial heart.
In 1973, a calf named "Tony" survived for 30 days on an early Kolff heart.
In 1975, the bull "Burk" survived 90 days on the artificial heart.
In 1976, a calf named "Abebe" lived for 184 days on the Jarvik 5 artificial heart.
In 1981, the calf "Alfred Lord Tennyson" lived for 268 days on the Jarvik 5.

Over the years, more than 200 physicians, engineers, students, and faculty members developed, tested and improved Dr. Kolff's artificial heart. To help manage his many endeavors, Dr. Kolff assigned project managers. Each project was named after its manager. Graduate student Robert Jarvik was the project manager for the artificial heart, which was subsequently renamed the Jarvik 7.

In 1981, Dr. William DeVries submitted a request to the FDA to implant the Jarvik 7 into a human being. On Dec. 2, 1982, Dr. Kolff's 35 years of dedication culminated in the first implant of the Jarvik 7 artificial heart into Dr. Barney Clark. Clark was hours from death prior to the surgery. He lived for 112 days with the artificial heart.

On Feb 11, 2009, Dr. Kolff died at the age of 97 in Philadelphia.

16.5 First Clinical Implantation of a Total Artificial Heart

In the morning of April 4, 1969, Domingo Liotta and Denton A. Cooley replaced a dying man's heart with a mechanical heart inside the chest at the Texas Heart Institute in Houston as a bridge for a transplant. The patient woke up and recovered well. After 64 h, the pneumatic powered artificial heart was removed and replaced by a donor heart. Replacing the artificial heart proved to be a bad decision, however; 32 h after transplantation, the patient died of what was later proved to be an acute pulmonary infection, extended to both lungs, caused by fungi, most likely caused by an immunosuppressive drug complication. If they had left the artificial heart in place, the patient may have lived longer [14].

The original prototype of Liotta–Cooley artificial heart used in this historic operation is prominently displayed in The Smithsonian Museum's "Treasures of American History" exhibit in Washington, DC.

16.6 First Clinical Applications of a Permanent Pneumatic Total Artificial Heart

The 85th clinical use of an artificial heart designed for permanent implantation rather than a bridge to transplant occurred in 1982 at the University of Utah. Artificial kidney design pioneer Dr. Willem Johan Kolff started the Utah artificial organs program in 1967 [15]. There, physician-engineer Dr. Clifford Kwan-Gett invented two components of an integrated pneumatic artificial heart system: a ventricle with hemispherical diaphragms that did not crush red blood cells (a problem with previous artificial hearts), and an external heart driver that inherently regulated blood flow without needing complex control systems [16]. Independently, American inventor and ventriloquist Paul Winchell designed and patented a similarly shaped ventricle and donated the patent to the Utah program [17]. Throughout the 1970s and early 1980s, veterinarian Dr. Donald Olsen led a series of calf experiments that refined the artificial heart and its surgical care.

During that time, as a student at the University of Utah, Dr. Robert Jarvik combined several modifications: an ovoid shape to fit inside the human chest, a more blood-compatible polyurethane developed by biomedical engineer Dr. Donald Lyman, and a fabrication method by Kwan-Gett that made the inside of the ventricles smooth and seamless to reduce dangerous stroke-causing blood clots [18]. On Dec. 2, 1982, Dr. William DeVries implanted the artificial heart into retired dentist Dr. Barney Bailey Clark (b. Jan. 21, 1921), who survived 112 days with the device, dying on March 23, 1983. Bill Schroeder became the second recipient and lived for a record 620 days.

Contrary to popular belief and erroneous articles in several periodicals, the Jarvik heart was not banned for permanent use. Since 1982, more than 350 people received the Jarvik heart as a bridge to transplantation (http://www.jarvikheart.com/basic.asp?id=69).

16.7 The Development of Permanent, Implantable, Electrically Powered Artificial Hearts

In the mid-1980s, artificial hearts were powered by dishwasher-sized pneumatic power sources whose lineage went back to Alpha-Laval milking machines. Moreover, two sizable catheters had to cross the body wall to carry the pneumatic pulses to the implanted heart, greatly increasing the risk of infection. To speed the development of a new generation of technologies, the National Heart, Lung, and Blood Institute opened a competition for implantable electrically powered artificial hearts. Three groups received funding: Cleveland Clinic in Cleveland, Ohio; the College of Medicine of Pennsylvania State University (Penn State Hershey Medical Center) in Hershey, Pennsylvania; and Abiomed, Inc. of Danvers, Massachusetts. Despite considerable progress, the Cleveland program was discontinued after the first 5 years.

Polymeric trileaflet valves ensure unidirectional blood flow with a low-pressure gradient and good longevity. State-of-the-art transcutaneous energy transfer eliminates the need for electric wires crossing the chest wall.

The first AbioCor to be surgically implanted in a patient was implanted on July 3, 2001 [19]. The AbioCor is made of titanium and plastic with a weight of 1 kg, and its internal battery can be recharged with a transduction device that sends power through the skin [19]. The internal battery lasts for half an hour, and a wearable external battery pack lasts for 4 h [20]. The FDA announced on Sept. 5, 2006, that the AbioCor could be implanted for humanitarian uses after the device had been tested on 15 patients [21]. It is intended for critically ill patients who cannot receive a heart transplant [21]. Some limitations of the current AbioCor are that its size makes it suitable for only about 50% of the male population, and its useful life is only 1–2 years [22]. By combining its valved ventricles with the control technology and roller screw developed at Penn State, Abiomed has designed a smaller, more stable heart, the AbioCor II. This pump, which should be implantable in most men and 50% of women and has a life span of up to 5 years [22], had animal trials in 2005, and the company hoped to get FDA approval for human use in 2008 [23].

16.8 First Clinical Application of an Intrathoracic Pump

E. Stanley Crawford and Domingo Liotta implanted the first clinical LVAD at the Methodist Hospital in Houston, Texas, on the evening of July 19, 1963, in a patient who had a cardiac arrest after surgery. The patient survived for 4 days under mechanical support but didn't recover from the complications of the cardiac arrest; finally, the pump was discontinued and the patient died.

16.9 First Clinical Application of a Paracorporeal Pump

On April 21, 1966, Michael DeBakey and Dr. Liotta implanted the first clinical LVAD in a paracorporeal position (where the external pump rests at the side of the patient) at the Methodist Hospital in Houston, in a patient experiencing cardiogenic shock after heart surgery. The patient developed neurological and pulmonary complications and died after a few days of LVAD mechanical support. In October 1966, DeBakey and Liotta implanted the paracorporeal Liotta–DeBakey LVAD in a new patient who recovered well and was discharged from the hospital after 10 days of mechanical support, thus constituting the first successful use of an LVAD for post-cardiotomy shock.

16.9.1 Recent Developments

In August 2006, an artificial heart was implanted into a 15-year-old girl at the Stollery Children's Hospital in Edmonton, Alberta. It was intended to act as a temporary fixture until a donor heart could be found. Instead, the artificial heart (called a Berlin Heart) allowed for natural processes to occur, and her heart healed on its own. After 146 days, the Berlin Heart was removed, and the girl's heart was able to function properly on its own [24].

With an increased understanding of the heart and continuing improvements in prosthetics engineering, computer science, electronics, battery technology, and fuel cells, a practical artificial heart may become a reality.

16.10 Total Artificial Heart

On Oct. 27, 2008, French professor and leading heart transplant specialist Alain F. Carpentier announced that a fully implantable artificial heart would be ready for clinical trial by 2011, and for alternative transplant in 2013. It was developed and was to be manufactured by Dr. Carpentier, the biomedical firm Carmat, and the venture capital firm Truffle. The prototype uses electronic sensors and is made from chemically treated animal tissues, called "biomaterials," or a "pseudo-skin" of biosynthetic, microporous materials. Another U.S. team with a prototype called 2005 MagScrew Total Artificial Heart, including Japanese and South Korean researchers, is competing to produce similar projects [25–27].

As of mid-2011, no information was available from Dr. Carpentier regarding his fully implantable manmade heart.

16.11 Heart Assist Devices

16.11.1 Ventricular Assist Device (VAD)

Patients who have some remaining heart function but who can no longer live normally may be candidates for ventricular assist devices (VAD), which do not replace the human heart but complement it by taking up much of the function.

The first left ventricular assist device (LVAD) system was created by Domingo Liotta at Baylor College of Medicine in Houston in 1962 [28].

Another VAD, the Kantrowitz CardioVad, designed by Dr. Adrian Kantrowitz, boosts the native heart by taking up over 50% of its function [29]. Additionally, the VAD can help patients on the waitlist for a heart transplant. In a young person, this device could delay the need for a transplant by 10–15 years [29].

The first heart assist device was FDA-approved in 1994, and two more received approval in 1998 [30]. While the original assist devices emulated the pulsating heart, newer versions, such as the Heartmate II [31], developed by the Texas Heart Institute of Houston, Texas, provide continuous flow. These pumps (which may be centrifugal or axial flow) are smaller and potentially more durable and longer-lasting than the current generation of total heart replacement pumps. Another major advantage of a VAD is that the patient can keep the natural heart, which can receive signals from the brain to increase and decrease the heart rate as needed. With the completely mechanical systems, the heart rate is fixed.

Several continuous-flow ventricular assist devices have been approved for use in the **European Union and as of August 2007 were undergoing clinical trials for FDA approval.**

16.12 Total Artificial Heart: Aortic Pumping System, Indian Initiative

Prof. Sujoy Guha, of IIT-Kharagpur, leading the artificial heart project team, indicated that they are developing a total artificial heart **(TAH):**

> The TAH will be of great help to patients whose heart muscles have become so weak that they need immediate transplantation. Angioplasty, stents and even bypass surgery are of no use for such patients because they cannot strengthen muscles. It is difficult to find donor organs and even if transplantation is done, the body develops auto rejection and severe medication is required to suppress immune reactions.

Several TAH are available and were developed in the U.S. Those are prohibitively expensive for India. Dr. Guha was kind enough to give this author the summary of his effort, to be included in this text. The prior efforts in this direction are also shown briefly. Readers may consult the patent documents for further information and clarification. He explained **the need:** "One step large pressure boost in existing systems lowers the device life. A stepwise pressure rise is required."

The stepwise rise in pressure in each diaphragm also reduces the chances of thrombus formation.

The valve opening is to be automatic and not pressure-/flow-driven, to reduce thrombosis.

To meet occasional high-flow demands, artificial heart action needs to be augmented by aortic pumping.

Most of the existing reciprocating systems exhibit excessively high

1. Noise characteristics
2. Vibration
3. Recoil (thrust) levels.

16.12.1 Novel Features

A miniaturized device consisting of an aortic pump coupled with a biventricular pump has been developed. Staged multi-actuated pump systems, with a series of interconnected diaphragm chambers, were used. The valves are magnetically controlled in a mode linked to the actuation of each section of the pump. The diaphragm membranes are compressed in succession by a motor having an electromagnetic arm, the length of which is controlled to increase the pressure from the outflow cannula.

The outflow pressure is also increased by increasing the stiffness of the diaphragm membrane from inflow toward outflow cannula.

Further, the aortic pump, consisting of a series of cuffs, creates a peristaltic action at multiple points in the ascending/descending aorta to generate a negative pressure, which will assist the biventricular pump in blood circulation.

16.13 Existing Prior Technologies

16.13.1 Diaphragm Pump

A pump housing is divided into a driving chamber and a pumping chamber by a flexible diaphragm. The driving chamber is coupled to a pump driver adapted to direct fluid to and from the driving chamber to drive the diaphragm. These are prior arts Prof. Guha used for his patent. These might be useful for the reader to have an overview of such prior efforts. All these line drawings are self-explanatory to some extent and marked as Fig. 16.A.

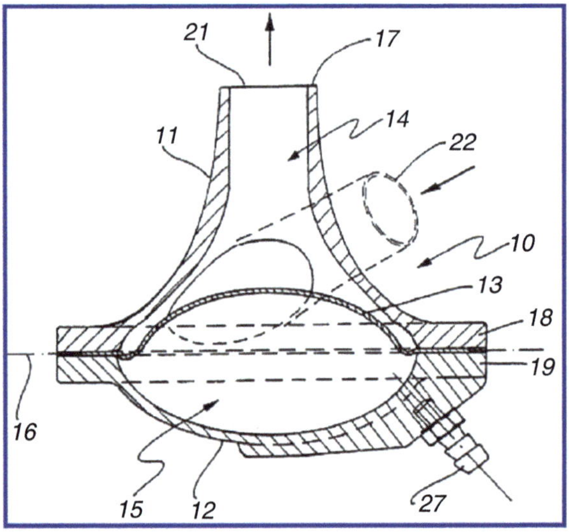

Advantages

The circular motion of the pump results in an effective washout of the diaphragm pump chamber junction, a possible area of stagnation.

Disadvantages

The blood gushes through the valves with a high force due to the sudden opening of the valves, thereby causing damage to the blood cells.

Development of flow instabilities.

16.13.2 Pulsatile Blood-Pumping System

A blood-pumping system is based on a spherical multivane and multichambered pump with an oscillating motion that delivers pulsatile flow.

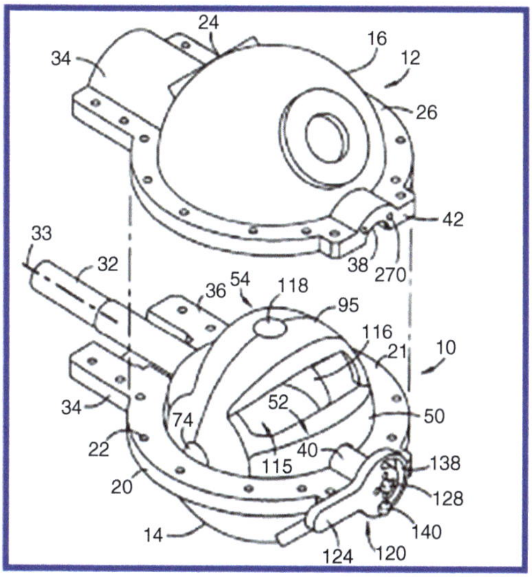

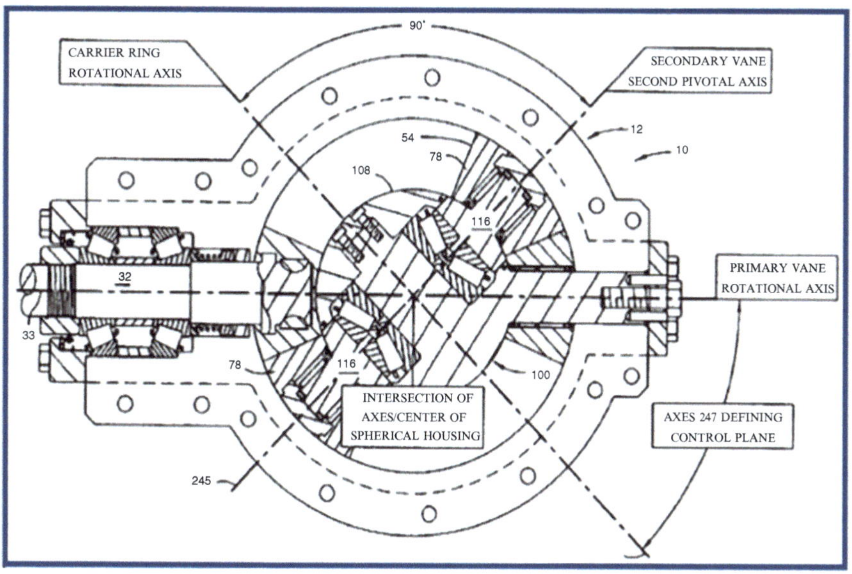

Advantages

Does not require valves and is relatively small in size.

Can match both normal flow rate and pulsatile flow to mimic the natural characteristics of the heart. Can be operated at relatively low RPM; simple motor requirements.

Disadvantages

The constant oscillating motion of the vanes can cause damage to the fragile blood cells. Increased chance of calcification due to pressure on the vanes.

Too many moving/rotating parts in contact with blood can cause trauma.

16.13.3 Orthotopic Total Artificial Heart

An orthotopic total artificial heart comprises a new design of two assembled chambers with an original layout between them and between their inlet and outlet ports. This enables better space utilization of the anterior mediastinum than that of prior art, realizing the required anatomical fit.

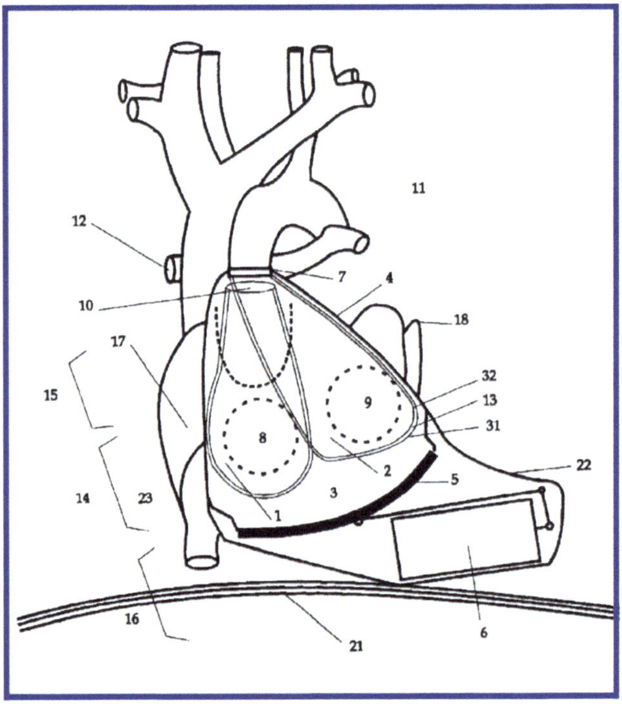

Advantages

Significantly better space utilization to place the driving mechanism inside the mediastinum and to increase the diastolic volume of each blood chamber

Directly connected to the vascular system through native blood vessels

Independent variation of the discharge volumes of each blood chamber

Nonthrombogenic walls of blood chambers.

Disadvantages

The large synthetic polymeric chambers can rupture due to repeated contractions.

16.13.4 Heart Assist Devices, Systems, and Methods

The apparatus comprises an aortic compression mechanism that may be fully implantable and a fluid reservoir and a pump mechanism adapted to pump a fluid from the reservoir to the aortic compression mechanism in order to actuate the aortic compression mechanism at least partly in counterpulsation with the heart.

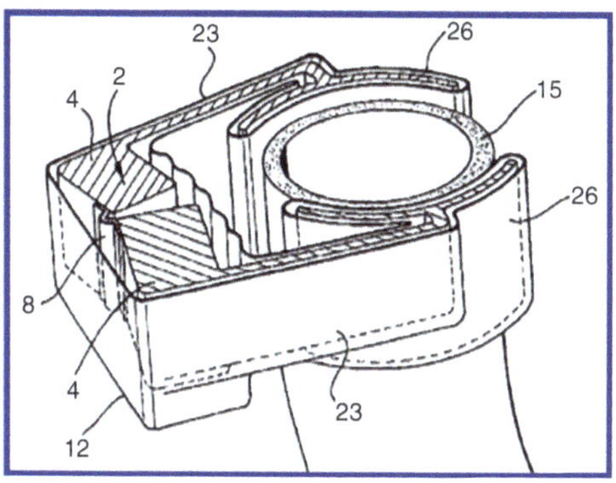

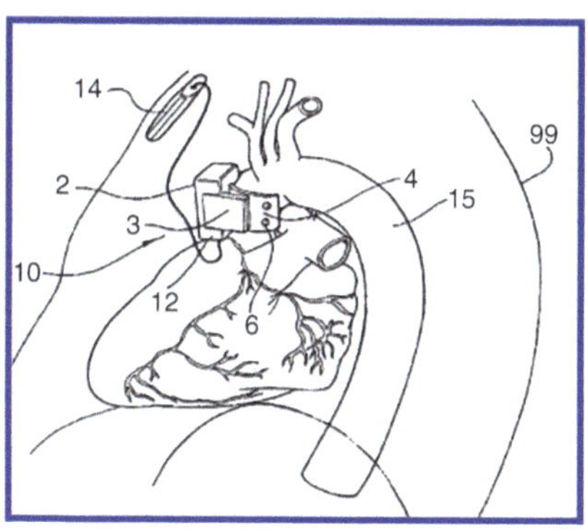

Advantages

Quickly and totally implanted in the heart, with a minimum of trauma to the patient using less invasive procedures.

The risk of limb ischemia in other IAB systems is avoided here because there is no contact with the blood supply whatsoever.

Disadvantages

Wear and fatigue of the enclosed region of aorta.

16.14 Disadvantages of Prior Developed Systems

Historically, blood-pumping mechanisms have presented many problems. For example, the pumping mechanism of the reciprocating (diaphragm) total artificial heart has been energized with gases (pneumatic systems), electricity (motors, solenoids, etc.), and skeletal muscles.

The energy sources and associated converter systems possess additional components that increase the complexity of the total system and thereby contribute to the overall unreliability.

Also, the size of the prior art systems for total artificial hearts is very restrictive to patient mobility and nonconducive to the recipient's quality of life.

The excessive size and complexity of energy conversion systems, as well as the overall pump design, exceed the available anatomical space.

Furthermore, most of these prior art reciprocating systems exhibit excessively high (1) noise characteristics, (2) vibration, and (3) recoil (thrust) levels.

16.15 Artificial Heart Pump

The artificial heart consists of two identical artificial ventricular pumps. Each consists of a staged multi-actuated pump system with a series of interconnected diaphragm chambers made of polyurethane inside a single housing.

The inflow and outflow cannula of these chambers is made of polyurethane, which will fuse distally with a Dacron graft to allow suturing with atrial ports and the ascending aorta and the pulmonary artery of the left and right ventricle, respectively.

The valves are magnetically controlled in a mode linked to the actuation of each section of the pump.

The diaphragm membranes are compressed in succession by a motor having an electromagnetic arm, the length of which is controlled to increase the pressure from the outflow cannula.

The outflow pressure also increases by increasing the stiffness of the diaphragm membrane from inflow toward outflow cannula. The motor is driven by electric power from a battery and is controlled by a controller.

16.16 Multistaged Multi-Actuated Pump System (Figs. 16.4 and 16.5)

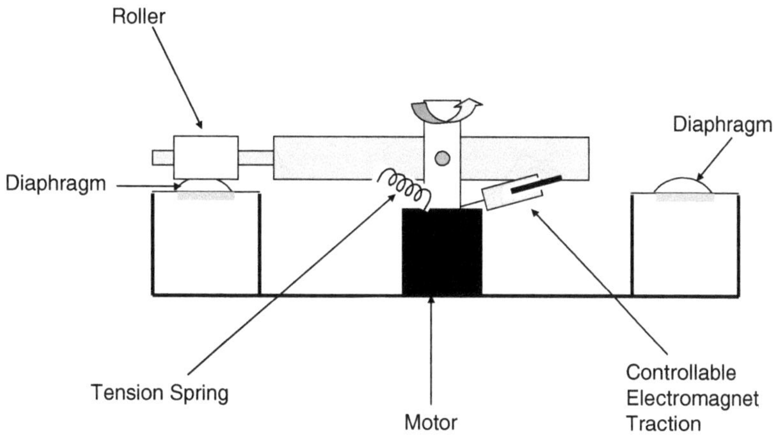

Fig. 16.4 Multistaged multi-actuated pump system

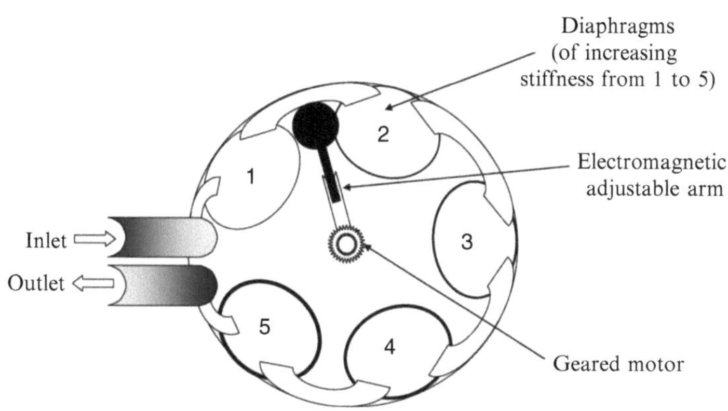

Fig. 16.5 Schematic diagram of a multi-actuated pump

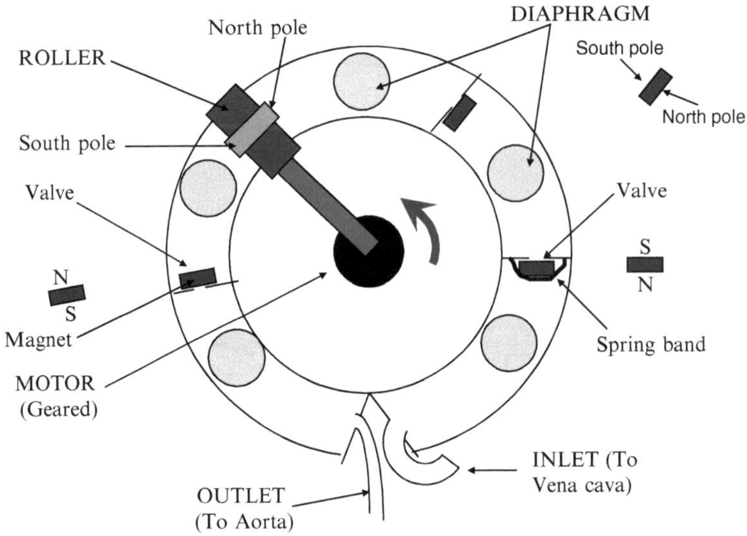

Multi-staged Programmable Pump

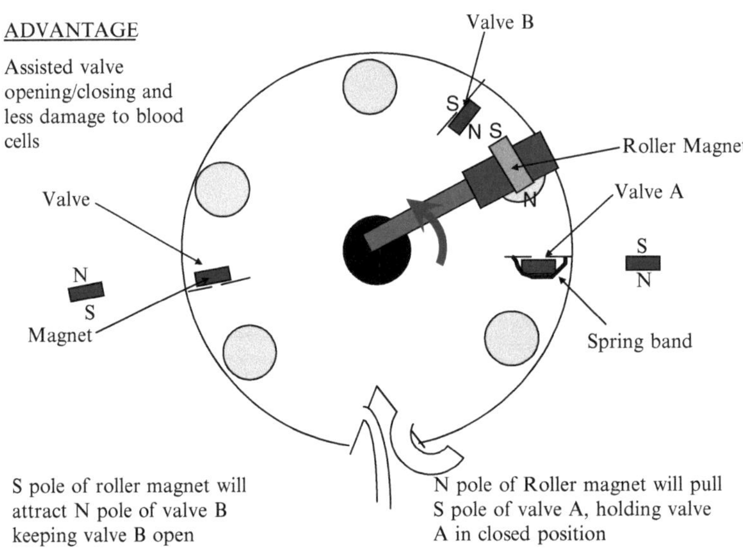

Multi - staged Programmable Pump

16.16.1 Heart Pump Prototype Model

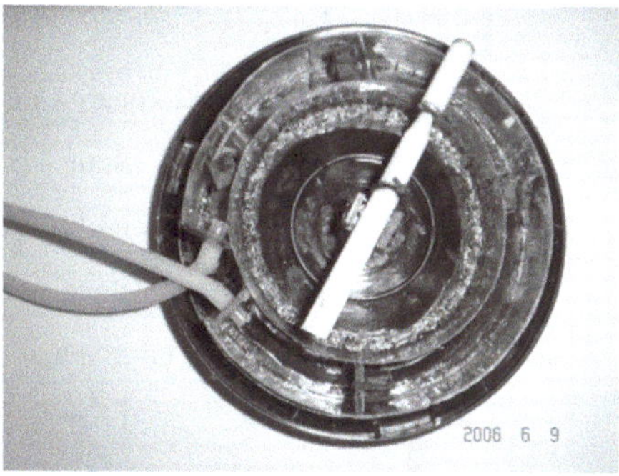

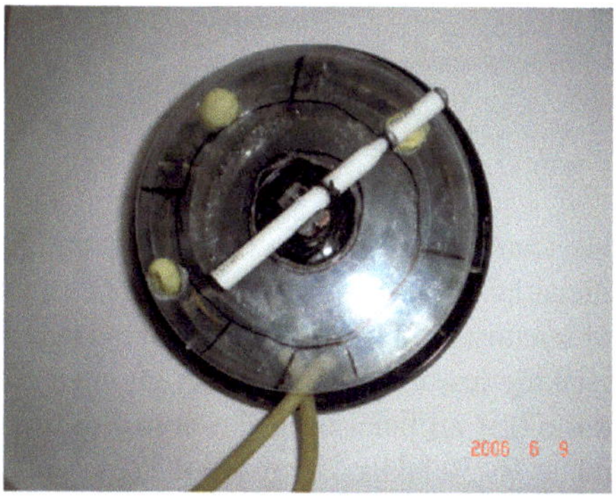

16.16.2 Advantages of the Artificial Heart

The entire system is compact and fit for implantation.

The series of diaphragms divides the pressure load of the pump and hence increases the longevity of the device.

The internal flow is designed to prevent excessive blood recirculation, stagnation, and mechanical trauma.

The increasing stiffness of the diaphragm and the adjustable electromagnetic arm increase the flow pressure gradually from the inflow to the outflow conduit.

The required blood flow and pressure are obtained.

The aortic pump functions in accordance with the heart pump to make the system more efficient.

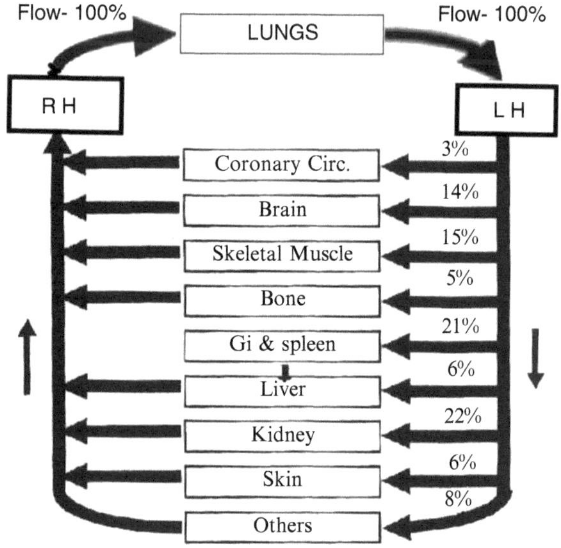

Fig. 16.6 Percentage of blood flow carried by different organs

Prof. Guha and his team are in the process of arranging animal trial in a goat model. The total artificial heart (TAH)—said to be the first in the country—has been developed by a team of scientists at IIT-KGP's school of medical science and technology. After 4 years of painstaking research, the scientists say their creation is better and far more affordable than the first artificial heart developed in the U.S., which showed a "high rate failure" and, at Rs.30 lakhs, is beyond the reach of the common man. The present one will cost Rs. 1 lakh and will be affordable to the general populace.

The inventors of the IIT-KGP-developed artificial heart that costs Rs.1 lakh hope to fit the heart into an ailing patient within a few months, once permissions from the Indian Council of Medical Research come through. The unique 13-chamber heart works fine in small animals, said a member of the team. Human tests are to be conducted at Medical College and Hospital (MCH), Kolkata. The TAH will be of great help to patients whose heart muscles have become so weak that they need immediate transplantation. Angioplasty, stents, and even bypass surgery are of no use for such patients because they cannot strengthen their muscles. It is difficult to find donor organs; even if transplantation is done, the body develops auto-rejection, and severe medication is required to suppress immune reactions, said Guha. As of mid-2011, the work was ongoing, and preclinical trials were being organized.

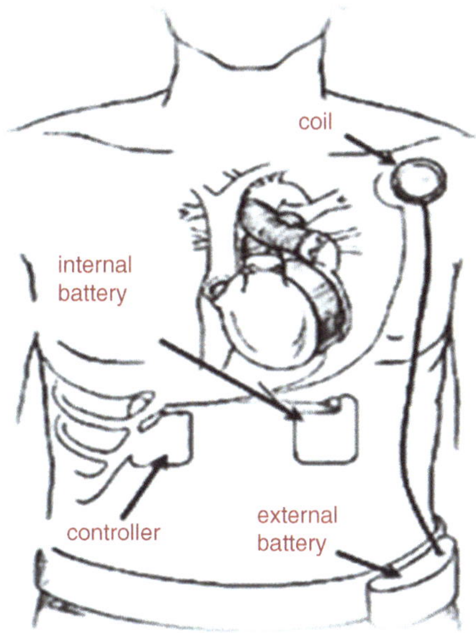

Fig. 16.7 A schema of the layout of artificial heart (Abicor)

Reading List

1. http://www.fda.gov/cdrh/mda/docs/p030011.html
2. http://www.syncardia.com
3. http://content.nejm.org/cgi/content/short/351/9/859
4. http://www.fda.gov/cdrh/ode/H040006sum.html
5. Abiomed claims its 1st implanted artificial heart. http://www.businessweek.com/ap/financial-news/D9917PE80.htm.
6. American Heart Association (2002) The Mechanical Heart celebrates 50 lifesaving years. 22 10 2002. 9 Feb 2008. http://www.americanheart.org/presenter.jhtml;jsessionid=EFNP3NSFU BXLICQFCXQCDSQ?identifier=3005888
7. Wayne State University School of Medicine
8. Stephenson LW et al (2002) The Michigan Heart: The world's first successful open heart operation? J Card Surg 17(3):238–246
9. Liotta D, Taliani T, Giffoniello AH, Sarria Deheza F, Liotta S, Lizarraga R et al (1961) Artificial heart in the chest: preliminary report. Trans Am Soc Artif Intern Organs 7:318
10. Liotta D, Taliani T, Giffoniello AH, Liotta S, Lizarraga R, Tolocka L, Pagano J (1961) Ablation experimentale et replacement du coeur par un coeur artificial intra-thoracique [in French]. Lyon Chirurgical 57:704
11. Jauhar S (2004) The artificial heart. N Engl J Med 350:542–544
12. Cooley DA, Liotta D, Hallman GL, Bloodwell RD, Leachman RD, Milam JD (1969) Orthotopic cardiac prosthesis for two-staged cardiac replacement. Am J Cardio 24:723–730
13. Fox RC, Swazey JP (1992) Spare parts: organ replacement in American society. Oxford University Press, New York, pp 102–104

14. Kwan-Gett CS, Van Kampen KR, Kawai J, Eastwood N, Kolff WJ (1971) Results of total artificial heart implantation in calves. J Thorac Cardiovasc Surg 62(6):880–889
15. Winchell's Heart—TIME
16. Patient gets first totally implanted artificial heart. CNN.com. 2001-07-03. http://archives.cnn.com/2001/HEALTH/conditions/07/03/artificial.heart/. Retrieved on 2008-07-13
17. AbioCor FAQs. abiomed.com. http://www.abiomed.com/products/faqs.cfm. Retrieved on 2008-07-13
18. FDA approves first totally implanted permanent artificial heart for humanitarian uses. FDA.gov. 2006-09-05.
19. Will we merge with machines? popsci.com. 2005-08-01. http://www.popsci.com/scitech/article/2005-08/will-we-merge-machines. Retrieved on 2008-07-13.
20. 14th Artificial heart patient dies: a newsmaker interview with Robert Kung, PhD. medscape.com. 2004-11-11. http://www.medscape.com/viewarticle/493622. Retrieved on 2008-07-13.
21. Scientists develop artificial heart that beats like the real thing. timesonline.co.uk
22. Total artificial heart to be ready by 2011: research team. afp.google.com
23. Liotta D, Hall CW, Henly WS, Beall AC, Cooley DA, DeBakey M (1963) Prolonged assisted circulation after cardiac or aortic surgery. Prolonged partial left ventricular bypass by means of intracorporeal circulation. This paper was finalist in: The Young Investigators Award Contest of the American College of Cardiology. Denver, May 1962. Am J Cardiol 12:399–404
24. Mitka M (2001) Midwest trials of heart-assist device. JAMA 286(21):2661
25. FDA approves two portable heart-assist devices. At Fda.Gov
26. Griffenhagen GB, Hughes CH. *The History of the Mechanical Heart*. Smithsonian Report for 1955 (Pub. 4241): 339–356, 1956.

Problems

1. Describe the human heart as four-chamber pump; find the volume of each chamber and flow rate showing the valves; also, indicate its destination. Label the name of the chamber and the valves. Figure 16.6 gives an indication of the percent flow of each organ. If the total blood in an adult human body is on the order of 5.5 l, find the quantity in circulation in each organ.
2. Briefly describe the workings of the AbioCor Total Artificial Valve. What is its current status? A line sketch of the system is shown in Fig. 16.7.
3. French professor and leading heart transplant specialist Alain F. Carpentier announced that a fully implantable artificial heart will be ready for clinical trial by 2011, and for alternative transplant in 2013. Find the current status of his project as you study this.
4. Discuss the Indian initiative in the design and fabrication of the total artificial heart at IIT-Kharagpur. What is the specialty of such a device?
5. A left ventricular assist device (LVAD) that pumps blood from the left ventricle to the aorta was developed. VADs need to be clearly distinguished from artificial hearts, which are designed to completely take over cardiac function and usually require the removal of the patient's heart.

VADs are designed to assist either the right (RVAD) or left (LVAD) ventricle, or both at once (BiVAD). Which of these types is used depends primarily on the underlying heart disease and the pulmonary arterial resistance that determines the load on the right ventricle. Figure 16.6 shows a schematic view. Try to explain how it works.

Chapter 17
The Design of Heart Valves

17.1 Introduction

Our cardiovascular system transports important substances, such as oxygen and nutrients, between tissues and organs. It also helps transport and eliminates waste products. Our heart, blood vessels, and blood form a sophisticated network that transports materials around our body. These materials are carried by the blood through the blood vessels and are kept in motion by the pumping action of the heart. The blood vessels of the cardiovascular system are divided into two main pathways. The blood vessels in the **pulmonary circuit** carry blood from the heart to the lungs and back to the heart. The **systemic circuit** consists of the pathways between the heart and all other areas of the body.

17.2 The Heart

The heart is a hollow muscular structure that contracts in a rhythmic pattern to pump approximately 5–6 l of blood per minute. The human heart has a pair of pumps that are divided into four chambers. The two top chambers are called *atria*, and the two lower chambers are called *ventricles* (right and left) (Figs. 17.1a, b).

The two types of chambers in the heart perform different functions: The atria collect the blood that enters the heart and push it to the ventricles, while the ventricles push blood out of the heart and into the arteries to go to the rest of the body.

The two atria are separated by an *interatrial septum*, while the *interventricular septum* divides the two ventricles. The atrium and ventricle of each side of the heart communicate with each other via an *atrioventricular orifice*. This orifice can be opened or closed off by an *atrioventricular valve*, also known as an *A–V valve*. The left A–V valve is known as the bicuspid (or mitral) valve, while the right A–V valve is termed the tricuspid valve.

S. Pal, *Design of Artificial Human Joints & Organs*,
DOI 10.1007/978-1-4614-6255-2_17,

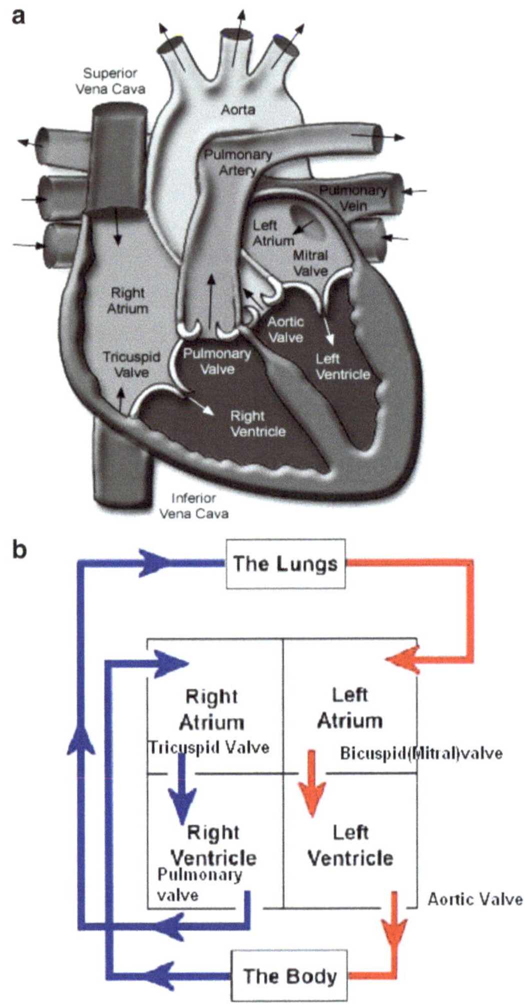

Fig. 17.1 (**a**) The cutaway view of a human heart showing the pair of pumps with four chambers (acknowledgment: Texas Heart Institute); (**b**) schematic of the heart and blood flow in the entire human body

In order to move blood through heart, our heart chambers undergo alternating periods of relaxation **(diastole)** and contraction **(systole),** allowing the chambers to fill up with and pump blood, respectively (Fig. 17.2).

The right atrium of the heart receives deoxygenated blood from two major veins: the *superior vena cava* and the *inferior vena cava*, as well as a smaller *coronary sinus* that drains blood from the heart wall. When this chamber contracts, blood moves out of the right atrium and goes into the right ventricle through the tricuspid valve.

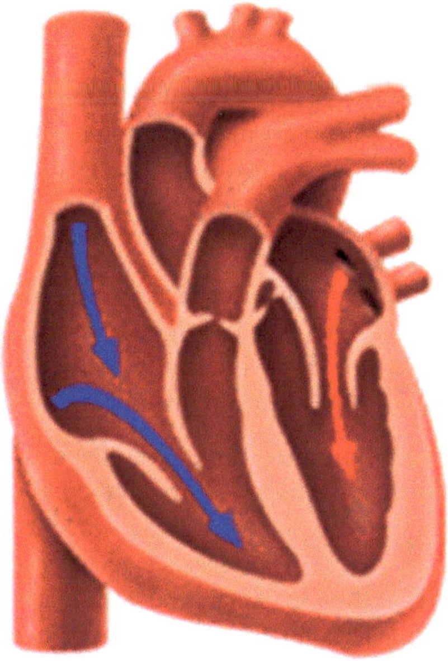

Fig. 17.2 Flow through the heart

Once the right ventricle is sufficiently filled with blood, it contracts, pumping blood via the pulmonary arteries into the pulmonary circuit of the cardiovascular system.

Newly oxygenated blood enters the left atrium of the heart through pulmonary veins. Once this chamber is filled with blood, the left atrial wall will contract, pushing blood into the left ventricle through the bicuspid valve. After the left ventricle is filled with blood, it contracts, forcing blood out of the ventricle and into the aorta. From the aorta, blood travels through the systemic circuit of the blood vessels, bringing oxygen to tissue cells throughout the body.

The heart is enclosed in a pericardial sac that is lined with the parietal layers of a serous membrane. Three layers of tissue form the heart wall:

1. Endocardium—innermost layer; epithelial tissue that lines the entire circulatory system
2. Myocardium—thickest layer; consists of cardiac muscle
3. Epicardium—thin, external membrane around the heart.

The internal cavity of the heart is divided into four chambers: **right atrium, right ventricle, left atrium, and left ventricle**. The two atria are thin-walled chambers that receive blood from the veins. The two ventricles are thick-walled chambers that forcefully pump blood out of the heart. Differences in thickness of the heart chamber walls are due to variations in the amount of myocardium present, which reflects the amount of pressure each chamber is required to generate. The right atrium receives deoxygenated blood from systemic veins; the left atrium receives oxygenated blood from the pulmonary veins.

17.3 Heart Valves

Heart valves act as one-way gates that allow blood to pass between heart chambers or from heart chambers to their associated blood vessels. They include the tricuspid and pulmonary valves for the right chambers and the bicuspid (or mitral) and aortic valves for the left chambers.

The tricuspid valve: The *tricuspid valve* is located between the atrium and ventricle on the right side of the heart. When this valve is open, blood passes from the right atrium into the right ventricle. The tricuspid valve prevents the reverse of blood flow back into the atrium by closing during ventricular contraction. As its name suggests, the tricuspid valve is made up of three leaves, or *cusps* (Fig. 17.3).

The pulmonary valve: With the tricuspid valve closed, the only outlet for blood in the right ventricle is through the *pulmonary trunk*. The pulmonary trunk splits into the left and right *pulmonary arteries*, which connect to the left and right lungs, respectively. The entrance to the pulmonary trunk is guarded by the *pulmonary valve*. The pulmonary valve is made up of three leaves that open when the right ventricle contracts and close when this chamber relaxes, allowing blood to flow from the right ventricle into the pulmonary arteries but not the reverse (Fig. 17.4).

The bicuspid valve: The *bicuspid* or *mitral valve* regulates the flow of blood from the left atrium to the left ventricle. Like the tricuspid valve, the bicuspid valve closes during ventricular contraction. The bicuspid valve is composed of two leaves (Fig. 17.5).

The aortic valve: The *aortic valve* consists of three leaves found at the entrance to the aorta. This valve lets blood out of the left ventricle as it contracts and blocks the pathway of blood from the aorta back into the left ventricle when this chamber relaxes (Fig. 17.6).

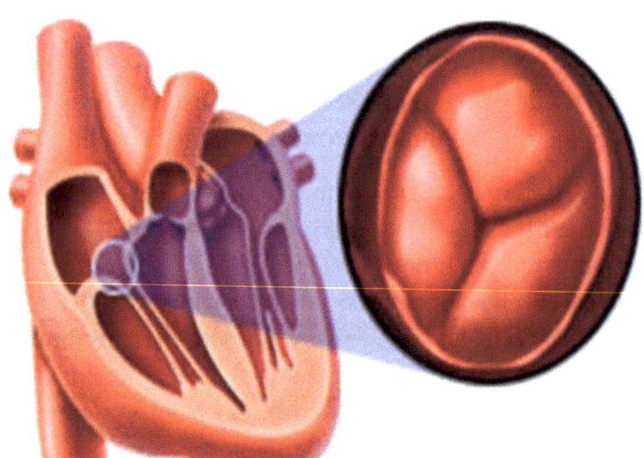

Fig. 17.3 The tricuspid valve

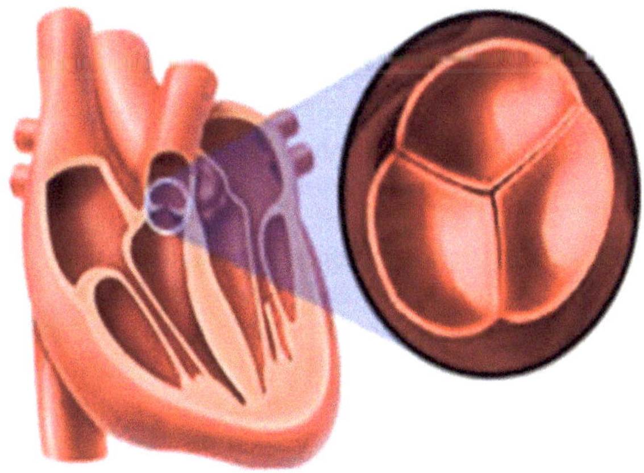

Fig. 17.4 The tricuspid valves closed

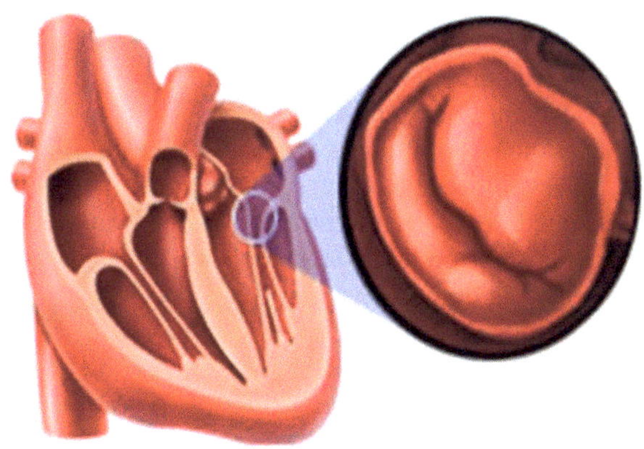

Fig. 17.5 The bicuspid valve

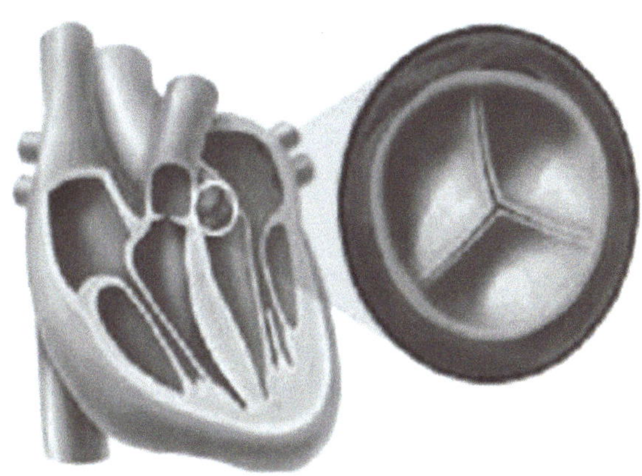

Fig. 17.6 The aortic valve

17.4 Cardiac Cycle

The cardiac cycle consists of two parts: systole (contraction of the heart muscle) and diastole (relaxation of the heart muscle). Atria contract while ventricles relax. The pulse is a wave of contraction transmitted along the arteries.

Valves in the heart open and close during the cardiac cycle. Heart muscle contraction is due to the presence of nodal tissue in two regions of the heart:

The SA node (sinoatrial node) initiates heartbeat.

The AV node (atrioventicular) causes ventricles to contract. The AV node is sometimes called the pacemaker since it keeps the heart beating regularly.

The heartbeat is also controlled by the autonomic nervous system (Fig. 17.7).

Blood flows through the heart from veins to atria to ventricles and goes out via arteries. Heart valves limit the blood flow to a single direction. One heartbeat, or cardiac cycle, includes atrial contraction and relaxation, ventricular contraction and relaxation, and a short pause. Normal cardiac cycles (at rest) take 0.8 s. Blood from the body flows into the vena cava, which empties into the right atrium. At the same time, oxygenated blood from the lungs flows from the pulmonary vein into the left atrium. The muscles of both atria contract, forcing blood downward through each AV valve into each ventricle.

Diastole is the filling of the ventricles with blood. Ventricular systole opens the SL valves, forcing blood out of the ventricles through the pulmonary artery or aorta. The sound of the heart contracting and the valves opening and closing produces a characteristic "lub-dub" sound. Lub is associated with closure of the AV valves; dub is the closing of the SL valves.

Human heartbeats originate from the sinoatrial node **(SA node)** near the right atrium. Modified muscle cells contract, sending a signal to other muscle cells in the heart to contract. The signal spreads to the atrioventricular node (**AV node**). Signals

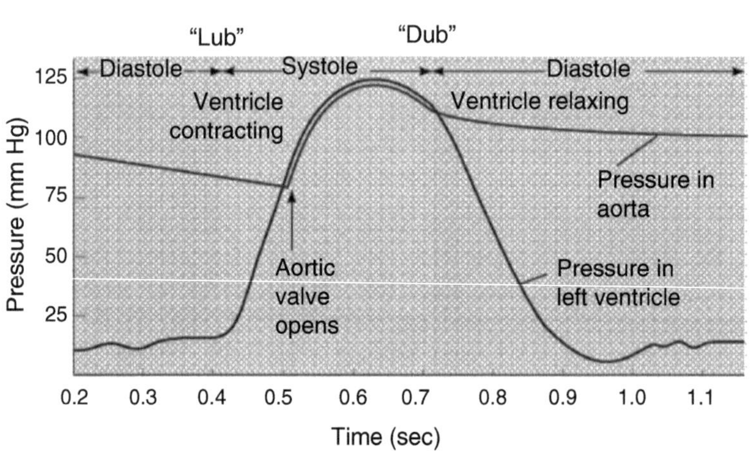

Fig. 17.7 The cardiac cycle

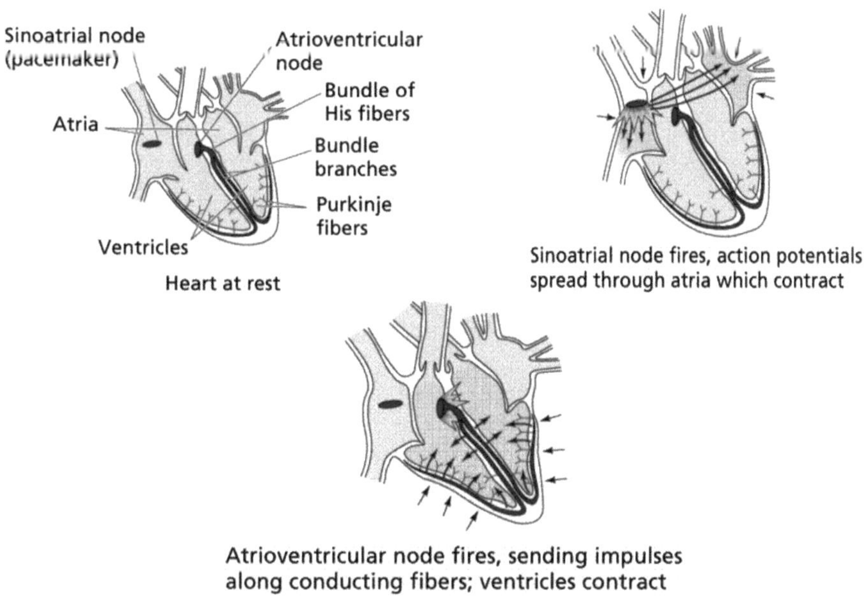

Fig. 17.8 The contraction of the heart and the action of the nerve nodes located on the heart

carried from the AV node, slightly delayed, through bundles of His fibers and Purkinje fibers cause the ventricles to contract simultaneously (Fig. 17.8).

An electrocardiogram (ECG) measures changes in electrical potential across the heart and can detect the contraction pulses that pass over the surface of the heart. There are three slow, negative changes, known as P, R, and T. Positive deflections are the Q and S waves. The P wave represents the contraction impulse of the atria, the T wave the ventricular contraction. ECGs are useful in diagnosing heart abnormalities (Fig. 17.9).

17.5 Cardiac Output

Cardiac output (**CO**) is the product of the heart rate (**HR**) and stroke volume (**SV**):

$$\mathbf{CO = HR \quad SV.}$$

For a 70-kg man, normal values are HR = 70/min and SV = 70 ml, giving a cardiac output of about 5 l/min. The cardiac index is the cardiac output per square meter of body surface area; normal values range between 2.5–4.0 l/min/m^2.

Heart rate is determined by the rate of spontaneous depolarization at the sinoatrial node (see above) but can be modified by the autonomic nervous system. The vagus nerve acts on muscarinic receptors to slow the heart, whereas the cardiac sympathetic fibers stimulate beta-adrenergic receptors and increase heart rate.

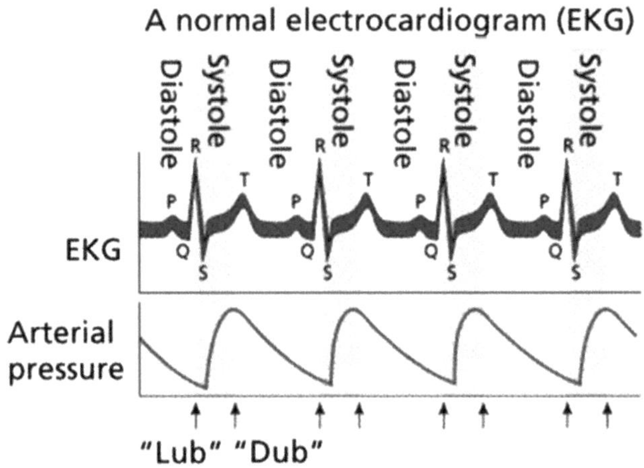

Fig. 17.9 Normal ECG/EKG pattern and arterial pressure variation

Stroke volume is determined by three main factors: preload, and contractility and afterload. These will be considered in turn:

Preload is the ventricular volume at the end of diastole. An increased preload leads to an increased stroke volume. Preload is mainly dependent on the return of venous blood from the body. Venous return is influenced by changes in position, intrathoracic pressure, blood volume, and the balance of constriction and dilatation (tone) in the venous system. The relationship between ventricular end-diastolic volume and stroke volume is known as **Starling's law of the heart**, which states that the energy of contraction of the muscle is related, or proportional, to the initial length of the muscle fiber. This is graphically illustrated in Fig. 17.10 by a series of "Starling curves."

Curves A and B illustrate the rise in cardiac output with increases in ventricular end-diastolic volume (preload) in the normal heart. Note that with an increase in contractility, there is a greater cardiac output for the same ventricular end-diastolic volume.

In the diseased heart (C and D), cardiac output is less and falls if ventricular end-diastolic volume rises to high levels, as in heart failure or overload.

As the volume at the end of diastole (**end-diastolic volume**) increases and stretches the muscle fiber, so the energy of contraction and the stroke volume increase, until a point of overstretching when the stroke volume may actually decrease, as in the failing heart. Cardiac output will also increase or decrease in parallel with stroke volume **if there is no change in heart rate**.

The curves show how the heart performs at different states of **contractility**, ranging from the normal heart to one in cardiogenic shock. This is a condition where the heart is so damaged by disease that cardiac output is unable to maintain tissue perfusion. Also shown are increasing levels of physical activity, which require a corresponding increase in cardiac output.

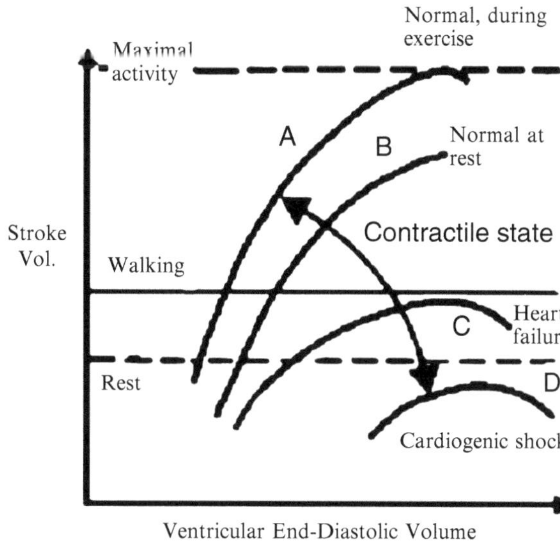

Fig. 17.10 Stroke volume vs. VED volume

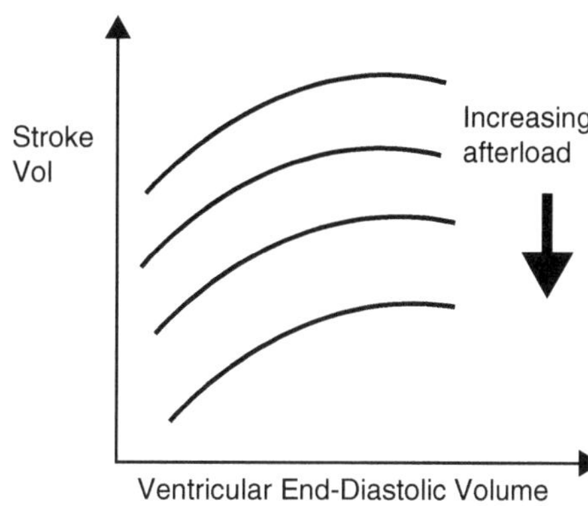

Fig. 17.11 Stroke volume vs. ventricular end-diastolic volume

Afterload is the resistance to ventricular ejection. This is caused by the resistance to flow in the systemic circulation and is the **systemic vascular resistance**. The resistance is determined by the diameter of the arterioles and precapillary sphincters; the narrower or more constricted this diameter is, the higher the resistance is. The level of systemic vascular resistance is controlled by the sympathetic system, which, in turn, controls the tone of the muscle in the wall of the arteriole, and hence the diameter. The resistance is measured in units of dyne·s/cm^5. A series of Starling curves with differing afterloads is shown in Fig. 17.11 demonstrating a fall in the stroke volume as the afterload increases.

A series of curves illustrates the effects of increasing the afterload on systemic vascular resistance. As the afterload increases, the patient moves to a lower curve, with a lower stroke volume for the same ventricular end-diastolic volume (preload).

The relationship between the systemic vascular resistance and the control of arterial pressure is discussed below.

Contractility describes the ability of the myocardium to contract in the absence of any changes in preload or afterload. In other words, it is the "power" of the cardiac muscle. The most important influence on contractility is the sympathetic nervous system. Beta-adrenergic receptors are stimulated by noradrenalin released from nerve endings, and contractility increases. A similar effect is seen with circulating adrenaline and drugs such as ephedrine, digoxin, and calcium. Contractility is reduced by acidosis, myocardial ischemia, and the use of beta-blocking and anti-arrhythmic agents.

Cardiac output will change to match changing metabolic demands of the body. The outputs of both ventricles must be identical and also equal the venous return of blood from the body. The balancing of cardiac output and venous return is illustrated during the response to exercise. Blood vessels dilate in exercising muscle groups because of increased metabolism, and the blood flow increases. This increases the venous return and right ventricular preload. Consequently, more blood is delivered to the left ventricle, and the cardiac output increases. There will also be increased contractility and heart rate from the sympathetic activity associated with exercise, further increasing cardiac output to meet tissue requirements.

Cardiac Reserve

- Cardiac reserve is the difference between cardiac output at rest and the maximum volume of blood the heart is capable of pumping per minute.
- It permits cardiac output to increase dramatically during periods of physical activity.

Flow Rate Through Blood Vessels

- directly proportional to the pressure gradient
- inversely proportional to vascular resistance

flow = difference in pressure/resistance

pressure gradient = difference in pressure between beginning and end of vessel (pressure = force exerted by blood against vessel wall and measured in millimeters of mercury).

Blood Flow

The relationship between the flow and the driving pressure is given by the **Hagen–Poisseuille** formula. This states that the flow rate Q in a tube is proportional to

$$Q \propto \text{Driving pressure} \times \text{Radius} / (\text{Length} \times \text{Viscosity}).$$

In blood vessels, the flow is pulsatile rather than continuous, and the viscosity varies with the flow rate, so the formula is not strictly applicable, but it illustrates an important point; small changes in radius result in large changes in flow rate. In both arterioles and capillaries, changes in flow rate are brought about by changes in tone and therefore in vessel radius.

Viscosity describes the tendency of a fluid to resist flow. At low flow rates, the red blood cells stick together, increasing viscosity, and remain in the center of the vessel. The blood closest to the vessel wall (which supplies side branches) therefore has a lower hematocrit. This process is known as **plasma skimming**. Viscosity is reduced in the presence of anemia, and the resulting increased flow rate helps maintain oxygen delivery to the tissues.

17.6 Artificial Heart Valves

An artificial (mechanical) heart valve is a manmade device that is used to replace one of a patient's own damaged or diseased **heart valves** that cannot be repaired. A biological valve, from either an animal (*xenograft*) or a deceased human donor (*allograft*), may also be used to replace the patient's damaged or diseased valve. In most cases, the use of an artificial heart valve can lengthen or even save a patient's life. The valves are durable and can last 30 years or longer. However, there is a risk of complications, and most patients will need to take **anticoagulants** for the rest of their lives to reduce the risk of **blood clot** formation. An artificial heart valve is inserted into the patient's heart as part of an **open-heart surgery** called **heart valve replacement** (Fig. 17.12).

Any of the patient's four heart valves (**aortic valve**, **mitral valve**, **pulmonic valve,** or **tricuspid valve**) may be replaced with an artificial heart valve. However, artificial heart valves tend to last about 30 years or even more, which is significantly longer than biological valves last. Usually, a porcine (pig) valve suits the human body. The weight of the pig has to be approximately 60–70 kg to attain a suitable size for a human valve. It is estimated that each year, almost 100,000 people will need a heart valve replacement. In India, at the start of this century, the current requirement was about 6,000–8,000 or more valves annually. It will increase at a much faster rate in the future due to the increase in aging patients, the adoption of a Western lifestyle, and the expectation for a good-quality life.

17.6.1 Potential Risks of Artificial Heart Valves

In most cases, replacing a diseased or damaged **heart valve** can lengthen or even save a life. Untreated, **valvular heart disease** can lead to **heart failure** and death.

There are a number of rare but possible complications that could arise from the surgery needed to insert the artificial heart valve (see **Heart Valve Replacement**). Rarely, there may be a problem with the artificial heart valve itself. Patients are encouraged to document the make, model, and serial number of their artificial heart valve in case any problems with it are announced in the future.

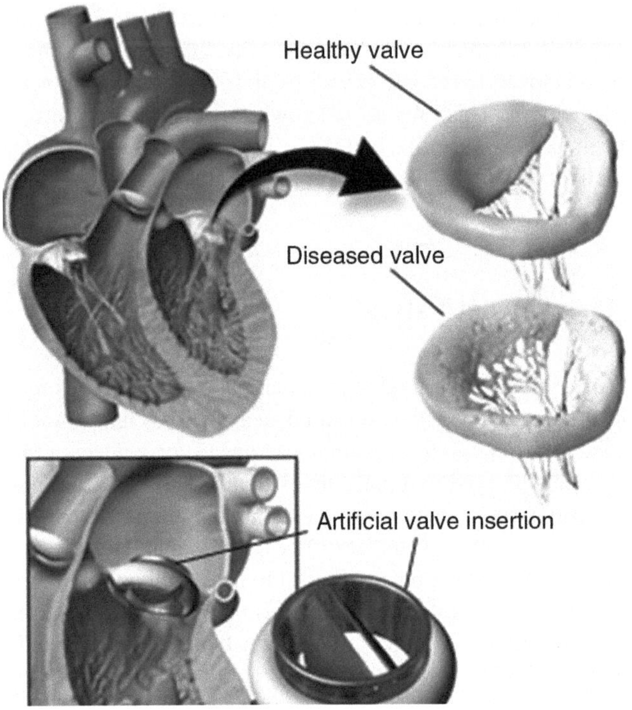

Fig. 17.12 Artificial heart valve and its placement

17.6.2 Lifestyle Considerations

People with artificial **heart valves** will need follow-up visits to have a painless **echocardiogram**, which can reveal any malfunction or leaking of the valve.People with artificial heart valves face an increased risk of developing a potentially dangerous inflammation called bacterial **endocarditis**. This risk is increased when bacteria enter the bloodstream, such as during a **dental**, medical, or surgical procedure, and infect the tissue surrounding the valve, or the valve itself. Therefore, people with artificial heart valves are generally advised by their physicians to take **antibiotics** before any of these procedures to minimize their risk of bacterial endocarditis. All people with artificial heart valves will also need to take medications called **anticoagulants** for the remainder of their lives. These medications help prevent the body's natural response of forming **blood clots** around a foreign object, such as an artificial heart valve. Anticoagulants reduce the risk of blood clot formation, thus reducing the patient's risk of valve malfunction, **stroke**, and other potentially dangerous complications. Patients taking anticoagulants may need to undergo regular **blood tests** to monitor their medication dosage. It is also wise for patients carry a form of identification (card, bracelet) stating that they have an artificial heart valve, in case of emergency.

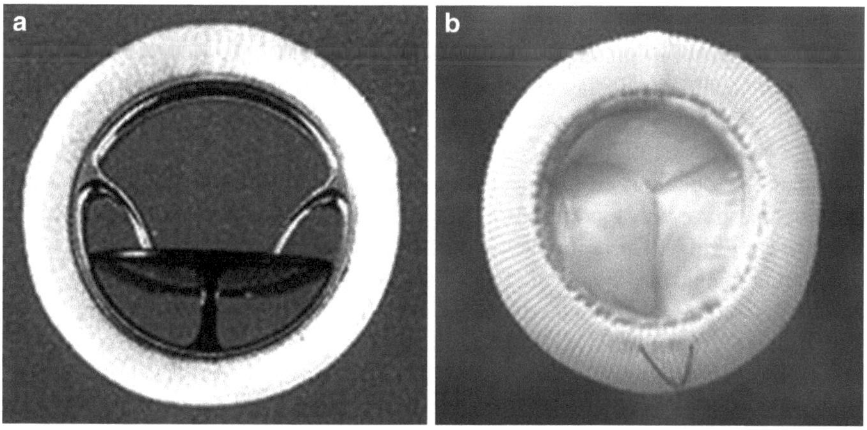

Fig. 17.13 Replacement heart valves: (**a**) tilting disk; (**b**) three-leaflet

17.6.3 Longevity and Replacements

On average, an artificial heart valve can last 30 years or more. The patient will need to see his or her physician for regular follow-up to detect any signs that the artificial heart valve is beginning to wear out. These signs include an unusual **heart murmur** or leakage that is growing more severe. As the leakage gets more severe, the patient may begin to notice symptoms such as **shortness of breath** or **chest pain**. Patients are urged to contact their physician if they begin to notice these types of symptoms (Fig. 17.13).

17.7 Design of Valves

By comparison with the revolutionary changes in many other fields of medicine, the development of artificial heart valves has progressed rather gradually since its beginning in 1960. The main problems recognized with the first designs, thromboembolism in mechanical valves and structural deterioration in tissue valves, were reduced to an acceptable level by 1965–1970 and were not substantially reduced further in the ensuing 25 years. However, the years since 1995 have seen a resurgence in the development cycle, and many new devices are currently under investigation

Heart valves can be classified into two broad categories according to the origin of their occluding mechanism: (1) **mechanical and** (2) **biological.** The introduction of the heart-lung machine made valve replacement a possibility in 1955; many replacement devices were attempted. Mechanical and biological valves of all kinds were tried, including designs with one, two, three, and four leaflets. The first valve replacements that led to long-term survivors were mechanical cage ball valves used by the Harken in the aortic position and Starr in the mitral position, both in 1960.

Once the early problems of valve fixation and durability were solved, the major concern that drove mechanical valve development was the reduction of thromboembolic complications. One favorite design was the tilting disk valve, several varieties of which exist. The original versions of the bileaflet valve did not endure, but this design was successfully implemented in 1977 on the basis of the transfer of pyrolytic carbon technology from spacecraft to heart valves. Thus, the mechanical valve designs that have prevailed until today are **the ball valve, tilting disk valve, and the bileaflet valve**. Current development of mechanical valves is concentrated in attempting to enhance the bileaflet valve design.

The first biological valves used successfully were transplants from human cadavers, called allografts, pioneered by Ross and Barratt-Boyes in 1962. The successful use of autologous grafts was begun with the pulmonary autograft in 1967. The goal of these biological valves was to reduce the complications associated with thromboembolism and the need for anticoagulation. Several homologous and heterologous materials were used to fabricate tissue valves but were eventually abandoned. During the 1960s, a major advance was the use of **glutaraldehyde** for the preservation of porcine valves pioneered by Carpentier and coworkers. **Glutaraldehyde-fixed** valves currently in use are aortic porcine valves and, in resurgence, valves fabricated of bovine pericardium. Current developments in tissue valve technology include improved methods of fixation, calcification mitigation treatments, and stentless designs.

17.7.1 Valve Descriptions

A heart valve functions as a check valve, opening to permit forward blood flow and closing to prevent retrograde flow, about 40 million times per year. Heart valve prostheses consist of an **orifice**, through which blood flows, and an **occluding mechanism** that closes and opens the orifice. There are two fundamental approaches to valve design: mechanical prostheses with rigid manufactured occluders, and biological prostheses (also called tissue valves) with flexible leaflet occluders of animal origin. The latter category includes replacement valves of human origin.

17.7.2 Mechanical Valves

The type of a mechanical valve is designated by its occluder: a ball, a circular disk, or two semicircular leaflets (bileaflet). For ball and disk valves, the occluder is guided and retained by structural members call struts attached to the orifice. The combination of orifice and struts is referred to as the valve housing. For bileaflet valves, the leaflets are retained and guided by a hinge or pivot mechanism: Projections of the leaflets fit into indentations or sockets in the housing, which serve to retain the leaflets and define their limits of travel.

It will not be out of place to quickly look into the materials that were used over the years in different heart valve designs (Table 17.1).

Table 17.1 Summary of the material composition of nine key designs of mechanical valves developed over the course of 30 years

Year	Name	Type	Poppet	Material
1959	Hufnagel	Ball	Polypropylene	Methacrylate
1964	Starr–Edwards 1000	Ball	Silastic	Stellite
1968	Wada–Cutter	Tilting disk	Teflon	Titanium
1969	Bjork–Shiley	Tilting disk	Delrin	Stellite
1970	Lillehei–Kaster	Tilting disk	Pyrolitic carbon	Titanium
1971	Bjork–Shiley	Tilting disk	Pyrolitic carbon	Stellite
1977	Medtronic–Hall	Tilting disk	Pyrolitic carbon	Titanium
1977	St Jude Medical	Bileaflet	Pyrolitic carbon	Pyrolitic carbon
1991	Jyros	Bileaflet	Vitreous carbon	Vitreous carbon

The most commonly used materials include

Stainless steel alloys (316L SMo)—for cage
Molybdenum alloys (Co-Cr-Mo)—for cage
Pyrolitic carbon for the valve housings and leaflets
Silicone, Teflon®
Polyester (Dacron®) for sewing rings.

17.7.3 Caged Ball Valve

The first clinically successful heart valve was the Starr–Edwards caged ball valve introduced in 1960. The valve underwent several slight design modifications for five years, resulting in the model currently used. The ball is a silicone rubber polymer impregnated with barium sulfate for radio-opacity. The cobalt-chromium alloy struts are joined at the apex to form a cage (Fig. 17a, upper left). When the ball opens by moving to the end of its cage, it creates a circular primary orifice and a ring-shaped secondary orifice between the ball and the housing. In the aortic position, there is a tertiary orifice between the equator of the ball and the aorta (Fig. 17.14).

17.7.4 Tilting Disk Valve

Tilting disk valves have separate projections into the orifice, either single arms or closed loops to retain and guide the disk-shaped occluder. Among the metals used for the housing are stainless steel and titanium. The disks are graphite with a coating of pyrolitic carbon. When the disk pivots to the open position, the primary orifice is separated into two areas, called the major and minor orifices (Fig. 17.15).

Fig. 17.14 Starr–Edwards mitral caged ball valve (courtesy of Baxter Edwards CVS)

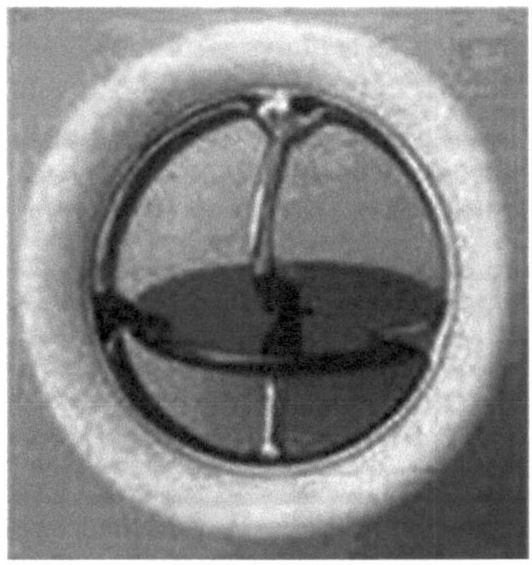

Fig. 17.15 Medtronic Hall tilting disk valve (courtesy of Medtronic Heart Valve Division)

The Bjork–Shiley valve was the first successful tilting valve. It became available in 1971 with a carbon-coated disk and both struts (inflow and outflow) welded to the chromium alloy orifice. The Convexo-Concave model introduced in 1979 had an integral inflow strut to eliminate the few inflow strut fractures that had occurred with previous models. Sorin manufactured a tilting disk valve patterned after the Shiley valve disk, but with both struts integral to the housing to avoid the possibility of strut fracture. It is currently available with a pyrolitic carbon-coated sewing ring and entire housing.

The Medtronic Hall valve has a titanium housing machined from a solid cylinder and a thin carbon-coated disk with flat parallel sides (Fig. 17.15). The disk opens to 75° in the aortic model and 70° in the mitral. The disk is retained and guided by an S-shaped guide strut that protrudes through a central hole in the disk. Four structural elements project perpendicularly from the annulus into to the orifice: a guide strut and three pivot struts (one inflow, two outflow). The two inflow pivot struts starts from near the top (inflow) edge of the orifice toward each other; the disk sits on the flat triangular bottom surfaces of these struts. The Omniscience valve is a streamlined, elegant-looking valve. It has a curved pyrolitic carbon disk with no indentations, a one- piece titanium cage, and a seamless polyester knit sewing ring. The disk opens to 80° and closes at an angle of 12° to the plane of the orifice. It has been in use since 1978 but underwent a design change in 1981–1982 involving a significant modification of the sewing ring.

17.7.5 Bileaflet Valves

The currently available bileaflet valves vary with regard to several design features. Although the features of the valves manufactured by St. Jude, Baxter, Carbomedics, Soren, ATS, and Medtronic are described here, clinical performance information is included only for the St. Jude valve, for which a large amount of long-term information is available. The two leaflets of a bileaflet valve swing apart during opening, resulting in three separate flow areas. The bileaflet valve housings are solid pyrolytic carbon over a graphite (titanium in the Sorin valve) substrate. All except the St. Jude and Sorin valves have stiffener rings to strengthen the housing, shield it from needles during implantation, and improve radiographic visualization (the Sorin valve has a titanium substrate, rather than graphite, which may confer some benefit). All except the St. Jude are rotatable after the plantation. Since its first implant in 1977, the St. Jude bileaflet valve has been used half a million times. Previous bileaflet valves were unsuccessful, but the results with this design, with pyrolitic carbon-coated housing and leaflets, introduced a new generation of mechanical prostheses. The housing of the valve included two rounded tabs, called pivot guards, that project out from the inflow side. The inside surfaces of these tabs contain the butterfly-shaped indentations that serve to retain the leaflets. The tabs containing the hinge sockets extend above the housing, whereas with the other bileaflet valves, the cavities in the housing containing the pivot recesses are located within the main body of the housing, approximately at the plane of the annulus. The leaflets open to 85° and swing through an arc of 55–60°, depending on the valve size, from fully closed to fully open (travel arc) (Fig. 17.16).

The TEKNA valve (Baxter Edwards), originally called Duromedics, introduced several modifications to the bileaflet configuration. The leaflets are curved and translate slightly in the direction of flow as they open. Unlike other bileaflet valves, the leaflets sit on a lip or shelf molded into the housing. This seating design may reduce regurgitation and the possibility for suture entrapment but do so at the expense of an increase gradient and higher closing impact. In the aortic model, the leaflets open to 77°, with a travel arc of 62°; these dimensions in the mitral model are 73° and 58°, respectively. The original Duromedics valve experienced some fractures of leaflets and housing; it was withdrawn from the market in 1988, reintroduced in 1990 as Carbomedics, the company that made pyrolitic carbon components, and in 1993 received the third marketing approval for a bileaflet valve in the United States. The Carbomedics bileaflet valve has flat leaflets that open 78–80°, with the resultant travel arc of 53–55°; it has a carbon-coated blood-contacting surface on the sewing ring.

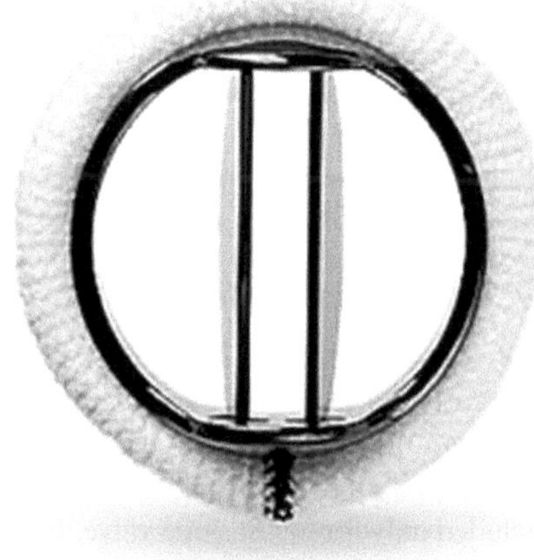

Fig. 17.16 St. Jude bileaflet valve (courtesy of St. Jude Medical, Inc.)

The Medtronic parallel bileaflet valve is unique in that the leaflets open to the maximum possible angle, 90°, to the plane of the housing with the travel arc of 50°. Unique design features include an active dual-mode pivotal washing and a housing profile optimized for flow.

17.8 Biological Valves

Biological valves include as wide a variety as mechanical valves.

1. **An autograft** is a valve that has been translocated within the same individual (e.g., the pulmonary valve in the aortic position).
2. **An autologous** tissue valve is a valve that has been fabricated from the patient's own nonvalvular tissue (e.g., pericardium).
3. **A homograft** valve is one that has been transplanted from a donor of the same species (a donor's aortic or pulmonary valve into a recipient's aortic or pulmonary position).
4. **A heterograft** is one that has been transplanted from another species; it may be either an intact valve [e.g., a valve constructed from heterologous tissue, e.g., bovine pericardium (Fig. 17.17)] or a porcine aortic valve (Fig. 17.18). The first successful biological valves were homografts. The homograft valve is not a homogeneous type of valve, but it has appeared in a range of subtypes according to many variable factors. Sterilization methods used include chemical (ethylene oxide, beta propiolactone), irradiation, and antibiotics, with antibiotics being favored today. Preservation for a short time (months) is accomplished with nutrient storage of 4 °C, but cryopreservation, which allows for indefinite storage, greatly increases the availability of homografts. Because of supply limitations with homografts, the most widely used valves are the partially manufactured heterograft valves. "Bioprosthesis" is a term Carpentier and Dubost introduced

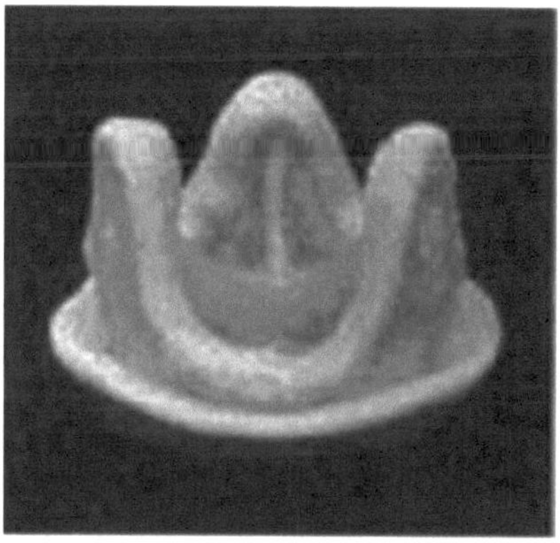

Fig. 17.17 Carpentier–Edwards bovine pericardial valve (courtesy of Medtronic Heart Valve Division)

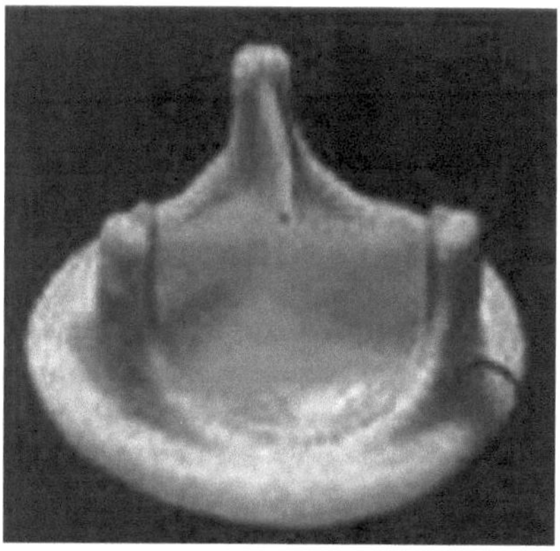

Fig. 17.18 Hancock porcine valve (courtesy of Baxter Edwards CVS)

for a biological tissue that has been treated to render it nonviable. Glutaraldehyde is used for fixing and preserving prosthetic heart valves because of three important biological actions: It sterilizes the tissue, renders it bioacceptable by destroying antigenicity, and stabilizes the molecular cross links between the collagen fibers to enhance durability.

17.8.1 Homograft, Autograft

The homograft valve is considered to be the preferred substitute for aortic valve replacement, especially for younger patients. It has excellent hemodynamics, no anticoagulant requirements, and low (or in some cases zero) thrombogenicity. The drawback is low availability and a more technically demanding operation. The pulmonary autograft procedure consists of an autotransplant of the pulmonary valve to the aortic position. The pulmonary valve is then replaced by an aortic or pulmonary homograft. The pulmonary autograft is perhaps the best aortic valve substitute for younger patients, as there is the potential for growth of the pulmonary valve in the aortic position, but this operation involves a double-valve replacement with the attendant early and late risks.

17.8.2 Autologous Pericardial Valve

An innovative valve concept has recently been developed and is being investigated. This is a new category of valve: an attempt to combine the reproducibility and the ease of insertion of the commercial stented heterograft valve and the benefits of autologous tissue. It is a frame-mounted autologous pericardial valve, which is assembled from a kit in the operating room. The kit consists of the tools to create the valve in a matter of minutes: a cookie-cutter-type tool for obtaining the correctly shaped piece of pericardium, a frame that snaps together around the tissue, and a holder to precisely align two pieces for assembly.

17.8.3 Porcine Heterograft Valves (Stented)

Most heterograft valves are mounted on rigid or flexible stents, to which are attached the leaflets and the sewing ring. Implantation involves fixing the sewing ring into place in (or above) the patient's annulus. The Hancock standard porcine valve (Medtronic) was the first commercially available porcine valve. The stent is made up of flexible polypropylene cylinder with the radio-opaque ring of cobalt-chromium alloy added for rigidity. The Hancock modified orifice was designed to overcome the undesirable hemodynamics caused by the muscular shelf of the porcine right coronary cusp by replacing that leaflet with one of the other two leaflets from another valve. The Hancock II valve incorporates second-generation features such as low-pressure fixation, calcification retardant treatment, and a thinner stent. The MO II valve is a modified Orifice valve with a modified scalloped sewing ring. The Medtronic's intact valve is distinctive in that the calcification-retardant treatment colors it blue. It is fixed in zero pressure, leaving the leaflets thinner and more flexible. Medtronic has recently introduced the Mosaicm valve, which incorporates features from both the intact valve and the Hancock II valve.

The Carpentier–Edwards standard porcine valve became available shortly after the Hancock valve and has been widely used. The frame of the valve is a flexible wire stent intended to reduce stresses on the leaflets and orifice yet retain its original contour over time. A one-piece, cylindrical, flexible mylar support surrounds a flexible wire frame. The annulus is asymmetrical rather than circular to incorporate the muscular septal ridge of the porcine right coronary cusp. In the Carpentier–Edwards supra-annular valve, the mounting structure of the aortic valve has been redesigned for the positioning above rather than within the annulus. The fixation treatment and the stent have been modified in an attempt to improve the leaflet's durability. The sewing ring was reconfigured to increase the effective orifice of the valve. The St. Jude BioImplant porcine valve is available internationally. Clinical investigation has begun on the X-Cell porcine valve, developed in conjunction with the St. Jude Medical and Hancock–Jaffe laboratories. The innovative design features of this valve include an extraction process to selectively remove calcification sites from the porcine tissue, sterilization by gamma irradiation treatment to reduce leaflet stiffness, and a clothless stent. It also features zero-pressure fixation and, in smaller (2.5-mm) sizes, a composite leaflet arrangement.

17.8.4 Porcine Heterograft Valves (Unstented)

The homograft is considered to have properties superior to those of the heterograft with regard to hemodynamics and thromboembolism. In an attempt to incorporate some of the advantages of a homograft into an easily available commercial product, several manufacturers have recently begun clinical testing of stentless porcine valves. This potential benefit is achieved at the expense of a more difficult implant technique. As with homografts, there are potentially three ways of implanting a stentless porcine valve:

1. as a replacement for the aortic root, with reimplantation of the coronary arteries
2. as a mini-root replacement, where the leaflets remain attached to the donor aortic wall, which is inserted within the host aorta
3. as a valve-only replacement, where the sides of the donor aorta are scalloped and the valve is sewn freehand into the subcoronary position in the host's aorta.

17.8.5 Bovine Pericardial Heterograft Valves

Pericardial valves are assembled using biological tissues as a fabric, rather than being harvested directly, as are porcine valves. The theoretical advantages include a more symmetrical and complete opening for optimal hemodynamics, the opportunity to allow extra tissue for eventual shrinkage, and a higher intrinsic percentage of collagen than in porcine valves. Since it is the collagen that is cross-linked during fixation with glutaraldehyde, a stronger and more durable tissue should result. The Ionescu–Shiley valve was the first commercially available pericardial valve, but it had an unacceptable rate of structural failure and was taken off the market. The Carpenter–Edwards pericardial bioprosthesis received FDA approval in 1991 and has become quite well accepted. It uses a sophisticated method of mounting the leaflets to the stent that does not depend on stitches passing through the leaflets. The leaflets are secured behind the stent pillar by a plastic plug, which serves as an anchor to prevent them from being pulled through the opening in the wire frame. An international model has a modified sewing ring, which is reinforced and more cone-shaped.

17.8.6 Complications

1. Structural deterioration refers to any change in valve function resulting from an intrinsic abnormality causing stenosis or regurgitation.
2. Nonstructural dysfunction: any abnormality resulting in stenosis or regurgitation that is not intrinsic to the valve itself. This includes inappropriate sizing, also called prosthesis–patient mismatch.
3. Thromboembolism includes any valve thrombosis or embolus except those secondary to infection or hemorrhage. This includes any neurological deficit and any peripheral arterial emboli unless proved to have resulted from another cause. Patients who do not awaken postoperatively or who awaken with a stroke or myocardial infarction are excluded. Valve thrombosis is listed as a subcategory of thromboembolism.
4. Anticoagulant-related hemorrhage includes any episode of internal or external bleeding (in patients taking anticoagulants or antiplatelets) that is fatal, causes a stroke, or is serious enough to require hospitalization or transfusion.
5. The diagnosis of prosthetic valve endocarditis is based on clinical criteria, including an appropriate combination of positive blood cultures and clinical signs or histological confirmation at reoperation or autopsy. Morbidity associated with active infection, such as thromboembolism or paravalvular leak, is included in this category only.

17.8.7 The Comparative Clinical Performance

The choice between valve types involves a tradeoff between an increased risk of thromboembolism-thrombosis-bleeding complex for mechanical valves versus the structural deterioration of tissue valves. In 1976, medical devices (including prosthetic heart valves) came under the jurisdiction of the FDA. The FDA then issued guidelines for Premarket Approval (PMA) applications for heart valves. In 1993, the FDA issued a guidance document based on objective performance criteria. This set the minimum amount of follow-up required for a PMA study at 800 valve-years.

The performance of mechanical valves has been noteworthy. The ball valve, in use for over 30 years, has had only a dozen structural problems that caused no major harm to the patient. The tilt valve had lower than a 1% rate of failures after 15 years of experience. The most popular type of bileaflet valve only reported several dozen failures to the FDA. However, in early 2000, one valve manufacturer recalled silver-coated valves because of a leaking problem in 2% of patients. In all, there have only been about 50 failures out of the approximately one million valves in service.

Approximately 265,000 prosthetic valves are now implanted worldwide each year, valued at over $700 million. About 60% of these are mechanical valves, with a market value of around $400 million. Over two million mechanical valves have been implanted in patients around the world during the last several decades.

17.9 Structural Deterioration of Biological Valves

Structural valve deterioration with tissue valves is not a constant risk event but increases with time. Thus, linearized rates are not appropriate and actuarial methods must be used to describe and compare them.

17.9.1 Porcine Valves

Stented porcine prostheses represent by far the most commonly used biological valves. Freedom from structural deterioration at 10 years ranges from about 60–90% for the aortic position and from about 60–80% for the mitral position. Conventional replacement therapies for heart valve disease are associated with significant drawbacks. The field of tissue engineering has emerged as an exciting alternative in the search for improved heart valve replacement structures. One of the principles behind this concept is the transplantation of living elements, embedded in a suitable scaffold material, to the diseased site where the structure becomes integrated with patients' tissue to restore natural function. Significant progress was made in the 1990s in the development of a living artificial heart valve alternative (LAHVA), with the identification of potential replacement sources for valve cells (Fig. 17.19).

17.9.2 Pericardial Valves

After an initial unsatisfactory experience with the Ionescu–Shiley valve, which did not perform exceptionally but was not equal to the standards of other contemporary bioprostheses, the Carpenter–Edwards valve seems to have revived the concept of bovine pericardium as an acceptable alternative for valve fabrication. There has now been over 20 years' experience with it, it has received FDA approval for marketing in the aortic position, and the results have been encouraging.

17.9.3 Homografts

Thromboembolism rates for homograft valves are considered quite low or even zero by some investigators. For structural deterioration, homografts should be evaluated according to the various methods of sterilization and preservation. It appears that the series that used a chemical or irradiation process for sterilization have the highest rates of failure, about 40% structural valve deterioration-free at 10 years. Those sterilized by antibiotics, whether stored in nutrient solution or cryopreserved, are more durable, with about 75% structural valve deterioration-free at 10 years.

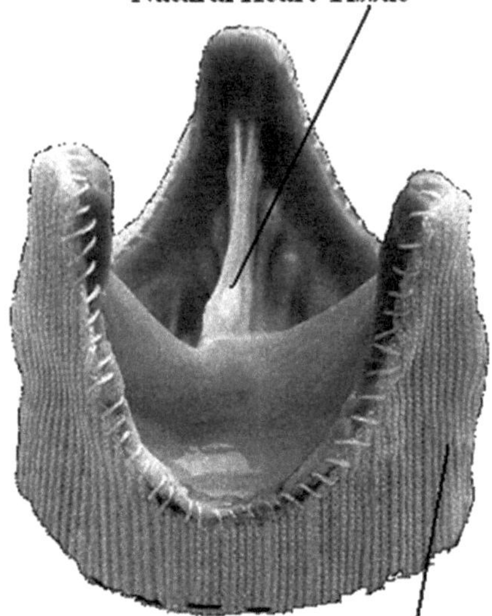

Fig. 17.19 Natural heart valve with polymer scaffold

17.9.4 Pulmonary Autograft

The pulmonary autograft used as the aortic valve replacement is considered a "permanent" valve and is especially appropriate for young patients. Excellent results have also been reported in treating endocarditis. Freedom from replacement for all reasons including endocarditis has been reported to be as high as 85% at 20 years but has also been found to be 48.5% at 19 years.

17.9.5 Materials of Construction and Manufacture

Most artificial valves are made of titanium, graphite, pyrolytic carbon, and polyester. The titanium is used for the housing or outer ring graphite-coated with pyrolytic carbon is used for the bileaflets, and 100% pyrolytic carbon is used for the inner ring. The pyrolytic carbon is sometimes impregnated with tungsten so that the valve can easily be seen following implantation. The sewing cuff, used to attach the valve to the heart, is made out of double-velour polyester.

Titanium is used for its strength, low density, and biocompatibility. The outer rings are made from machined bar stock. Lock rings and wire, used to hold the cuff in place, are also made from titanium. The polyester comes in the form of tubes. All plastic components are deburred by the supplier, which involves removing any bumps from the surface. Occasionally, the valve manufacturer may have to deburr some parts.

The pyrolytic carbon coating is produced by depositing gaseous hydrocarbons (usually methane) onto a heated graphite substrate at temperatures of 1,800–2,300 °C

(3,272–4,172 °F) in a chamber. These gases break down into carbon. The inner rings are made from 100% pyrolytic carbon using a fluidized bed process at another manufacturer. This material's atomic microstructure helps resist cracking, making it ductile. However, the processing method can still introduce micro cracks that must be detected.

The polyester cuffs are made by a sewing process that includes various looping, folding, and stitching steps. The manufacturing process therefore consists mainly of various assembly and inspection steps.

17.9.6 Assembly

The assembly of parts takes place in a clean room to avoid contamination. The leaflets are attached to the inner rings, which are then placed in the housing or outer ring.

While this is being done, the sewing cuffs are being made. A special pressurized heating process is then used to form the cuffs around the valve, which is done at several hundred degrees. The valves are then mounted into a rotator assembly, which the surgeon uses for implanting.

17.9.7 Sterilization and Packaging

After the valves are assembled and tested, they are sterilized in a double-plastic container. Steam sterilization is used, which involves temperatures up to 132 °C (270 °F) for 15 min or more. To make sure the sterilization process is in order, a biological indicator is placed inside. If the indicator shows no growth of bacteria or other viable organisms, the valves and its packaging have been properly sterilized. Each plastic-encased valve is then packaged in a box for shipping.

17.9.8 Quality Control

Usually, all components are inspected visually, dimensionally, and functionally prior to assembly to make sure they meet specifications. The diameter of each ring is measured and assigned a size, which is then matched to the appropriate bileaflet to make sure they will fit together. High-power microscopic analysis is used to check components for scratches. In total, up to 50 inspections are made during the assembly process.

Proof testing is used to determine the structural quality of potentially flawed heart valves. In this method, a valve is loaded to a certain stress level using a special pressurization fixture to see if it will fail at this stress. During the stress test, acoustic emission technology is used to detect minute cracks that might go undetected so that these valves can be rejected. Then the valves are sterilized and packaged.

17.9.9 The Future

Blood clotting is still a problem with mechanical valves, and manufacturers continue to improve designs, sometimes using super-computing modeling tools, as well as surgical procedures. The shape of the orifice is being improved to reduce pressure losses, turbulence, and shear stresses. The flow area is maximized by using stronger materials, which minimizes the wall thickness. Tapering the sides of the valve pumps blood more efficiently. Operations are also being developed that only require an 8–10-cm (3–4-in.) incision instead of 12 in. (30 cm). Manufacturing efficiencies will continue to improve.

Researchers are looking at making heart valves out of plastic material that are flexible and strong enough to perform satisfactorily over a long time.

17.10 Failure Modes in Mechanical Valves

Problems that interfere with the successful performance of valves can be grouped as listed below:

1. Degradation of valve components.
2. Structural failure.
3. Clinical complications associated with the valve.
4. Clinically, valve failure has been considered to be present if any of the following events occur:
   a) Require reoperation and/or cause death
   b) Anticoagulant-related hemorrhage (ACH)
   c) Prosthetic valve occlusion (thrombosis or tissue growth)
   d) Thromboembolism
   e) Prosthetic valve endocarditis (PVE)
   f) Hemodynamic prosthetic dysfunction, including structural failure of prosthetic components (strut failure, poppet escape, ball variance)
   g) Reoperation for any other reason (e.g., hemolysis, noise, incidental)

The performance of mechanical valves is in several ways related to valve design and structural mechanics. The design configuration affects the load distribution and dynamics of the valve components, which in conjunction with the material properties determine the durability and successful performance of the valve. The flow engendered by the geometry of the components determines the extent of flow separation and high-shear regions. The hinges in the bileaflet and tilting disk valves can produce regions of flow stagnation, which may cause localized thrombosis, which may in turn restrict occluder movement.

17.11 Choice of Valves

The embolic-thrombosis-bleeding complex with mechanical valves and structural failure with biological valves serve to distinguish between the two valve types. But, judging from the wide variation in reported results with each model of valve, patient-specific factors must influence the results more than valve-specific ones, and it is impossible to rank valves, within valve types, on the basis of complication rates. However, some general recommendations can be made with regard to valve selection. Though not covered in this section, valve repair, when practical, should be considered preferable to replacement, especially in the mitral position, but also in the aortic positions. When replacement is necessary, an argument can be made for a particular class of valve under certain circumstances. A biological valve should be used when the patient cannot or will not take anticoagulants, desires pregnancy, or has a short life expectancy. A mechanical valve should be used if the patient will be on anticoagulants anyway (because of a trial fibrillation or mechanical valve in another position), is in renal failure or on dialysis, or has a long life expectancy. Mechanical valves should also be considered first for double-valve replacement, because the thromboembolic risk is not an additive with two valves, but the risk of structural deterioration is additive.

17.12 Future Developments

The current trend in mechanical valves is toward further development of the bileaflet valve principle enhancing the very successful St. Jude valve design. New directions include the search for better hemodynamics, for example, parallel leaflets and lower thrombogenicity. Better anticoagulant management has the potential to reduce bleeding complications and also to reduce thromboembolic events. If markers of thrombogenicity can be identified and measured preoperatively, tissue valves can be preferentially used in high-risk patients and mechanical valves in low-risk patients, even older ages. The primary advantage of biological valves is a reduced need for anticoagulant, but the offsetting disadvantage is poorer durability. The search for improved durability will define the newer generation of biological valves. A major factor is low- or zero-pressure fixations, which allow the leaflets to be fixed in the neutral position and to retain more of the natural flexibility. Anticalcification treatments may improve durability, and the elimination of the stent should provide improved hemodynamics.

Recent trends in the choice of materials indicate a preference toward soft occluder materials. One team in Germany is working toward bileaflet valves with soft occluders. Medtronic–Hall also has announced that it will be looking for a valve with a soft occluder in the near future. The advantages of using soft occluder material are many. They absorb the impact forces generated during valve closure, thereby reducing the chance of suture dehiscence. The reduction in the impact forces also reduces the load that needs to be transferred to the surrounding tissues through the suture ring, reducing the irritation caused by the continuous movement at the cloth–metal interface. Another improvement caused by the soft occluder is the reduction in the probability of occurrence of cavitation and damage related to it.

List for Further Study

1. Hufnagel CA, Harvey WP, Rabil PJ, McDermott TF (1954) Surgical correction of aortic insufficiency. Surgery 35:673–683
2. Campbell JM (1950) An artificial aortic valve. J Thorac Surg 19:312–318
3. Roberts WC (1976) Choosing a substitute cardiac valve; type, size, surgeon. Am J Cardiol 38:633
4. Ajit P Yoganathan (1995) Cardiac valve prosthesis. In: Handbook of biomedical engineering. CRC Press, New York, 123: 1847–1870
5. Bain WH, Nashef SAM (1991) Tilting disk valves. In: Bodnar E, Frater R (eds), Replacement cardiac valves. Toronto, Pergamon, 8, 187–200
6. Horstkotte D, Bodnar E (1991) Bileaflet valves. In: Bodnar E, Frater R (eds) Replacement cardiac valves. Pergamon, Toronto, 9, 201–228
7. Bhuvaneshwar GS, Muraleedharan CV, Ramani AV, Valiathan MS (1991) Evaluation of materials for artificial heart valves. Bull Mater Sci 14:1361–1374, 8
8. Flanagan TC, Pandit A (2003) Living artificial heart valve alternatives: a review. Eur Cell Mater 6:28–45

Problems

1. Show in a schematic diagram the mechanical equivalent of the human heart, indicating the approximate volume of each chamber and the location of valves and their features. What conditions of the heart valve necessitate its replacement?
2. How do you propose to design a mechanical heart valve for replacement of the mitral valve? What materials are to be used for each part, and how will you fix it with the heart tissue? What tests are necessary to predict the life expectancy of such a valve? Could you suggest the manufacturing aspect of the product?
3. Discuss the differences between different types of mechanical valves and tissue valves. Indicate their advantages and disadvantages, life expectancy, blood coagulation, complications, and other relevant aspects.

Example from the Literature for Valve Area Calculation

Aortic Valve Area Calculation

Aortic valve area calculation is an indirect method of determining the area of the aortic valve. The calculated aortic valve orifice area is currently one of the measures for evaluating the severity of aortic stenosis. A valve area of less than 0.8 cm^2 is considered to be severe aortic stenosis [1, 2]. There are many ways to calculate the valve area of aortic stenosis. The most commonly used methods involve measurements taken during echocardiography. For interpretation of these values, the area is generally divided by the body surface area, to arrive at the patient's optimal aortic valve orifice area.

Planimetry is the tracing of the opening of the aortic valve in a still image obtained during echocardiographic acquisition during ventricular systole, when the valve is supposed to be open. While this method directly measures the valve area, the image may be difficult to obtain due to artifacts during echocardiography, and the measurements are dependent on the technician, who has to manually trace the perimeter of the open aortic valve. Because of these reasons, planimetry of the aortic valve is not routinely performed.

The continuity equation states that the flow in one area must equal the flow in a second area if there are no shunts in between the two areas. In practical terms, the flow from the left ventricular outflow tract (LVOT) is compared to the flow at the level of the aortic valve. Using echocardiography, the aortic valve area is calculated using the time–velocity integral, which is the most accurate and preferred method.

The Gorlin equation states that the aortic valve area is equal to the flow through the aortic valve during ventricular systole divided by the systolic pressure gradient across the valve times a constant. The flow across the aortic valve is calculated by taking the cardiac output (measured in ml/min) and dividing it by the heart rate (to give output per cardiac cycle) and then dividing it by the systolic ejection period measured in seconds per beat (to give the flow per ventricular contraction):

$$\text{Valve area (cm}^2\text{)} = \frac{\text{Cardiac output}\left(\frac{\text{ml}}{\text{min}}\right)}{\text{Heart rate}\left(\frac{\text{beats}}{\text{min}}\right)\ \text{Systolic ejection period (s)}\ 44.3\ \sqrt{\text{mean gradient (mmHg)}}}.$$

The Gorlin equation is related to flow across the valve. Because of this, the valve area may be erroneously calculated as stenotic if the flow across the valve is low (i.e., if the cardiac output is low). The measurement of the true gradient is accomplished by temporarily increasing the cardiac output by the infusion of positive inotropic agents, such as dobutamine.

Example: An individual undergoes left and right heart cardiac catheterization as part of the evaluation of aortic stenosis. The following hemodynamic parameters were measured. With a heart rate of 80 beats/min and a systolic ejection period of 0.33 s, the cardiac output was 5 l/min. During simultaneous measurement of pressures in the left ventricle and aorta (with the use of one catheter in the left ventricle and a second in the ascending aorta), the mean systolic pressure gradient was measured at 50 mmHg. What is the valve area as measured by the Gorlin equation?

Answer:

$$\text{Aortic valve area} = \frac{5{,}000\,\frac{\text{ml}}{\text{min}}}{80\,\frac{\text{beats}}{\text{min}} \bullet 0.33\,\text{s} \bullet 44.3 \bullet \sqrt{50\,\text{mmHg}}} \approx 0.6\,\text{cm}^2$$

The Hakki equation given below is a simplification of the Gorlin equation, relying on the observation that in most cases the numerical value of heart rate (bpm) •systolic ejection period (s)•44.3 1,000.

The resulting simplified formula is

$$\text{Aortic valve area } (\text{cm}^2) \approx \frac{\text{Cardiac output}\left(\frac{1}{\text{min}}\right)}{\sqrt{\text{mean gradient (mmHg)}}}.$$

Example: An individual undergoes left and right cardiac catheterization for the evaluation of aortic stenosis. Measurements include an aortic pressure of 120/60, an LV pressure of 170/15, and a cardiac output of 3.5 l/min. What is the aortic valve area?

Answer: The peak gradient between the LV and aorta is (170–120) 50 mmHg. This gives

$$\text{Aortic valve area} \approx \frac{3.5}{\sqrt{50}} \approx 0.5\ \text{cm}^2.$$

Some relevant references for further study on this methodolgy:

1. Varadarajan P, Kapoor N, Bansal RC, Pai RG (2006) Survival in elderly patients with severe aortic stenosis is dramatically improved by aortic valve replacement: results from a cohort of 277 patients aged >/=80 years. Eur J Cardiothorac Surg 30(5): 722–728
2. Hakki A, Iskandrian A, Bemis C, Kimbiris D, Mintz G, Segal B, Brice C (1981) A simplified valve formula for the calculation of stenotic cardiac valve areas. Circulation 63(5): 1050–1055.

Chapter 18
The Kidney and Its Artificial Replacement

18.1 Introduction

The main function of the pair of kidneys in the human body is to form urine out of blood plasma, a function that basically consists of two processes:

1. Removing waste products from blood plasma
2. Regulating the composition of the blood plasma.

These activities not only lead to the excretion of nonvolatile metabolic waste products but also are responsible for the remarkable constancy of the volume, osmotic pressure, pH, and electrolyte composition of the extracellular body fluids.

The kidneys lie in the back of the abdominal cavity just below the diaphragm, one on each side of the vertebral column. Each kidney consists of about a million individual units, all similar in structure and function. These tiny units are called nephrons, whose structure is shown in Fig. 18.1. A nephron is composed of two parts—a cluster of capillary loops called the glomerulus and a tubule. The tubule runs a tortuous course and ultimately drains via a collecting duct into the funnel-shaped expansion of the upper head of the urethra.

The kidney works only on plasma. The erythrocytes (RBC) supply oxygen to the kidneys but serve no other function in urine formation. Each substance in plasma is handled in a characteristic manner by the nephron, involving particular combinations of filtration, reabsorption, and secretion.

The renal arteries carry blood at a very high pressure from the aorta into the glomerular capillary tuft, which is controlled by the contraction state of the muscle of the arterioles leading to the tuft. The fluid pressure within the tuft forces some of the fluid part of the blood, by filtration, through the thin walls of the capillary into the glomerulus and on into the tubule of the nephron. The glomerulus filtration consists of blood plasma without proteins. The total amount of glomerular filtration is about 180 l per day, whereas the amount of urine formed from it is only 1–1.5 l. This means that very large amounts of water and other substances are reabsorbed by the kidney tubules. The reabsorption is partly an automatic process, because the absorption of

S. Pal, *Design of Artificial Human Joints & Organs*,
DOI 10.1007/978-1-4614-6255-2_18,

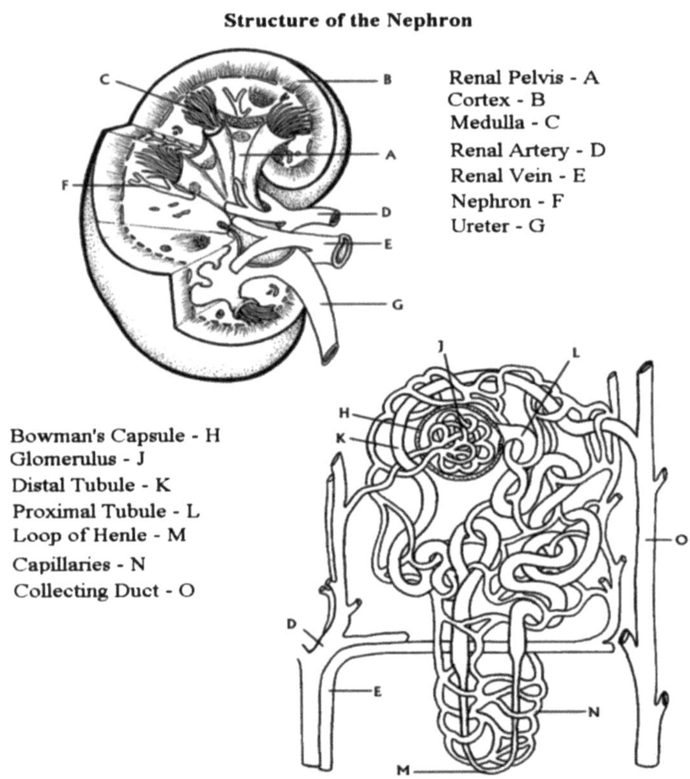

Fig. 18.1 The schematic of a kidney, indicating its structure

water is accurately controlled by the antidiuretic hormones of the pituitary gland, in relation to the body fluids' need for water. The absorption of electrolytes, such as sodium and phosphate, is partly controlled by the suprarenal gland; the connection of others, like chloride and bicarbonate, is related to the acid–base balance. Some of the reabsorption from the glomerular filtration is also a passive, automatic process of diffusions depending upon pressure gradients. This applies to water itself and to electrolytes, including sodium potassium, chloride, calcium, and bicarbonate.

There are other substances, such as urea, phosphate, and sulfates, that are the waste products of metabolism. They are unwanted by the body. The tubules are selectively porous to substances of importance to the body and impermeable to the unwanted. Therefore, the unwanted substances cannot diffuse back into the plasma, and a large proportion is thus execreted in the urine. As the filtrate passes down the tubules, the concentration of waste products rises steadily; the specific gravity of normal urine varies between 1.015 and 1.030 as compared with 1.010 for glomerular filtrate. Other substances in the glomerular filtrate such as glucose and amino acid show little tendency to diffuse through the tubular walls and are returned to the plasma by a process of active reabsorption.

The total blood flow through the kidney is about 1.2 l/min. The total extracellular fluid amounts to about 15 l. The blood plasma and the extracellular fluids are in equilibrium with each other. Therefore, an amount of blood equivalent to all the cellular fluids can pass through the kidneys once every 15 min. The water and electrolyte content of the blood plasma that is the extracellular fluid are closely controlled by the kidneys. Kidneys also play an important role in maintaining the acid–base balance.

18.2 How Do Kidneys Help Maintain Health?

In addition to removing wastes and fluid from body, kidneys perform these other important functions:

- Regulate body water and other chemicals in blood such as sodium, potassium, phosphorus, and calcium.
- Remove drugs and toxins introduced into body.
- Release hormones into blood to help the body

  1. Regulate blood pressure.
  2. Make red blood cells.
  3. Promote strong bones.

18.3 Diseases

18.3.1 Chronic Kidney Disease (CKD) and Related Facts

- Twenty six million Americans—1 in 9 U.S. adults—have CKD, and another 20 million more are at increased risk [1]. The data for India are not available, but CKD is on the rise due to the adoption of Western-type eating and drinking habits and many other reasons, including a sedentary lifestyle.
- Early detection can help prevent the progression of kidney disease to kidney failure.
- Heart disease is the major cause of death for all people with CKD.
- Glomerular filtration rate (GFR) is the best estimate of kidney function.
- Hypertension causes CKD, and CKD causes hypertension.
- Persistent proteinuria means CKD.
- High-risk groups include those with diabetes, hypertension, and a family history of kidney disease.
- Three simple tests can detect CKD: **blood pressure, urine albumin, and serum creatinine**.

18.3.2 What Is Chronic Kidney Disease?

Chronic kidney disease (CKD) includes conditions that damage the kidneys and decrease their ability to keep us healthy by doing the jobs listed. If kidney disease gets worse, wastes can build to high levels in our blood and make us feel sick. We may develop complications such as high blood pressure, anemia (low blood count), weak bones, poor nutritional health, and nerve damage. Also, kidney disease increases the risk of having heart and blood vessel disease. These problems may happen slowly over a long period of time. Chronic kidney disease may be caused by diabetes, high blood pressure, and other disorders. Early detection and treatment can often keep chronic kidney disease from getting worse. When kidney disease progresses, it may eventually lead to kidney failure, which requires dialysis or a kidney transplant to maintain life.

18.3.3 What Causes CKD?

The two main causes of chronic kidney disease are diabetes and high blood pressure, which are responsible for up to two thirds of the cases. Diabetes happens when blood sugar is too high, causing damage to many organs, including the kidneys and heart, as well as blood vessels, nerves, and eyes. High blood pressure, or hypertension, occurs when the pressure of blood against the walls of the blood vessels increases. If uncontrolled, or poorly controlled, high blood pressure can be a leading cause of heart attacks, strokes, and chronic kidney disease. Also, chronic kidney disease can cause high blood pressure.

Other conditions that affect the kidneys are

- Glomerulonephritis, a group of diseases that cause inflammation and damage to the kidney's filtering units. These disorders are the third most common type of kidney disease.
- Inherited diseases, such as polycystic kidney disease, which causes large cysts to form in the kidneys and damage the surrounding tissue.
- Malformations that occur as a baby develops in its mother's womb. For example, a narrowing may occur that prevents normal outflow of urine and causes urine to flow back up to the kidney. This causes infections and may damage the kidneys.
- Lupus and other diseases that affect the body's immune system.
- Obstructions caused by problems such as kidney stones, tumors, or an enlarged prostate gland in men.
- Repeated urinary infections.

18.3.4 The Symptoms of CKD

Most people may not have any severe symptoms until their kidney disease is advanced. However, a person may notice that he or she

- Feels more tired and has less energy
- Has trouble concentrating
- Has a poor appetite
- Has trouble sleeping
- Has muscle cramping at night
- Has swollen feet and ankles
- Has puffiness around eyes, especially in the morning
- Has dry, itchy skin
- Needs to urinate more often, especially at night.

Anyone can get chronic kidney disease at any age. However, some people are more likely than others to develop kidney disease. The risk for kidney disease is increased in patients who

- Have diabetes
- Have high blood pressure
- Have a family history of chronic kidney disease
- Are of an advanced age
- Belong to a population group that has a high rate of diabetes or high blood pressure, such as African Americans, Hispanic Americans, Asians, Pacific Islanders, and American Indians. Currently, in India, there is a high rate of increase in diabetes and high BP and to some extent CKD as well.

18.4 Dialysis

Dialysis is a treatment that does some of the things that healthy kidneys do. It is needed when a patient's own kidneys can no longer take care of his or her body's needs. Patients need dialysis when they develop end-stage kidney failure—usually by the time they lose about 85–90% of their kidney function.

Like healthy kidneys, dialysis keeps the body in balance, by doing the following:

- **Removing waste, salt, and extra water** to prevent them from building up in the body
- Keeping a safe level of certain chemicals in the blood, such as potassium, sodium, and bicarbonate
- Helping to control blood pressure.

18.4.1 Kidney Failure

Kidney failure is not always permanent. Some kinds of acute kidney failure get better after treatment. In some cases of acute kidney failure, dialysis may only be needed for a short time until the kidneys get better.

In chronic or end-stage kidney failure, the patient's kidneys do not get better, and he or she will need dialysis for the rest of his or her life. Suitable candidates may choose to be placed on a waiting list for a new kidney. Dialysis can be done in a hospital, in a dialysis unit outside the hospital, or at home.

There are two types of dialysis—**hemodialysis and peritoneal dialysis**.

18.4.2 Hemodialysis

In hemodialysis, an artificial kidney (hemodialyzer) is used to remove waste and extra chemicals and fluid from the blood. To get blood into the artificial kidney, the doctor needs to make an access (entrance) into blood vessels. This is done by minor surgery to the arm or leg.

Sometimes, an access is made by joining an artery to a vein under the skin to make a bigger blood vessel, called a fistula. However, if the blood vessels are not suitable for a fistula, the doctor may use a soft plastic tube to join an artery and a vein under the skin. This is called a graft.

Occasionally, an access is made by means of a narrow Silastic tube, called a catheter, which is inserted into a large vein in the neck. This type of access may be temporary but is sometimes used for long-term treatment.

The time needed for dialysis depends on several factors:

- How well the kidneys work
- How much fluid waste is gained between treatments
- How much waste is in the body
- How big the patient is
- The type of artificial kidney used.

Usually, each hemodialysis treatment lasts **about 4 h and is done three times per week**. A type of hemodialysis called high-flux dialysis may take less time.

18.4.3 Peritoneal Dialysis

In peritoneal dialysis, blood is cleansed inside the body. The doctor will do surgery to place a plastic tube called a catheter into the abdomen (belly) to make an access. During the treatment, the abdominal area (called the peritoneal cavity) is slowly filled with dialysate through the catheter. The blood stays in the arteries and veins that line the peritoneal cavity. Extra fluid and waste products are drawn out of the blood and into the dialysate.

There are several kinds of peritoneal dialysis, but the two major ones are **continuous ambulatory peritoneal dialysis (CAPD) and continuous cycling peritoneal dialysis (CCPD).**

Continuous ambulatory peritoneal dialysis (CAPD) is the only type of peritoneal dialysis that is done without machines. A patient usually does this four or five times a day at home and/or at work. A bag of dialysate (about two quarts) is put into peritoneal cavity through a catheter. The dialysate stays there for about 4 or 5 h before it is drained back into the bag and thrown away. This is called an exchange. A new bag of dialysate is used with each exchange. While the dialysate is in the peritoneal cavity, the patient can go about his or her usual activities at work, school, or home.

Continuous cycling peritoneal dialysis (CCPD) usually is done at home using a special machine called a cycler. This is similar to CAPD except that a number of cycles (exchanges) occur. Each cycle usually lasts 1 1/2 h and exchanges are done throughout the night, while the patient is asleep.

18.5 The Artificial Kidney

18.5.1 Parallel Flow Dialyzers

The parallel flow dialyzer has a low internal resistance, which allows adequate blood flow through the dialyzer with the patient's arterial blood pressure, eliminating the need for a blood pump. The dialyzing surface area of a parallel flow dialyzer is about 1 m^2 at a blood flow rate of 200 ml/min and a dialyzer flow of 500 ml/min. The urea and creatinine clearance are about 80 and 64 ml/min, respectively. The rigid supports used in parallel flow dialyzers permit negative pressure to be created on the dialysate side of the membrane for ultrafiltration. The water is ultrafiltered at a rate of 9.2 ml/min, with a negative pressure of 130 mmHg. The rate is 1.8 ml/min without negative pressure. The dialysate flows continuously at 500 ml/min in a direction countercurrent to the blood, permitting exchange to take place throughout the dialyzer (Figs. 18.2 and 18.3).

18.6 Coil Hemodialyzer

A coil hemodialyzer comprises a tubular membrane placed between flexible supports wrapped around a rigid cylindrical core. The coil is immersed in a dialyzing bath. The tubular membrane can be of cellophane or cuprophane. The average wall thickness of the cellophane membrane is 20–30 μm and that of cuprophane in the range of 18–75 μm. The coil membrane support is woven screens or unwoven lattice usually. The twin coil is made with three layers of woven polyvinyl chloride-coated fiber glass screen separated by four narrow strips of the same material, which are sewn into place with cotton thread. Coil dialyzers are available with several design variations:

1. The type of membrane
2. The membrane support

3. The number of blood channels (1, 2, or 4)
4. The width of the blood channel (38–100 mm)
5. The surface area (0.7–1.9 m^2).

They are characterized by high dialysis flow rates and a high resistance to blood.

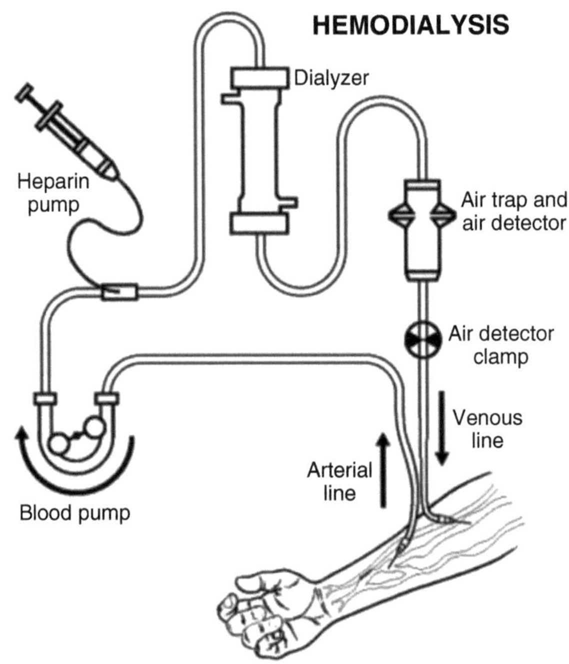

Fig. 18.2 Schema of an artificial kidney machine

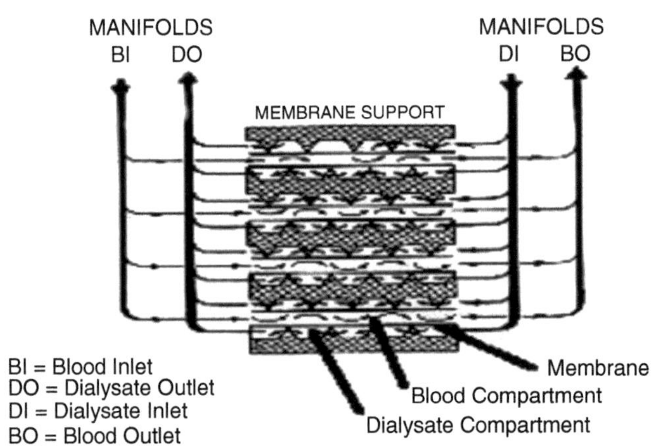

Fig. 18.3 Schema of membrane dialyzer

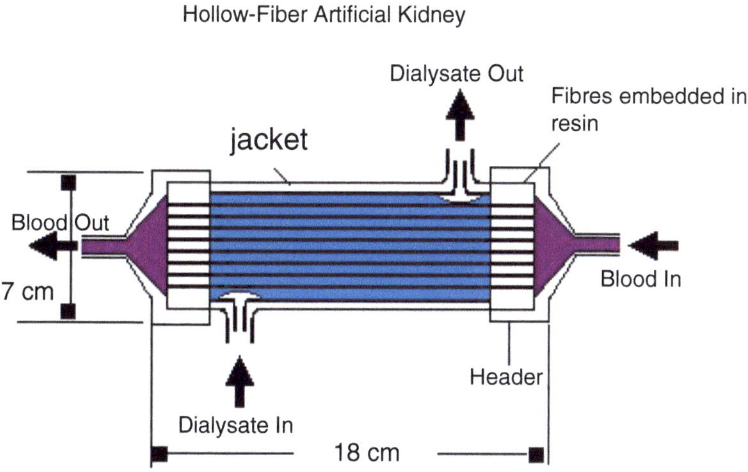

Fig. 18.4 Schema of hollow fiber artificial kidney

18.6.1 Hollow Fiber Hemodialyzer

Hollow fibers are (Fig. 18.4) most commonly used in a hemodialyzer. A hollow fiber hemodialyzer consists of about 10,000 hollow de-acetylated cellulose dilacerate capillaries. The capillaries are jacked in a plastic cylinder 18 cm in length and 7 cm in diameter. The capillaries are seated on each end into a tube sheet with an elastometer. The capillary have 200–300-μm internal diameters and a wall thickness of 25–30 μm. The dialyzing area is approximately 9,000 cm^2/unit. The primary volume with blood manifolds exclusive of tubing is approx 130 ml. The blood is introduced and removed from the hemodialyzer through manifold headers. The dialysate is drawn through the jacket under a negative pressure around the outside of the capillaries countercurrent to the blood flow. The dialyzers are disposable. Disposal dialyzers offer the advantage of reduced risk of infection and reduced operator setup time. The dialyzer sterilization procedure is also eliminated. However, the use of disposable dialyzers is an expensive procedure. This has necessitated the development of a **method of cleaning dialyzer cartridges so that they may be reused**. However, there are several difficulties associated with the practice of reusing dialyzers (Fig. 18.4).

18.7 Performance Analysis of Dialyzers

The dialyzers' performance can be compared in terms of their clearance of urea and creatinine, priming volume, residual blood volume, ultrafiltration rate, convenience of handling, and cost.

Fig. 18.5 Dialysance vs. blood flow

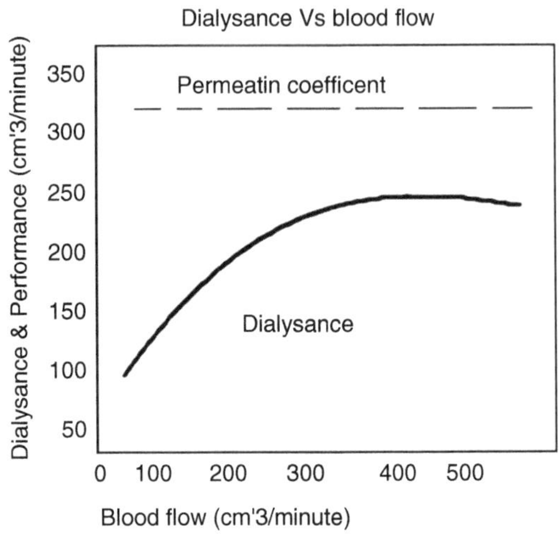

Clearance: The overall performance of a dialyzer is expressed as the clearance, analogous to that of a natural kidney. It represents the part of the total blood flow rate through the dialyzer that is completely clear of solute urea and the number of toxic solutions in their blood, which are generated daily. Despite the uncertainty as to which solutes and how much should be removed, the performance of dialyzers is generally as for a spectrum of molecular weight solution. **The molecular weight of urea generally is 60, that of creatinine 113,** that of **vitamin B12 1355, and that of insulin 5200.**

The clearance of urea and creatinine is measured at clinically useful blood flow rates and standard dialysis fluid addition or flow rates. It is calculated as

$$\text{Clearance} = \text{blood flow rate} / (\text{A} + \text{Bx blood flow rate}).$$

A and B are constants with 95% confidence limits of the mean by least squares approximation. The blood flow rates are measured by bubble transmit time over a 2-m track using the mean of three measurements. Urea and creatinine concentrations are measured in the plasma from a 1-ml sample of heparinized blood (Fig. 18.5). Usually, blood flow is maintained between 75–300 ml/min and dialysate fluid flow rates at 500 ml/min.

The performance of the hemodialyzer is usually given by dialysance, D**:**

$$D = Q_b (C_{bi} - C_{bo}) / (C_{di} - C_{bo}),$$

where

Q_b = blood flow rate

C_{bi} = blood solute concentration at the dialyzer inlet

C_{bo} = blood solute concentration at the dialyzer outlet

C_{di} = dialysate solute concentration at the inlet.

The dialysance is also calculated from the clearance, by the following relationship:

$$\text{Clearance} = \text{dialysance} / (1 + \text{dialysance} / \text{dialysis fluid addition rate}).$$

It may be noted that the clearance may vary with time despite a quasi-steady-state condition. If the dialysate is recirculated, its solute concentration increases, which effectively reduces the concentration driving force. For a given blood flow rate, the clearance is greater for the smaller-molecular-size constituents. This is due to less membrane resistance and a higher liquid diffusion coefficient for smaller-molecular-weight solutes. On the other hand, the contribution of the membrane resistance value becomes greater with the increase in solute molecular weight. Keeping these facts in mind, scientists have tried to design a dialyzer with a large surface area and develop more permeability membranes.

$$\text{Performance capacity} = \text{K}.A = A / \text{R}$$

where
A = surface area
K = permeability coefficient
R = mass-transfer resistance.

18.7.1 *Ultrafiltration Rate*

The fluid removal during dialysis takes place due to hydrostatic and osmotic **trans-membrane pressure gradients**. The rate of fluid removal due to hydrostatic pressure effects depends upon the specification of the dialyzer in terms of mass-transfer coefficient and surface area. However, it has a linear function of the transmembrane pressure gradient.

$$\text{Mean transmembrane pressure} = 1/2\left[P_{bi} + P_{bo}\right] - 1/2\left[P_{di} + P_{do}\right]$$

where
P_{bi} = blood inlet pressure
P_{bo} = blood outlet pressure
P_{di} = dialysate inlet pressure
Pdo = dialysate outlet pressure.

The pressure loss generated by blood and dialysate flows in their respective flow paths should be small. This ensures that the local transmembrane pressure (ΔPm) will not vary excessively from the mean pressure. High values of ΔPm can result in deformation of the membrane and possible rupture. The pressure drop ΔPm across a dialyzer is directly proportional to the length of the passage and the viscosity of the fluid and inversely proportional to the number of blood passages and some function of their cross-sectional area. The relationship of the pressure drop to blood flow is not linear at increased flow, which is accompanied by an increase in pressure that can cause a widening of the blood pressure passage and a decrease in ΔPm/Qb.

18.7.2 Residual Blood Volume

Residual blood volume measured after an 800-ml saline wash in the fluid remaining in the dialyzer and flow lines is circulated through a 1 l bottle of 0.04% ammonium solution for 10 min. The residual blood volume is calculated from the formula

$$\textbf{Residual blood volume} = U\left(\textbf{1,000} + \textbf{volume of dialyzer and lines in ml}\right) / \textbf{200}S$$

where
U = the hemoglobin concentration of the recirculated fluid
S = the hemoglobin concentration of a sample of arterial blood taken at the end of dialysis and diluted 1:200 with 0.04% ammonia.

Residual blood volumes of hemoglobin of 1.8–6.3 ml are quoted in the literature depending upon the dialyzer type and wash-back volume. Here line means flow lines through dialyzer tubings.

18.7.3 Priming Volume

The volume of the blood within the dialyzer is known as the priming volume. It is desirable that this should be minimal. The priming volume of a present-day dialyzer ranges between 75–200 ml, depending on the membrane area geometry and operating conditions. The requirement of a low priming volume permits the use of the patient's own blood to prime the circuit without serious hypovolemic effects. This is particularly significant in the case of long-term dialysis therapy.

Extracorporeal blood volumes become important with those dialyzers requiring priming. Priming is usually accompanied at relatively low pressures. Recent innovations have largely required the extracorporeal volume, and saline priming is frequently used.

18.7.4 Pyrogenicity

Pyrogen (body temperature riser) reactions are rare with all disposable dialyzers. However, they are known to exist with dialyzers, but at rates well lower than 1%.

18.7.5 Leakage Rate

Blood-to-dialysis fluid leak with the dialyzer is found to be 3%, but it varies with the dialyzer, the batch of membrane, and the skill of the operator. The leak rate from all curophane coils is high, however.

18.8 The Hemodialysis Machine and Its Parts

The artificial kidney, or dialyzer, is part of an overall hemodialysis machine. Other parts in this machine include

- A **blood pump** moves blood through plastic tubing to the dialyzer to be cleansed and returns it to the body.
- An **inflow dialysate line** delivers dialysate (the cleansing fluid) to the dialyzer.
- An **outflow hose** carries used dialysate away from the dialyzer to a drain.
- A **heparin pump** provides the right amount of heparin to keep the blood from clotting.
- **Monitors** are used to detect any potential problems.
  - An **air bubble detector** prevents air bubbles from getting into the bloodstream.
  - A **blood pressure monitor** makes sure the patient's blood pressure does not become too high or too low.

  Alarms are used to alert the dialysis team of any potential problem.

18.9 Conclusions

18.9.1 Dialysis Does Not Help Cure Kidney Disease

Dialysis does some of the work of healthy kidneys, but it does not cure kidney disease. It is a palliative treatment. Patients on dialysis will need to have dialysis treatment the remainder of their life unless they are able to receive a kidney transplant. Patients may have some discomfort when the needles are put into fistula or graft, but most patients have no other problems. The dialysis treatment itself is painless. However, some patients may have a drop in their blood pressure. If this happens, patients may feel sick to their stomach, vomit, have a headache, or have muscle cramps. With frequent treatments, those problems usually go away.

Hemodialysis and peritoneal dialysis have been performed since the mid-1940s. The use of dialysis as a regular treatment was begun in 1960 and is now standard around the world. CAPD began in 1976. Thousands of patients have been helped by these treatments.

Dialysis is very costly. However, in the United States, the federal government pays high percentage of all dialysis costs for most patients. Private health insurance or state medical aid also helps with the costs.

Many patients live normal lives except for the time needed for treatments. Dialysis usually makes patients feel better because it helps many of the problems caused by kidney failure. Patients and their families will need time to get used to dialysis. Dialysis patients have to control their diets and may be on a special diet. Patients may not be able to eat everything and may need to limit the volume of liquid consumed. The diet may vary according to the type of dialysis.

18.10 Recent Developments in Artificial Kidneys

The kidneys serve a tremendously important role in humans, as 180 l of blood plasma are cleaned and filtered every single day. Damage to the kidneys is common for a wide variety of reasons and can be quite lethal. As with all organs, the availability of healthy donor organs for transplantation is very low compared to the number of people who need them. The insight of an astute researcher led to the development of the artificial kidney. In 1939, while in Groningen, the Netherlands, Dr. Willem Kolff (1911–2009) took a piece of cellophane tubing (50 cm long), put blood in it with 400 mg urea (metabolic waste in blood plasma), and shook it up and down in a bath with saline. Surprisingly, within 5 min, nearly all of the urea was removed by dialysis because the surface area was large and both the blood and the dialyzing fluid were in continuous movement. To make an artificial kidney, we could simply multiply this 50-cm length of cellophane tubing by 20, and we would have enough cellophane to make clinical dialysis worthwhile.

After Kolff's discovery, the first clinically useful artificial kidney was made. It was made of a long cellophane tubing wrapped around a steel drum that was spun in a saline bath. Blood was taken from the patient, circulated through the tubing, and then flowed back into the vein. Modifications on this design have been made over time, as there are now cheaper, smaller, more efficient, computer-controlled machines. These artificial kidneys are used in hospitals and healthcare facilities all over the world to help inpatients and outpatients with failing kidneys.

Artificial kidneys are also being used in novel contexts. Take, for instance, the "Dialysis at Sea" cruise line. The cruise company has dialysis machines available that are capable of doing standard, high-efficiency, and high-flux treatments. These machines are transported to the ship, set up, and calibrated on the cruise's departure day (Fig. 18.6).They meet with the patient on the day of sailing, introduce the medical staff, and provide a general orientation. Whenever possible, dialysis is done at

Fig. 18.6 The Kolff (1998) dialyzer

sea to allow as much free time aboard and in port as possible. Travelers can receive the same number of treatments as they do at home.

Unfortunately, an artificial kidney that can be an internal implant has not yet been made. The best devices we have for such an implant are not devices at all; they are actual kidney transplants. However, there have been some at least partially successful attempts to make the artificial kidney machine small enough so that severe chronic dialysis patients can carry it around externally and lead active lives.

Artificial kidneys have certainly served individuals who need help filtering their blood due to kidney failure in order to live their lives. For that reason alone, we should realize the great importance of such an invention. As Kolff pointed out, though, sometimes such inventions can have a bigger impact than on just one individual.

An artificial organ should not be used unless there is a reasonable hope that it will restore the recipient to a happy existence. There are a few exceptions to that rule. At times, an artificial organ has played an important role in history. Marshall Tito of Yugoslavia was kept alive with an artificial kidney until a smooth transfer of government took place.

Yuri Andropov, the former head of the KGB who became president of the Soviet Union, was maintained with dialysis for nine months. Then Mikhail Gorbachev was appointed.

18.10.1 Nanotechnology

Conventional dialysis, in which a patient's blood is pumped through an external filter to drain out accumulating toxins, is far from ideal for the 1.4 million people with kidney disease worldwide whose lives depend on it. The common regimen of three half-day blood-cleansing sessions per week removes, on average, just 17% of the toxins that a healthy kidney would clear, so that only one third of all dialysis patients survive for more than 5 years of treatment.

Nanotechnology could offer an alternative, according to nephrologist William Fissell at the University of Michigan. He and colleagues are working on nanopore membranes that could enable dialysis to be miniaturized into implantable devices that provide round-the-clock clearance of toxins, untethering dialysis patients from bulky pumps and clinics. "This is a fundamentally liberating technology," says Fissell (2007).

Fissell and colleague Shuvo Roy, a biomedical engineer at the Cleveland Clinic Foundation, claim to have solved half of the challenge: engineering nano-membranes that are efficient enough to support a compact, low-power implant. The team secured a patent for the concept recently. However, engineering pores with the required selectivity—pores that drain away the worst toxins without robbing the body of critical proteins such as albumin, blood clotting factors, and antibodies—is proving to be tougher than expected.

As currently practiced, dialysis is a crude procedure. Patients are hooked up intravenously to a powerful pump that circulates their blood through a cartridge of porous plastic fibers. Fluids, dissolved toxins, and salts pass through the fibers and are discarded, while the proteins and blood cells caught in the sieve are supplemented with electrolyte before returning to the patient. The filter's poor fluid dynamics are a function of their imprecision: Filter manufacturing produces a wide range of pores, so to avoid having too many large pores, which would suck out valuable proteins, the fibers must be manufactured with a preponderance of very small pores. The machine's pump makes up the difference, forcing blood through these inefficient sieves.

In contrast, Fissell and Roy converted pores into ultrathin wafers of silicon with lithographic precision. The result is a homogeneous array of pores, each capable of flow rates several orders of magnitude higher than the average pore in a conventional filter. The pores mimic the exquisitely precise yet efficient diaphragms that filter blood in a human kidney, resembling a panel of Venetian blinds, says Fissell. Ron Adams 541 737 310 OSU Portland firm develops portable kidney dialysis machine.

Current prototypes contain roughly 10,000 pores per square millimeter, according to Fissell. Next-generation membranes, now being engineered, will have more than 100,000 pores or slits per square millimeter and provide more than 10 times the flow. An implanted device carrying several hundred square centimeters of this next-generation membrane should, Fissell estimates, filter at least 30 ml of blood per minute at average blood pressures about one third of normal kidney function. The implant would be tucked under the skin; small fluid bags worn externally could receive the ultrafiltrate and supply replacement electrolytes.

18.10.2 Portable Dialysis Machine

"Current dialysis machines are based on 30-year-old technology and employ filter systems that are only about 28% efficient," said Michael Baker, chief executive officer of Home Dialysis Plus, the firm developing the device.

"By employing the micro technology being developed at OSU's College of Engineering, the filter efficiency grew tremendously to about 90%," M. Baker said. "And we're able to reduce the dimensions of a dialysis machine from the size of a refrigerator to the size of a piece of carry-on luggage, which makes treatment portable." Prof. M. Baker was from OSU, College of Engineering, USA.

OSU researchers and Home Dialysis Plus officials say the reduced size and increased efficiency will improve the lives of dialysis patients because treatments will be done in the home while patients are asleep at night. Many dialysis patients now have to limit travel and other opportunities due to time-consuming treatments that can take up to 4 hours.

"A growing number of studies indicate that longer and/or more frequent in-home dialysis offers not only superior therapy, but also the opportunity to reduce costs," Baker said. "These revelations are prompting more and more dialysis patients to demand the health benefits, the convenience, and the quality of life that in-home dialysis has to offer. We launched Home Dialysis Plus to give dialysis patients the life-improving technology they've been asking for."

Home Dialysis Plus was founded by Altman Browning and Company, a Portland product development firm that has been working closely with researchers at the

OSU College of Engineering to develop the new microtechnology for use in the dialysis application. Researchers say the technology, known as **multiscale materials and devices**, or MMD, could eventually enable the development of a wearable dialysis device, or even an implantable version.

"MMD technology is allowing us to reduce a filter that is more than 20 cm tall by 7.6 cm in diameter to the size of about four sugar cubes," Baker said. "As this technology develops, the possibility of an implantable dialysis device becomes very real."

Jim Curtis, a certified hemodialysis technologist, worked as area manager for Fresenius Medical Care North America, a company that operate 11 dialysis clinics throughout Oregon. Currently he is the director, technical services at Home Dialysis Plus Ltd. He said, "MMD technology has the potential to revolutionize the artificial kidney because it enables such good surface area in such a small space. Many studies show that patients are better off dialyzing less rapidly, but over a longer period of time."

Citing a recent Canadian study that compared patients who underwent traditional dialysis treatments (three times a week for approximately 4 h) to patients who underwent nocturnal dialysis (8 h of dialysis nightly), Curtis said recovery time following treatment was reduced from approximately 6 h to about 7 min.

"Nocturnal dialysis gives patients more time in their lives because they spend this time sleeping anyway," Curtis said. "And they don't have to spend hours every week feeling washed out or nauseous before and after treatments."

The leading causes of kidney failure are diabetes and hypertension, Curtis said. In 2011, the number of people suffering kidney failure was expected to be 600,000 in the U.S. alone.

In addition to being a fraction of the size of current dialysis machines, the Home Dialysis Plus device will operate much more quietly and is designed not to look like a typical piece of medical equipment, but to blend into a home environment, Baker said.

During the 1960s, Dr. Richard Drake developed the Drake Willock dialysis said machine, which rapidly became the best-selling dialysis system in the world. Drake's Portland-based company, DWS, Inc., employed 250 employees before it was sold in the 1970s. Drake, now retired, says the use of MMD technology in the Home Dialysis Plus filter is an important advance.

"That is a major breakthrough, because filters have always been a problem," he said. "Reducing the size and increasing the efficiency is very important."

Although other companies, like Hewlett-Packard, are partnering with OSU researchers to employ MMD technology in new devices, Altman Browning is the first company on the brink of commercializing a viable product.

Researchers at OSU are partnering with researchers at the University of Oregon, Portland State University, Pacific Northwest National Laboratory, and other institutions to develop MMD technology, which U.S. Senator Ron Wyden has predicted will be an "economic sparkplug" for the Pacific Northwest.

"MMD technology has the potential to spawn an entire industry, one that could rival the microelectronics boom of the 1950s and 1960s," Ron Adams Dean Engg. OSU, USA said "The Home Dialysis Plus product is just the tip of a very large iceberg."

MMD technology includes the use of microchannels, which enable high rates of heat and mass transfer and are the reason for the high rates of efficiency in the Home Dialysis Plus filter.

A new high tech membrane may soon improve the effectiveness of dialysis and might someday lead to implantable kidney device.

List for Further Reading

1. Coresh J, Selvin E, Stevens LA et al (2007) Prevalence of kidney disease in US. JAMA 298:2038–2097
2. Fissell WH, Humes HD, Fleischman AJ, Roy S (2007) Dialysis and nanotechnology: now, 10 years, or never? Blood Purif 25(25):12–17
3. Lindsay RM, Leitch R, Heidenham AP, Kortas C (2003) The London daily/nocturnal hemodialysis study: study design, morbidity, and mortality results. Am J Kidney Dis 42(Supp 1): S5–S12
4. Fissell W, Manley S, Westover A, Humes HD, Fleischman AJ, Roy S (2006) Differentiated growth of human renal tubule cells on thin-film and nanostructured materials. ASAIO J 52(52):221–227
5. Saito A, Aung T, Sekiguchi K, Sato Y, Vu D, Inagaki M, Kanai G, Tanaka R, Suzuki H, Kakuta T (2006) Present status and perspectives of bioartificial kidneys. J Artif Organs 9(3):130–135
6. Saito A, Aung T, Sekiguchi K, Sato Y (2006) Present status and perspective of the development of a bioartificial kidney for chronic renal failure patients. Ther Apher Dial 10(4):342–347
7. Wang P, Takezawa T (2005) Reconstruction of renal glomerular tissue using collagen vitrigel scaffold. J Biosci Bioeng 99(6):529–540
8. Fissell W, Fleischman AJ, Roy S, Humes HD (2007) Development of continuous implantable renal replacement: past and future. Transl Res 150(6):327–336
9. www.dialysiswhilecrusing.com
10. Curtis, J. (2008) Dialysis & Transplant action vol. 37(8), 332–333

Exercises

1. Explain with a neat sketch the basic structure and function of a human kidney. How does it separate waste product from blood plasma? Estimate how much fluid the pair of kidneys handles. Give an idea about the composition of plasma and urine.
2. What are the common modes of kidney failure? How can hemodialysis improve the situation? Describe the two types of dialysis procedures being used. Is there any need for dialysis patients to modify their lifestyle?
3. How can nanotechnology help in devising a compact dialyzer? Use some recent advances in the area to explain your answer.
4. What is an artificial kidney? Explain its construction and function. How has its function improved in recent years? What is a wearable kidney machine? Develop a mathematical model for a kidney machine.
5. What is the new MMD technology that will reduce the cost and size of the artificial kidney dialyzer?

Chapter 19
Skin and the Design of Artificial Skin

19.1 Anatomy of the Skin

The skin is the body's largest organ, covering the entire outside of the body, weighing approximately 3 kg, and covering about 5–6 m^2 of the body's surface. In addition to serving as a protective shield against heat and cold, light, injury, pollution, and infection, the skin also

- **Regulates body temperature**
- **Stores water, fat, and vitamin D**
- **Can sense painful and pleasant stimulation**.

Throughout the body, the skin's characteristics vary (i.e., thickness, color, and texture). For instance, the head contains more hair follicles than anywhere else, while the soles of the feet and the surface of the palm contain none. In addition, the soles of the feet and the palms of the hands have much thicker layers (Fig. 19.1).

The skin is made up of the following layers, with each layer performing specific functions:

- **Epidermis**
- **Dermis**
- **Fat layer**.

Epidermis: The epidermis is the thin outer layer of the skin. The epidermis itself is made up of three sublayers:

Stratum corneum (horny layer): This layer contains continually shedding, dead keratinocytes (the primary cell type of the epidermis). The keratin, a protein formed from the dead cells, protects the skin from harmful substances.

Keratinocytes (squamous cells): This layer contains living keratinocytes (squamous cells), which help provide the skin with what it needs to protect the rest of the body.

Basal layer: The basal layer is the inner layer of the epidermis, containing basal cells. Basal cells continually divide, forming new keratinocytes and replacing the old

S. Pal, *Design of Artificial Human Joints & Organs*,
DOI 10.1007/978-1-4614-6255-2_19,

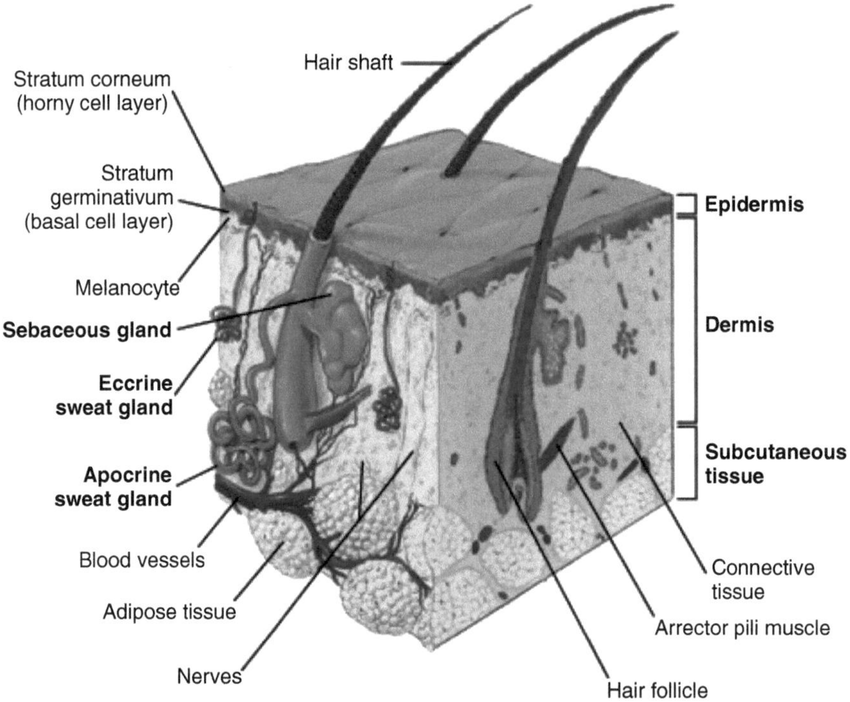

Fig. 19.1 A vertical three-dimensional total view of skin tissue

ones that are shed from the skin's surface. The epidermis also contains melanocytes, which are cells that produce **melanin** (skin pigment).

The **dermis** is the middle layer of the skin. The dermis is made up of the following:

Blood vessels, lymph vessels, hair follicles, and sweat glands.

The dermis is held together by a protein called **collagen**, made by fibroblasts (skin cells that give the skin its strength and resilience). This layer also contains pain and touch receptors.

The **subcutis** is the deepest layer of skin and is also known as the subcutaneous layer. The subcutis, consisting of a network of collagen and fat cells, helps conserve the body's heat while protecting other organs from injury by acting as a "shock absorber."

19.2 Modeling of Soft Biological Tissues

The diversity of mechanical properties encountered in soft biological tissues is huge. Soft organic tissues are in general characterized by very complex mechanical behavior. They show nonlinear, anisotropic, viscoelastic, and, in some cases, also

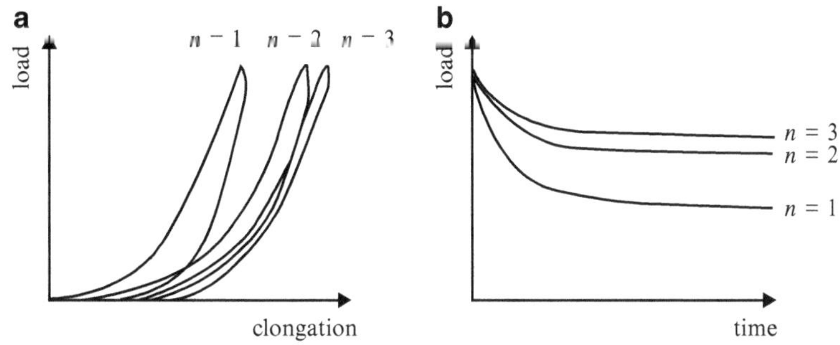

Fig. 19.2 (**a**) Load-deformation (cyclic) and (**b**) stress-relaxation behavior of soft tissue

viscoplastic behavior. They often have a layered or an even more complicated structure. The mechanical properties are inhomogeneous; that is, they depend on the position in the material. The perfusion of the organs and their constituting tissues also plays an important role regarding the elastic properties.

19.2.1 The Stress–Strain Relationship

There are mainly two sources of elasticity in soft biological tissues. The first source of elasticity is due to changes in internal energy, whereas the second one is due to changes in entropy. Change in entropy occurs in tissues whenever changes in orientation or waviness of fibers during loading or unloading occur. A typical load-elongation and load–time diagram for soft tissue is shown in Fig. 19.2.

With repeated loading cycles, the load-deformation curves shift to the right in a load-elongation diagram and the hysteretic effects diminish. In a load–time diagram, the load–time curves shift upward with increasing repetition number. By repeated cycling, eventually a steady state is reached at which no further change will occur unless the cycling routine is changed. In this state, the tissue is said to be preconditioned. Any change of the lower or upper limits of the cycling process requires new preconditioning of the tissue. Preconditioning occurs due to internal changes in the structure of the tissue. Hysteresis, nonlinearity, relaxation, and preconditioning are common properties of all soft tissues, although their observed degrees vary.

The hysteresis in the stress–strain relationship clearly shows the viscoelastic behavior of soft biological tissue. In a viscoelastic material, the history of strain affects the actual observed stress. As well, loading and unloading occur on different stress–strain paths. The hysteresis of most biological tissues is assumed to show only little dependence on the strain rate within several decades of strain-rate variation. This insensitivity to strain rate over several decades is not compatible with simple viscoelastic models consisting, e.g., of a single spring and dashpot element. With such a simple viscoelastic approach, the material model will show a maximum

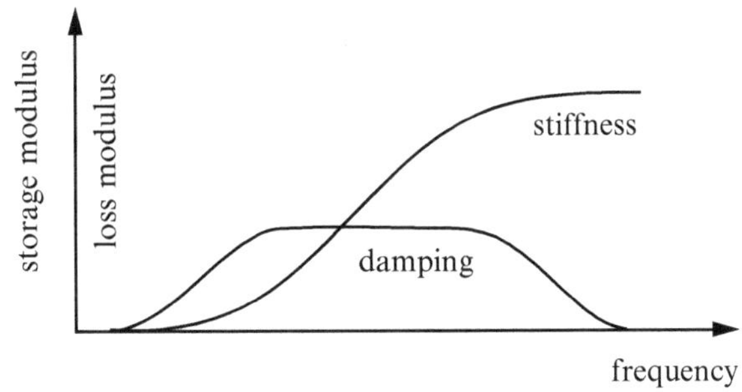

Fig. 19.3 Storage and loss modulus of soft tissue with frequency variation

hysteresis loop at a certain strain rate, whereas all other strain rates will show a smaller hysteresis loop. A model consisting of a discrete number of spring-dashpot elements therefore produces a discrete hysteresis spectrum with maximum dissipation at discrete strain rates. If the relaxation times of the different elements are chosen adequately, a series of spring-dashpot elements might be used as an approximation to a continuous relaxation spectrum. Living tissues often show a viscoelastic behavior, as shown qualitatively in Fig. 19.3.

In Fig. 19.2, the viscoelastic material properties are characterized by storage and loss modulus, which are concepts only valid for linear elasticity.

With a series of spring-dashpot elements, arbitrary viscoelastic material properties can be modeled.

19.3 Mechanical Properties of Soft Tissue

Mechanical characterization is a very important criterion for understanding soft tissue behavior. Apart from skin, there are other tissues, such as muscles, ligaments, tendons, and fascia.

Material tests were performed in tension and compression on swine, lamb, and human tissues with different experimental setups by many authors, including the present author.

Three experimental setups were used: a custom-made **micro testing machine (MTM)** with ±2-mN load resolution and ±1-mm stroke capacity; the **ELF-3100 testing instrument**, with ±22-N and ±2.5-mm ranges; and a 100L dynamic material testing machine with a load of ±50 N and a displacement of ±25-mm ranges (Figs. 19.4, 19.5 and 19.6).

19.4 Fitting Hyperelastic Material Models to the Experimental Data

The strain energy potentials were fitted to the experimental data. The material was assumed to be incompressible. The tested material models included Arruda-Boyce, Marlow, van der Waals, reduced polynomial ($N=1$: neo-Hookean; 2 and 3: Yeoh), polynomial ($N=1$: Mooney–Rivlin, and 2), and Ogden ($N=1$ and 3).

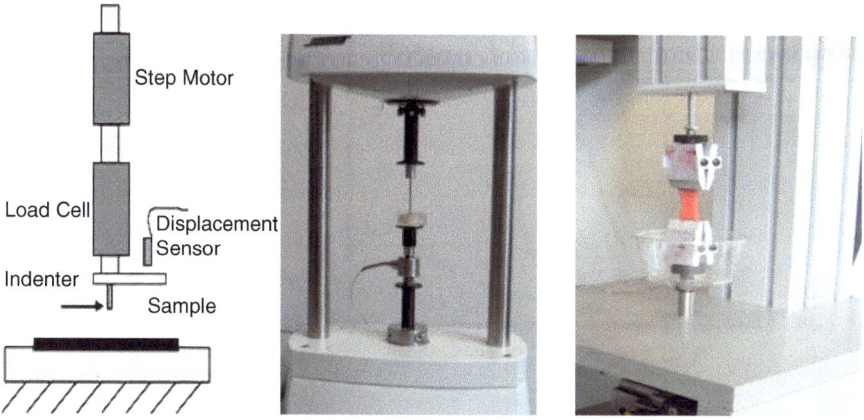

Fig. 19.4 Experimental setup for in vitro soft tissue test. Compression mechanism 3 (prestretched)

Fig. 19.5 A standard proportionate tensile testing sample as per ASTM protocol. It may be increased or decreased proportionately according to the material's availability. The scaling can be done proportionately. The *smaller part* is the failure area

Findings Generally, material behaviors for skin and fat were favorably represented by the fitted strain energy potentials. However, the material behavior for muscle was relatively less represented by the strain energy potential. It was found that for the tested soft tissues, the polynomial ($N=2$) and Ogden ($N=1$) models gave the best fitted results, with R^2 higher than 0.99. However, since the fitting results were for a limited amount of tissue samples, the selection of material models for the proposed study should not be limited to these two options.

19.5 Artificial Skin

The term "artificial skin" refers to skin grown in a laboratory that can be used as skin replacement for people who have suffered severe burns or skin diseases.

The skin is the largest organ in the human body. Severe damage to large areas of skin exposes the human body to dehydration and to infections, which can result in

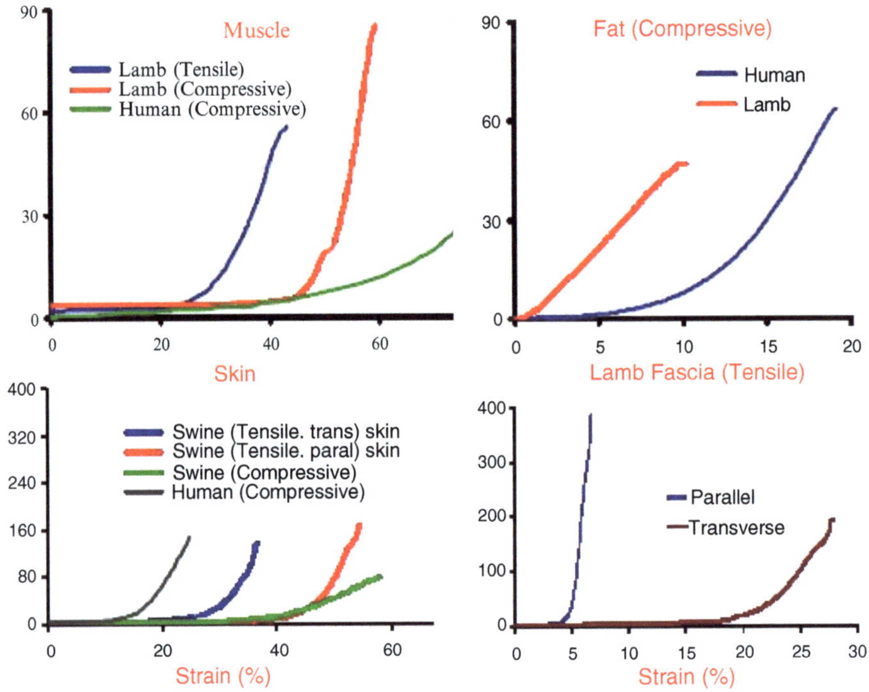

Fig. 19.6 Stress–strain results for muscle, fat, skin, and fascia from tensile and compression material tests

death. Traditional ways of dealing with large losses of skin have been to use skin from other parts of a patient's body (such as the thighs) or from a different person/cadaver. The former approach has the disadvantage that there may not be enough skin available, while the latter suffers from the possibility of rejection due to immunogenic reaction or infection.

Recently, there has been progress toward creating artificial skin. Typically, a collagen scaffold is used (the protein that underlies the structure of skin), which can additionally be seeded with the patient's own cells [1–3] or with foreskin from newborns that was removed during circumcision [4].

The Fraunhofer Institute for Interfacial Engineering and Biotechnology is working toward a fully automated process for producing artificial skin. Their goal for the moment is a simple two-layer skin without blood vessels that can be used to study how skin interacts with consumer products, such as creams, medicines, etc. Eventually, they hope to produce more complex skin that can be used in transplants [5].

Individuals who suffer extensive loss of skin, commonly due to fire, are seriously ill and in danger of succumbing either to massive infection or to severe fluid loss. Subjects who survive these early threats must often cope with problems of rehabilitation arising from deep, disfiguring scars and crippling contractures. In this chapter, we describe the physicochemical, biochemical, and mechanical considerations that form the basis for a two-stage design of a membrane useful as an experimental wound closure. Stage 1 of the design, applicable to short-term acute use, calls for a membrane that efficiently displaces air pockets from a carefully prepared wound bed, free of weak boundary layers, and maintains the moisture flux through the wound at an optimal level. Optimization of the surface energy, modulus of elasticity, energy to fracture ratio, and moisture permeability of the membrane are among the essential attributes of Stage 1 design. Stage 2 of the design, applicable to long-term, chronic use, focuses on a nonantigenic membrane, which acts as a biodegradable template for the synthesis of neodermal tissue. A survey of candidate materials suggests reasons for the selection of a porous, cross-linked collagen—glycosaminoglycan coprecipitate—as the chemical basis for an evolving design that was initiated 10 years ago. Over the past several years, a set of membranes has been iteratively designed on this basis and has been used satisfactorily to cover large experimental full-thickness skin wounds in guinea pigs. Such membranes have effectively protected these wounds from infection and fluid loss for over 25 days without rejection and without requiring change or other invasive manipulation. When appropriately designed for the purpose, the membranes have also strongly retarded wound contraction and have become replaced by newly synthesized, stable connective tissue. Several rules relating the molecular structure and morphology of these membranes to the cellular response of adjacent tissue have also been derived. This report is the first in a series that details the methodology of preparation and the record of performance. (These research results were from the Fibers and Polymers Laboratories, Department of Mechanical Engineering, Massachusetts Institute of Technology, Cambridge, MA, USA, and Shriners Burn Institute, Harvard Medical School, Boston, MA, USA).

The present author also did work on a chitosan–PVA blended membrane. After a successful animal trial, Dr. N. Bhattachryya, a Kolkata-based surgeon, undertook a clinical trial on bedsore patients with successful results. Figure 19.7 shows some results.

19.6 Tissue Expanders

Tissue expanders are of great value in maxillofacial, plastic, and reconstructive surgery. Tissue expansion was developed for a specific indication; however, within a very short time, the concept of tissue expansion found wide applicability. The indications for tissue expansion were burns, trauma, and sequelae of previous surgery resulting in massive contraction/contracture/scarring or disfigurement of the local tissue. An expander actually helps in growing the local tissue as required so that the disfigurement can be corrected easily with the same type of tissue. Tissue expansion is a good and safe technique (Fig. 19.8).

Tissue expansion is a reconstructive surgical technique that allows the body to "grow" extra skin where there has been tissue loss due to trauma or disease. The

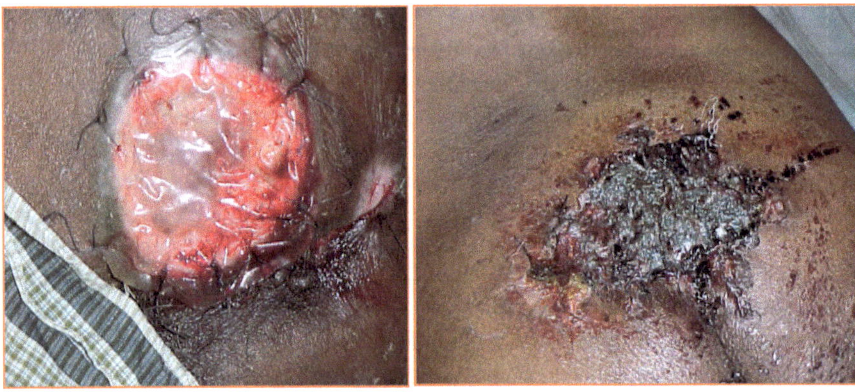

Fig. 19.7 The clinical results from bedsore patients using chitosan–PVA membrane

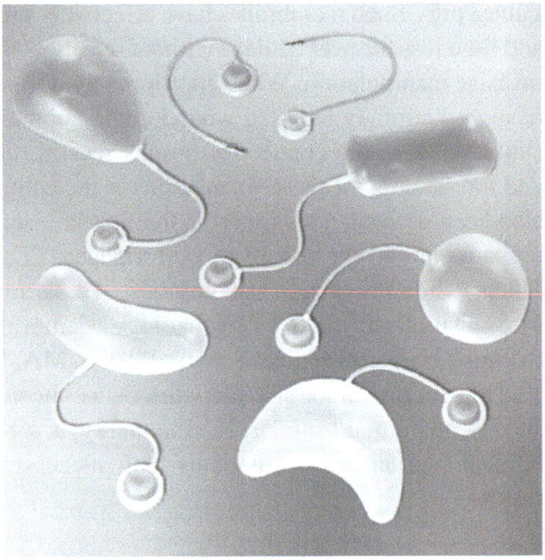

Fig. 19.8 Tissue expanders of various shapes are placed under the skin and expanded by inserting saline solution under pressure in the small, shaped vessel made of silicone

most common application is in postmastectomy breast reconstruction, but tissue expansion can be used in almost any part of the body following tissue loss due to injury. It is also quite advantageous for reconstruction of the scalp, because the "new" skin created contains matching hair follicles (as opposed to using skin grafts or flaps from other parts of the body, which may leave bald spots on the scalp).

How Tissue Expansion Works: A balloon-like expander with a silicone shell is inserted under the skin near the area in need of repair. Over time, the shell is gradually filled with saline (saltwater), causing the skin to stretch and grow. Once sufficiently stretched, the expander is removed with a second surgery, and the excess new skin is then placed over the defect and sutured in place. The expanders are inserted most frequently in the scalp, trunk, back, and neck area, avoiding the important anatomical structures.

Uses

- Revisions of facial scars, blemishes, and moles
- Burn wound and contracture correction
- Postsurgery defects
- Hair loss replacement
- Cleft lip and palate repair
- Correction of an underdeveloped breast
- Postmastectomy breast reconstruction.

Advantages

Expanders offer a near-perfect match of skin color, texture, and sensation.

There is less risk of tissue loss because the skin remains connected to its blood and nerve supply.

Scars are less noticeable than with a skin graft, and morbidity of the donor area is not a factor.

Disadvantages

Expansion can take as long as 3 weeks to 4 months. An expander creates what may be considered an unsightly bulge while in place (fine for breast reconstruction, but undesirable in facial reconstruction), and discomfort sometimes becomes unbearable to the patient, who requires multiple visits for saline injections to further inflate the expander. However, recently we developed a silicone expander where the patient himself will be able to inject additional saline according to the surgeon's advice.

References

1. http://www.technologyreview.com/biotech/18059/
2. http://web.mit.edu/newsoffice/1996/artificialskin.html
3. http://www.sciencedaily.com/releases/2008/01/080104140344.htm
4. http://www.answers.com/topic/artificial-skin-2
5. "Artificial Skin Manufactured In Fully Automated Process".
6. Fung Y.C. (1993) Biomechanics & Mechanical properties of living tissues, Springer Verlag N.Y.

Exercises

1. What is the storage and loss modulus of elasticity of polymeric material? Depict in a neat sketch the load-deformation diagram of human skin.
2. Dr. N. Bhattacharyya of a Kolkata Hospital did a clinical trial with bedsore patients after animal experimentation and ethical clearance using chitosan–PVA blended composite artificial skin. He observed excellent results in his trial. He wanted a large-sized chitosan–PVA biomembrane to heal severe burn injuries for ladies from the South 24-Pgs district of West Bengal. We used the artificial skin

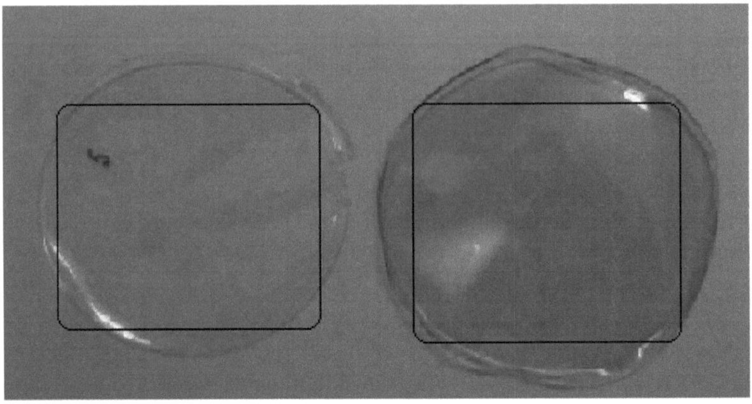

Fig. 19.9 Chitosan–PVA membrane cast in our laboratory. To facilitate visualization of the transparent sheet, rectangular borders are given

shown in Fig. 19.9. How do you propose to make a 400-mm × 300-mm membrane and sterilize that membrane?

3. How do you propose to test a skin sample from a goat to estimate the requirement of the artificial skin although goat skin and human skin do not function similarly? What is a tissue expander and how does it work? Illustrate some examples in cosmetic surgery. The ASTM has suggested a specific dog bone–shaped sample for tensile testing. Find it from this text.

Chapter 20
The Artificial Pancreas

20.1 Endocrine Physiology

Before we talk about the artificial pancreas, we have to understand its structure and function. The pancreas, which is shown in Fig. 20.1, is located below the stomach and above the duodenum. It releases endocrine hormones (**insulin, amylin, and glucagon**) into the portal vein, where it flows directly to the liver.

The pancreas produces three hormones that are important to glycemic control: **insulin**, which lowers blood glucose; **amylin**, which slows digestion, slows the rate of glucose entering the bloodstream, and temporarily suppresses the release of glucagon; and **glucagon**, which raises the blood glucose.

Upon digestion of carbohydrates, glucose levels in the blood will begin to rise. As the blood and glucose flow into the pancreas, insulin and amylin are secreted by the pancreatic beta cells directly into the bloodstream in response to elevated blood glucose levels. Insulin causes blood glucose to be removed from the bloodstream and stored in the liver and muscle cells. Notice that as the blood sugar goes higher, additional insulin will bring the blood sugar back down in a classic negative-feedback loop. As insulin is released from the beta cells, amylin is also released into the bloodstream. Amylin slows gastric emptying and also inhibits the release of glucagon from the pancreatic alpha cells. The effect of amylin is to spread out the blood glucose peak after eating, reducing the quantity of insulin needed. As the blood sugar level comes back toward normal, the beta cells will stop spurting insulin and amylin. As the glucose level approaches a low mark, the pancreatic alpha cells will release glucagon directly into the bloodstream. Glucagon causes the liver to release stored glucose back into the bloodstream. Notice that increased glucagon will increase blood glucose levels to produce a positive error in the negative-feedback loop. Together, the three endocrine hormones work as a system to maintain the blood glucose level between high and low boundaries.

When the beta cell produces insulin from proinsulin, a connecting peptide (or C-peptide) is also manufactured and released into the bloodstream. The **absence of C-peptide in the blood indicates that insulin has not been released from the**

S. Pal, *Design of Artificial Human Joints & Organs*,

DOI 10.1007/978-1-4614-6255-2_20,

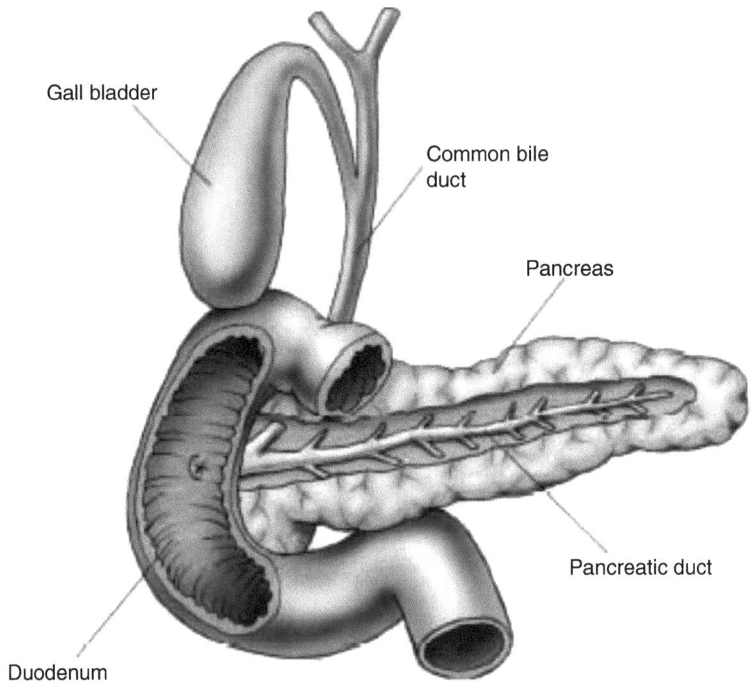

Fig. 20.1 The cutaway view of the pancreas and its associated organs

pancreas, and this fact confirms the diagnosis of diabetes type 1. C-peptide was believed to be only a byproduct of natural insulin production, but recent studies suggest that C-peptide exerts a beneficial therapeutic effects on diabetic nociceptive neuropathy [1].

Ideally, to replicate the natural function of the pancreas as closely as possible, an artificial pancreas might someday replace all of the beneficial endocrine functions lost, including the delivery of **insulin, amylin, glucagon, and C-peptide**.

20.2 Artificial Pancreas

The **artificial pancreas** is a technology in development to help people with diabetes control their blood glucose level automatically by providing the substitute endocrine functionality of a healthy pancreas.

There are several important exocrine (digestive) and endocrine (hormonal) functions of the pancreas, but it is the lack of insulin production that is the motivation to develop a substitute. While the current state of insulin replacement therapy is popular for its life-saving capability, the task of manually managing the blood sugar level with insulin alone is arduous and inadequate for the patient.

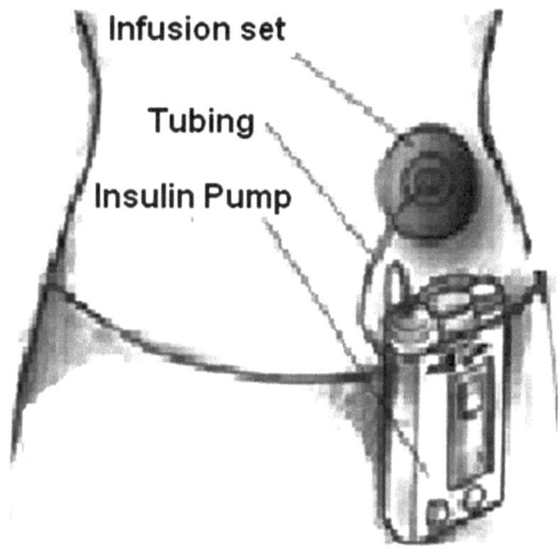

Fig. 20.2 Artificial insulin therapy setup in place

The goal of the artificial pancreas is twofold:

1. To improve insulin replacement therapy until glycemic control is practically normal, as evidenced by the avoidance of the complications of hyperglycemia.
2. To ease the burden of therapy for the insulin dependency.

Different approaches under consideration include

The medical equipment approach—using an insulin pump under closed-loop control using real-time data from a continuous blood glucose sensor. This is an emerging technology and will be discussed in detail.

The bioengineering approach—The development of a bioartificial pancreas consisting of a biocompatible sheet of encapsulated beta cells. When surgically implanted, the islet sheet will behave as the endocrine pancreas and will be viable for years.

The gene therapy approach—The therapeutic infection of a diabetic person by a genetically engineered virus that causes a DNA change of intestinal cells to become insulin-producing cells.

20.3 Intensive Insulin Therapy and Insulin Pump

The insulin pump is used to automatically deliver basal insulin continuously, and bolus insulin at meal times, by pressing the buttons. Before meals, a blood glucose value is entered into the pump to calculate the correction bolus to bring the blood glucose level back to the target value.

Insulin pump therapy, shown in Fig. 20.2, is used by tens of thousands of people of all ages. Many studies have shown improved glucose management outcomes for those using insulin pumps. While it does allow for more flexibility in lifestyle and the potential to even out the wide blood sugar fluctuations that are often experienced when injecting insulin, it may not be the right choice for every person.

In insulin-dependent persons, blood glucose levels have been roughly controlled using insulin alone. The carbohydrate in the amount of food is estimated by weighing foods, and the measurement is used to estimate the amount of insulin necessary to cover the meal. The calculation is based on a simple **open-loop model**: The insulin-to-carbohydrate ratio (adjusted based on past success) is multiplied by the grams of carbohydrate to calculate the units of insulin needed. That quantity of insulin is then adjusted based on a premeal blood glucose measurement (insulin bolus increased for a high blood sugar or insulin bolus delayed and reduced for a low blood sugar). Insulin is injected or infused under the skin and enters the bloodstream in approximately 15 min. After the insulin has acted in the bloodstream, the blood glucose level can be tested again and then adjusted with the injection of more insulin, or by eating more carbohydrates, until balance is restored.

There are notable differences with insulin replacement compared to the function of pancreatic insulin delivery:

1. The insulin dose is predicted based on measured food (where an accurate measurement of carbohydrate is difficult), whereas pancreatic insulin is released in proportional response to actual blood glucose levels.
2. Pancreatic insulin is released into the portal vein, where it flows almost directly to the liver, which is the major organ for storing glycogen (50% of insulin produced is used by the liver).
3. Pancreatic insulin is pulsatile, which helps maintain the insulin sensitivity of hepatic tissues.
4. Injected insulin is delivered subcutaneously (under the skin) but not directly to the bloodstream, so there is a delay before injected insulin begins to reduce blood glucose (although this can be compensated for by injecting insulin 15 min before eating).
5. Replacement insulin therapy does not include amylin (although Symlin is now available for use), which can reduce the insulin need by 50%.
6. Replacement insulin is dosed as a best compromise between an aggressive use for lowering the blood sugar when eating but also a conservative use to avoid a postprandial low blood sugar due to excess insulin, whereas pancreatic function releases insulin aggressively and later includes automatic release of glucagon at the end of an insulin cycle to manage the blood sugar level and avoid hypoglycemia.

An insulin pump to infuse rapid-acting insulin is the first step in simulating the function of the pancreas. The pump can accurately deliver small increments of insulin compared to an injection, and its electronic controls permit shaping a bolus over time to match the insulin profile required for a given situation. The insulin pump is controlled by the pump user to bolus manually based on a recent blood glucose measurement and an estimate of the grams of carbohydrate consumed. This predictive approach is said to be *open-loop*. Once a bolus has been calculated and delivered, the pump

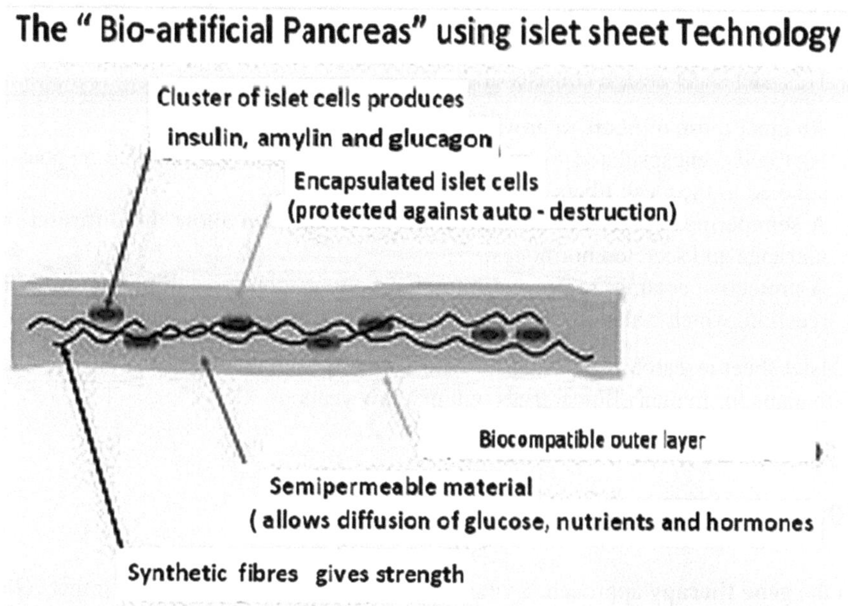

Fig. 20.3 The bioartificial pancreas

continues to deliver its basal-rate insulin in the manner that has been programmed into the pump controls based on the predicted insulin requirements of its user.

While insulin replacement is appreciated as a life-saving therapy, its practical use in controlling blood glucose levels sufficiently to avoid the long-term complications associated with hyperglycemia is not ideal.

20.4 Bioengineering Approach to an Artificial Pancreas

The bioartificial pancreas: Figure 20.3 shows a cross section of bioengineered tissue with encapsulated islet cells, which deliver endocrine hormones in response to glucose.

A biological approach to the artificial pancreas is to implant bioengineered tissue containing islet cells, which would secrete the amount of insulin, amylin, and glucagon needed in response to sensed glucose.

When islet cells have been transplanted via the Edmonton protocol, insulin production (and glycemic control) was restored at the expense of immunosuppression. Encapsulation of the islet cells in a protective coating has been developed to block the immune response to transplanted cells, which relieves the burden of immunosuppression and improves the longevity of the transplant [2].

One concept of the bioartificial pancreas uses encapsulated islet cells to build an *islet sheet*, which can be surgically implanted to function as an artificial pancreas [3].

This islet sheet design consists of

1. An inner mesh of fibers to provide strength for the islet sheet.
2. Islet cells, encapsulated to avoid triggering a proliferating immune response, adhered to the mesh fibers.
3. A semipermeable protective layer around the sheet, to allow the diffusion of nutrients and secreted hormones.
4. A protective coating, to prevent a foreign-body response resulting in a fibrotic reaction, which walls off the sheet and causes failure of the islet cells.

Islet sheet research is pressing forward with large animal studies at the present, with plans for human clinical trials within a few years.

20.5 Gene Therapy Approach

In the **gene therapy** approach, a viral vector is designed to deliberately infect cells with DNA to carry on the viral production of insulin in response to the blood sugar level.

Technology for gene therapy is advancing rapidly such that there are multiple pathways possible to support endocrine function, with the potential to practically cure diabetes [4] (Fig. 20.4).

Gene therapy can be used to **manufacture insulin directly**: An oral medication, consisting of viral vectors containing the insulin sequence, is digested and delivers its genes to the upper intestines. Those intestinal cells will then behave like any viral infected cell and will reproduce the insulin protein. The virus can be controlled to infect only the cells that respond to the presence of glucose, such that insulin is produced only in the presence of high glucose levels. Due to the limited numbers of vectors delivered, very few intestinal cells would actually be impacted, and they would die off naturally in a few days. Therefore, by varying the amount of oral medication used, the amount of insulin created by gene therapy can be increased or decreased as needed. As the insulin-producing intestinal cells die off, they are boosted by additional oral medications [5].

Gene therapy might eventually be used to **cure the cause of beta cell destruction**, thereby curing the new diabetes patient before the beta cell destruction is complete and irreversible.

Gene therapy can be used to **turn duodenum cells and duodenum adult stem cells into beta cells**, which produce insulin and amylin naturally. By delivering beta cell DNA to the intestine cells in the duodenum, a few intestine cells will turn into beta cells, and subsequently adult stem cells will develop into beta cells. This makes the supply of beta cells in the duodenum self-replenishing, and the beta cells will produce insulin in proportional response to the carbohydrates consumed [6].

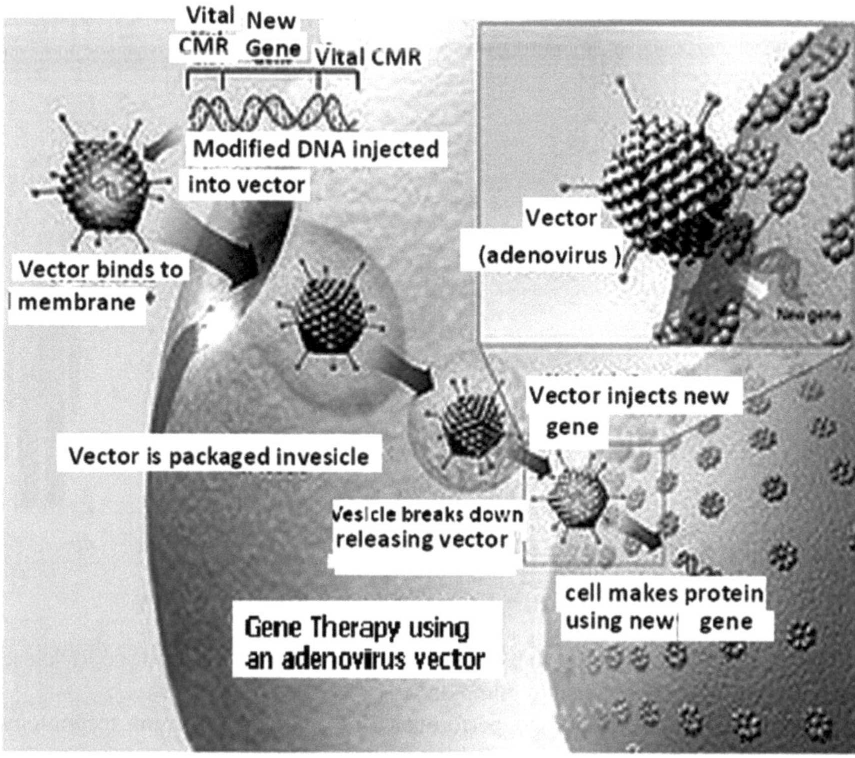

Fig. 20.4 Gene therapy for control of blood sugar level

20.6 Medical Equipment Approach

20.6.1 Development of Continuous Blood Glucose Monitoring

Technology for continuous blood glucose monitoring supports the mission of the artificial pancreas by

1. Automatically providing a blood glucose reading every few minutes without finger sticks from the user.
2. Monitoring trends pertaining to rising and falling blood sugars, which is helpful in the prediction of blood glucose levels in the immediate future.
3. Comparing blood sugar levels and predictions against a high blood sugar threshold, and then prompting the user that a correction bolus from an insulin pump is needed immediately.
4. Comparing blood sugar levels and predictions against a low blood sugar threshold, and then prompting the user to reduce the basal insulin from the pump or to eat something.

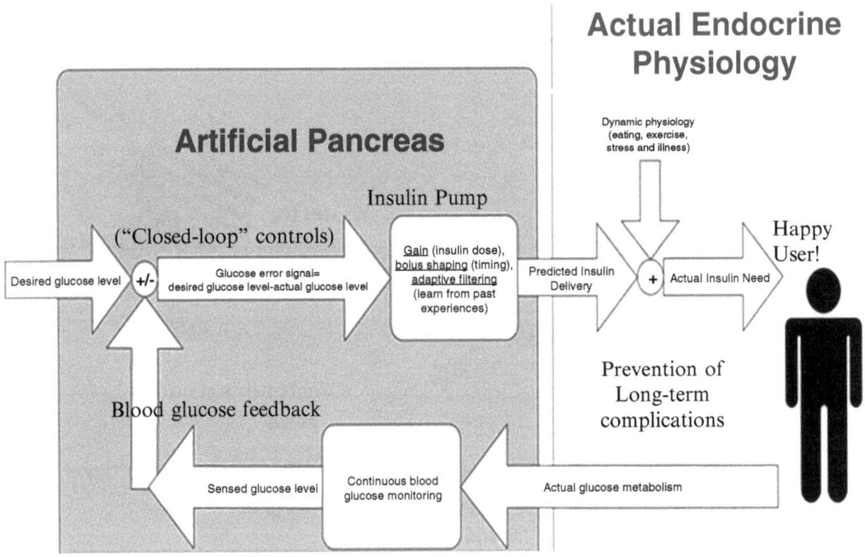

Fig. 20.5 Artificial pancreas using continuous blood glucose monitoring

These capabilities suggest that a stream of real-time data can be used to "close the loop" and control the insulin pump directly.

Some issues with the present performance of continuous sensing technology suggest that additional study is needed for application to the artificial pancreas:

1. Continuous sensors require calibration a few times a day, by performing a manual blood glucose test with a finger stick and then entering the blood glucose data into the continuous system for a sensor correction.
2. Continuous sensors are measuring interstitial glucose, so there is a time delay between the sensor data and the true blood glucose.
3. Automatic control removes the intellect of the user, which can be an additional safeguard when the data are subject to error and must be verified before taking action.

As the state of the art for blood glucose monitoring continues to advance, so does the promise of the artificial pancreas.

20.6.2 Feedback of Real-Time Blood Glucose Data to an Insulin Pump for Basal Control

The **medical equipment approach** to an artificial pancreas involves the automatic control of an insulin pump with feedback from a continuous blood glucose sensor (Fig. 20.5).

The first step in controlling an insulin pump based on continuous blood glucose data is to automatically control the basal rate of the insulin pump. When a bolus has not recently been performed, the pump can manage the blood glucose level by adjusting the basal rate as needed:

When the blood sugar is increasing, a small correction bolus can be automatically delivered and a higher basal rate can be set.
When the blood sugar is decreasing, the basal rate can be halted to deny the quantity of insulin needed to bring the blood glucose level back up until the basal rate can be continued at a new lower rate.
With adaptive filtering techniques, the pump can "learn" the unique basal rates for the person as a function of the time of day.

When controlling the basal rate alone, the closed loop can still correct a meal bolus error that was too large or small for the food consumed by

Recognizing an imbalance between the bolus "insulin on board" and the level of blood glucose.
Automatically bolusing to correct a shortage of insulin.
Automatically reducing or interrupting the basal rate to correct an abundance of insulin.
Using adaptive filtering techniques to "learn" the carbohydrate-to-insulin ratios for each meal bolus.

20.7 First Clinical Tests: Implantable Insulin Pumps and Continuous Glucose Sensors

In France, a human clinical trial of an artificial pancreas is underway. The system is fully automated by combining Medtronic MiniMed's long-term glucose sensor and its implantable insulin pump [7]. A summary of the project shows promise as well as some present limitations:

The implantable sensor is inserted into a neck vein leading to the heart.
The sensor is connected, via an electrical-type wire under the skin, to the implantable insulin pump: As blood sugar levels fluctuate, a signal tells the pump how much insulin to deliver.
The sensor accurately measured glucose in 95% of cases when compared with values obtained by finger sticks.
The blood glucose levels were maintained in the normal range more than 50% of the time in the patients using the pump connected to the sensor. Events of hypoglycemia dropped to less than 5% are alarming. While implantable insulin pumps work for an average of eight years before they have to be changed, the sensors stop working after an average of nine months.
The mathematical programs that calculate just how much insulin should be delivered at different parts of the day also need to be refined.

20.8 Insulin and Amylin Combination

When **pramlintide** (brand name Symlin or synthetic amylin) is used in combination with insulin, the benefits for postprandial glycemic control are substantial [8].

Pramlintide is a relatively new treatment for diabetes. The treatment involves

A separate injection of pramlintide before a meal.
A reduction in insulin bolus by 50% for that meal [9].

Pramlintide can be infused using an insulin pump. At the present time, the mixing of pramlintide and insulin in the same cartridge is not an approved practice, so two infusion pumps are used simultaneously. Since insulin and amylin are co-secreted by the pancreatic beta cells in response to raising blood glucose levels, using pramlintide and insulin together more closely duplicates the function of the pancreas.

Symlin has the potential to support the artificial pancreas project because (1) in the future, insulin and pramlintide may be automatically infused together at a mixture from a single automatic insulin pump, or two infusion pumps could be used automatically with the insulin pump acting as master and the symlin pump acting as slave, or a dual system in one pump machine (two cartridges, a dual-infusion-set tube, and two subcutaneous insertions), (2) it improves postprandial glycemic excursions relative to insulin alone, this supports the possible use of an automatic bolus with less impact due to the delay of the insulin bolus, and (3) it simply duplicates the natural pancreas function, the full benefits of which are not fully understood.

20.9 Feedback of Real-Time Blood Glucose Data to an Insulin Pump for Bolus Control

The ability of the electronic controls of the infusion pump, particularly in the bolus-shaping capability, suggests that the control algorithm may replicate the function of the healthy pancreas in a more copycat fashion. At present, the insulin bolus is a predictive dose based on what is about to be eaten, and then infused completely. Even with the benefit of the closed-loop control of the basal insulin, the standard bolus is still a "guess and then fix it later" approach, compared to the pancreatic physiology, where insulin and amylin are released from the beta cells in pulses almost directly to the liver in response to the immediate blood glucose level. The natural release from the beta cells is a closed-loop response to sensed glucose, and the shape of the insulin delivery is adaptable and appropriate to the food eaten and the body's present metabolic capability.

As technology for continuous blood glucose monitoring improves, the integrated components will support a typical application of control theory by employing the proportional, integral, and derivative control algorithms [10]. This will make it feasible to infuse an *adaptive bolus* that changes its shape and integral dose based on the measured performance of the bolus in progress, depending on

The rate of glucose increase (i.e., the derivative function would deliver more insulin for a rapid increase in blood sugar).
The peak of the glucose curve (i.e., the proportional function would deliver more insulin for a higher peak in blood sugar).
The duration of elevated glucose (i.e., the integral function would deliver more insulin for a long duration of high blood sugar).

The adaptive bolus could start with an assumption of typical proportions and a bolus shape like **the combination bolus**. This could include

1. A prebolus of pramlintide (optional perhaps, but resolves the issue with insulin timing).
2. Initiation of a combination bolus with the initial spike sized in proportion to the present blood glucose level and trends in the change of blood glucose level.
3. Modification to the square-wave portion of the bolus, increasing or extending if blood sugar is increasing, and decreasing or limiting in duration when blood sugar is decreasing.

The benefits of an automatic bolus delivery might include

Increased accuracy in the total insulin delivered relative to what was needed.
Freedom to the user of the artificial pancreas.
Elimination of glycemic excursions due to user error (such as forgetting to bolus in conventional pump therapy).
Adaptability to changes in digestion of carbohydrates based on food choices.
Adaptability to variable metabolic needs due to stress, illness, or exercise.

20.10 Glucagon Combination

The purpose of glucagon is to raise blood sugar, primarily by promoting the release of stored glucose in the liver. Human glucagon has been synthesized by recombinant DNA technology and is available in a dry powder form in the glucagon rescue kit. Glucagon injection pens are also sometimes provided to diabetics in the UK along with insulin. This is useful for the rescue of unconscious diabetics from a severe state of hypoglycemia [11].

In healthy pancreatic function, glucagon production is initially suppressed by beta cell production of insulin and amylin when blood sugar is high, and then is later produced by low or falling blood sugar. The natural pancreatic function uses glucagon at the end of an insulin cycle to release glucose from the liver, with two advantages:

1. To prevent low blood sugar.
2. To speed the overall insulin action by cancelling the insulin tail.

If an artificial pancreas was to simulate the natural endocrine pancreas to the maximum extent, then insulin and amylin would be used at the beginning of an insulin cycle and glucagon would be used at the end of the insulin cycle. Research with diabetic pigs given an insulin–glucagon combination via separate subcutaneous infusion pumps demonstrated closed-loop control without incidence of hypoglycemia. While the copycat endocrine function including glucagon seems desirable, the benefits relative to the cost and complexity of an artificial pancreas without glucagon are not yet known.

20.11 Research Around the World

In the United States in 2006, the Juvenile Diabetes Research Foundation (JDRF) launched a multiyear initiative to help accelerate the availability of an artificial pancreas to people with diabetes. The overall goal of the Artificial Pancreas Project is to accelerate the development, regulatory approval, and acceptance of continuous glucose monitoring and artificial pancreas technology in the shortest possible timeframe. The long-term goal is for broad patient access and a thriving competitive market for these devices and products.

The JDRF's role in quickening the development and availability of the artificial pancreas consists of funding research in order to look over the outcomes of patients using the artificial pancreas, keeping close contact with the Food and Drug Administration so that the standards of the patient are met, advocating for healthcare coverage of technologies such as the artificial pancreas, and working to ensure clinical acceptance of technologies such as the artificial pancreas. In January of 2012, Medtronic said it received regulatory approval for the first remote glucose monitor that will let parents check the blood sugar of a diabetic child sleeping in another room.

Reading List

1. Kamiya H et al (2006) C-peptide reverses nociceptive neuropathy in Type 1 diabetes. Diabetes 55(12):3581–3587
2. Cerco Medical: Science: Methods
3. Cerco Medical: Company: Islet Sheet Research
4. Gene Therapy Approaches to Diabetes
5. http://www.hopkinsbayview.org/healthcarenews06/060605diabetes.html
6. Artificial pancreas information on MedicineNet.com
7. Ratner RE, Dickey R, Fineman M, Maggs DG, Shen L, Strobel SA, Weyer C, Kolterman OG (2004) Amylin replacement with pramlintide as an adjunct to insulin therapy improves long-term glycaemic and weight control in Type 1 diabetes mellitus: a 1-year, randomized controlled trial. Diabet Med 21(11):1204–1212
8. SYMLIN.com – Home

9. http://www.journalofdst.org/pdf/REVIEW/VOL-1-1-REV1-KLONOFF.pdf
10. Bowen, R. Glucagon may add another dimension to artificial pancreas. JDRF Res Frontline March 2007, 1–2.
11. Artificial Pancreas Project: Juvenile Diabetes Research Foundation International

Problems

1. The pancreas is a vital organ in the human body. Show in a neat sketch its structure, and briefly describe its function. How does it control blood sugar? What are the normal limits of blood sugar in humans? How does hyperglycemia affect other organs? People say blood sugar is the mother of all diseases; is such a statement justifiable?
2. Describe a modern wearable blood sugar controller that can replace the function of the pancreas. Explain with scientific reasoning.
3. Discuss the various approaches for control of blood sugar automatically.
4. Can you estimate the number of people suffering from hyperglycemia in India? Is it more prevalent in the Western world? Is it steadily increasing? Why it is affecting the younger population in our country? Suggest a possible reason.

Chapter 21
The Liver and Its Artificial Replacement

21.1 Introduction

The **liver** is a vital organ present in vertebrates and some other animals that is necessary for survival. It has a wide range of functions, including detoxification, protein synthesis, and production of biochemicals necessary for digestion. Currently, there is no way to compensate for the absence of liver function in sustaining life.

The liver plays a major role in metabolism, that is, digestion, and has a number of functions in the body, including glycogen storage, decomposition of red blood cells, plasma protein synthesis, hormone production, and detoxification. It lies below the diaphragm in the thoracic region of the abdomen. It produces bile, an alkaline compound that aids in digestion, via the emulsification of lipids. The liver's highly specialized tissues regulate a wide variety of high-volume biochemical reactions, including the synthesis and breakdown of small and complex molecules, many of which are necessary for normal vital functions. Medical terms related to the liver often start with **hepato- or hepatic** from the Greek word for liver.

It is a reddish-brown organ with four lobes of unequal size and shape. A human liver normally weighs between 1.4–1.6 kg (3.1–3.5 lb) and is a soft, pinkish-brown, triangular organ. It is both the largest internal organ (the skin being the largest of all organs) and the largest gland in our body.

It is located in the right upper quadrant of the abdominal cavity, resting just below the diaphragm. The liver lies to the right of the stomach and overlies the gallbladder. It is connected to two large blood vessels: the hepatic artery and the portal vein. The hepatic artery carries blood from the aorta, whereas the portal vein carries blood containing digested nutrients from the small intestine and the descending colon. These blood vessels subdivide into capillaries, which then lead to a lobule. Each lobule is made up of millions of hepatic cells, which are the basic metabolic cells.

S. Pal, *Design of Artificial Human Joints & Organs*,

DOI 10.1007/978-1-4614-6255-2_21, © Springer Science+Business Media New York 2014

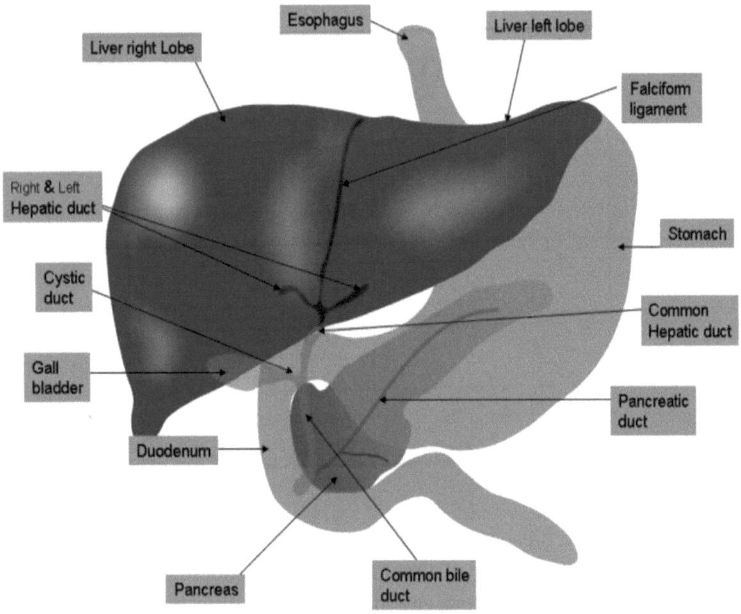

Fig. 21.1 Anatomical features of a human liver

21.2 Blood Flow

The liver receives a dual blood supply from the hepatic portal vein and hepatic arteries. Supplying approximately 75 % of the liver's blood supply, the hepatic portal vein carries venous blood drained from the spleen, gastrointestinal tract, and its associated organs. The hepatic arteries supply arterial blood to the liver, accounting for the remainder of its blood flow. Oxygen is provided from both sources; approximately half of the liver's oxygen demand is met by the hepatic portal vein, and half is met by the hepatic arteries. Blood flows through the sinusoids and empties into the central vein of each lobule. The central veins coalesce into hepatic veins, which leave the liver and empty into the inferior vena cava (Fig. 21.1).

21.3 Biliary Flow and the Biliary Tree

The term **biliary tree** is derived from the arboreal branches of the bile ducts. The bile produced in the liver is collected in bile canaliculi, which merge to form bile ducts. Within the liver, these ducts are called **intrahepatic** (within the liver) bile ducts, and once they exit the liver, they are considered **extrahepatic** (outside the liver). The intrahepatic ducts eventually drain into the right and left hepatic ducts, which merge to form the common hepatic duct. The cystic duct from the gallbladder joins with the common hepatic duct to form the common bile duct.

Bile can either drain directly into the duodenum via the common bile duct or be temporarily stored in the gallbladder via the cystic duct. The common bile duct and the pancreatic duct enter the second part of the duodenum together at the ampulla of Vater.

21.4 Lobes

Traditional gross anatomy divided the liver into four lobes based on surface features. The falciform ligament is visible on the front (anterior side) of the liver. This divides the liver into a left anatomical lobe and a right anatomical lobe.

If the liver is turned upside down, to look at it from behind (the visceral surface), there are two additional lobes between the right and left. These are the caudate lobe (the more superior), and below this the quadrate lobe. From behind, the lobes are divided by the ligamentum venosum and ligamentum teres (anything left of these is the left lobe); the transverse fissure (**or porta hepatis**) divides the caudate from the quadrate lobe; and the right sagittal fossa, which the inferior vena cava runs over, separates these two lobes from the right lobe.

Each of the lobes is made up of lobules; a vein goes from the center of each lobule and then joins to the hepatic vein to carry blood out from the liver.

On the surface of the lobules there are ducts, veins, and arteries that carry fluids to and from them. The various functions of the liver are carried out by the liver cells, or hepatocytes. Currently, there is no artificial organ or device capable of emulating all the functions of the liver. Some functions can be emulated by liver dialysis, an experimental treatment for liver failure.

21.5 Synthesis

The liver performs several roles in carbohydrate metabolism, gluconeogenesis (the synthesis of glucose from certain amino acids, lactate or glycerol). Note that humans and some other mammals cannot synthesize glucose from glycerol:

Glycogenolysis (the breakdown of glycogen into glucose)
Glycogenesis (the formation of glycogen from glucose; muscle tissues can also do this).

The liver is responsible for the mainstay of protein metabolism, synthesis as well as degradation. The liver also performs several roles in lipid metabolism:

Cholesterol synthesis
Lipogenesis
The production of triglycerides (fats).

The liver produces coagulation factors I (fibrinogen), II (prothrombin), V, VII, IX, X, and XI, as well as protein C, protein S, and antithrombin.

In the first-trimester fetus, the liver is the main site of red blood cell production. By the 32nd week of gestation, the bone marrow usually completely takes over that task.

The liver produces and excretes bile (a yellowish liquid), which is required for emulsifying fats. Some of the bile drains directly into the duodenum, and some is stored in the gallbladder and released when required.

The liver also produces insulin-like growth factor 1 (IGF-1), a polypeptide protein hormone that plays an important role in childhood growth and continues to have anabolic effects in adults. The liver is a major site of thromboprotein production. Thromboprotein is a glycoprotein hormone that regulates the production of platelets by the bone marrow.

21.6 Breakdown

The breakdown of insulin and other hormones: The liver breaks down hemoglobin, creating metabolites that are added to bile as pigment (bilirubin and biliverdin). The liver breaks down or modifies toxic substances (e.g., methylation) and most medicinal products in a process called drug metabolism. This sometimes results in toxication, when the metabolite is more toxic than its precursor. Preferably, the toxins are conjugated to avail excretion in bile or urine. The liver converts ammonia to urea.

21.7 Other Functions

The liver stores a multitude of substances, including glucose (in the form of glycogen), vitamin A (1–2 years' supply), vitamin D (1–4 months' supply), vitamin B12 (1–3 years' supply), iron, and copper.

The liver is responsible for immunological effects: The reticuloendothelial system of the liver contains many immunologically active cells, acting as a "sieve" for antigens carried to it via the portal system. The liver produces albumin, the major osmolar component of blood serum. The liver synthesizes angiotensinogen, a hormone that is responsible for raising the blood pressure when activated by renin, an enzyme that is released when the kidney senses low blood pressure.

21.8 Bioartificial Liver Device

A **bioartificial liver device (BAL)** is an artificial extracorporeal supportive device for an individual who is suffering from acute liver failure.

Currently, the purpose of BAL-type devices is not to permanently replace liver functions, but to serve as a supportive device [1], either allowing the liver to regenerate properly upon acute liver failure, or to bridge the individual's liver functions until a suitable transplant is obtained.

21.8.1 Function

BALs are essentially bioreactors, with embedded hepatocytes (liver cells) that perform the functions of a normal liver. They process oxygenated blood plasma, which is separated from the other blood constituents [2]. Several types of BALs are being developed, including hollow fiber systems and flat membrane sheet systems [3].

21.8.2 Hollow Fiber System

One type of BAL is similar to kidney dialysis systems that employ a hollow fiber cartridge. Hepatocytes are suspended in a gel solution, such as collagen, which is injected into a series of hollow fibers. In the case of collagen, the suspension is then gelled within the fibers, usually by a temperature change. The hepatocytes then contract the gel by their attachment to the collagen matrix, reducing the volume of the suspension and creating a flow space within the fibers. Nutrient media is circulated through the fibers to sustain the cells. During use, plasma is removed from the patient's blood. The patient's plasma is fed into the space surrounding the fibers. The fibers, which are composed of a semipermeable membrane, facilitate the transfer of toxins, nutrients, and other chemicals between the blood and the suspended cells. The membrane also keeps immune bodies, such as immunoglobulins, from passing to the cells to prevent an immune system rejection [4].

21.8.3 Comparison to Liver Dialysis

The advantages of using a BAL over other dialysis-type devices (e.g., liver dialysis) is that metabolic functions (such as lipid and plasma lipoprotein synthesis, regulation of carbohydrate homeostasis, production of serum albumin and clotting factors, etc.), in addition to detoxification, can be replicated without the use of multiple devices. There are several BAL devices currently in clinical trials.

A series of studies in 2004 showed that a BAL device reduced mortality by about half in acute liver failure cases [5]. The studies, which covered 171 patients in the U.S. and Europe, compared standard supportive care to the use of a bioreactor device using pig liver cells.

On June 28, 2010, Harrison Wein wrote in *Nature Medicine* about recent advances in BAL.

21.9 Progress Toward an Artificial Liver Transplant

Liver transplantation is currently the only available treatment for severe liver failure, but there aren't enough donors to fill the need. Researchers have now made transplantable liver grafts for rats that may point the way toward a successful liver transplant substitute for humans.

Our liver's job is to help fight infections and clean our blood. It also helps digest food and stores energy for when we need it. People needing a liver transplant in the U.S. are placed on a national waiting list kept at the United Network for Organ Sharing. Their blood type, body size, and severity of sickness all play a role in when they'll receive a liver. Whole livers can be donated only from people who have just died. Currently, there's an estimated shortfall of about 4,000 livers per year. It may be much higher in India.

Liver cell transplantation has shown some promise but has limited uses. To be successful, an artificial transplant must be sufficiently large to provide enough liver function. That requires a network of small blood vessels—called a microvascular network—to transport oxygen and nutrients throughout the structure.

Decellularization is the process of removing cells from a structure but leaving a scaffold with the architecture of the original tissue. It has shown some success in other organs. One group of scientists reported the decellularization of an entire heart that preserved the original architecture and microvascular network. A research team led by Dr. Korkut Uygun at Massachusetts General Hospital tried a similar approach for the liver; the work was supported by NIH's National Institute of Diabetes and Digestive and Kidney Diseases (NIDDK) and others.

In the June 13, 2010, online edition of *Nature Medicine*, Dr. Uygun and his team explained the process they developed. They used a gentle detergent over three days to decellularize the liver while preserving its structure. A matrix of proteins remained behind to hold the liver's shape. Using a dye, the researchers showed that the microvascular network in each eerily translucent liver was intact.

The researchers were then able to successfully introduce functional hepatocytes, i.e., liver cells, back into the matrix. When they tested the recellularized matrix, they found that it carried out liver-specific functions at levels comparable to a normal liver. Grafts transplanted into rats maintained their functional hepatocytes as well, for a few hours. The researchers noted, however, that successful engineering of an entire functional liver will require other types of cells.

Researcher Dr. Colin McGucklin and co-workers at Newcastle University say that pieces of artificial liver could be used to repair livers injured by disease, alcohol abuse, or other causes. These artificial livers could also be used outside the body in a manner analogous to the dialysis process used to keep alive patients whose kidneys have failed.

In next 10 years' time, entire livers could be grown in the lab and then be transplanted into human beings. The stem cells used by Drs. McGucklin and his team in their research are gathered from umbilical cords ("cord blood"), seen by some as a more ethical alternative to stem cells created from human embryos.

The cells are then placed in a **Bioreactor**, a device developed by NASA to simulate the weightless environment of space. The cells are situated in a growth medium that is constantly rotated, putting the cells in an endless state of free-fall. Ordinary cell growth in a nutrient medium in a dish does not provide a culture environment that supports three-dimensional tissue assembly. Epithelial cells without a three-dimensional assembly environment lack the proper clues for growing into the variety of cells that make up a particular tissue. Epithelial cells are the basic cells that differentiate tissue into specific organ functions. In a rotating Bioreactor, scientists can fool cells into behaving as though they are in a body.

References and for Further Study

1. Allen J, Hassanein T, Bhatia S (2001) Advances in bioartificial liver devices. Hepatology 34(3):447–455
2. Strain A, Neuberger J (2002) A bioartificial liver—state of the art. Science 295(5557): 1005–1009
3. Current work on the bioartificial liver. http://biomed.brown.edu/Courses/BI108/BI108_2002_Groups/liver/webpage/Current_Work.htm
4. University of Minnesota bioartificial liver: how it works. http://hugroup.cems.umn.edu/Research/bal/BAL-howitworks.htm
5. Bioartificial liver device – Wikipedia the free encyclopedia, originally published in 2007

Problems

1. Describe the structure and the various important functions of a normal human liver with a nice sketch. When it is necessary to replace it?
2. A **bioartificial liver device (BAL)** is an artificial extracorporeal supportive device for an individual who is suffering from acute liver failure. Describe the design and its function and limitations.
3. What is decellularization? How may the concept be utilized to understand the function of the liver?

Chapter 22
Female Organs and Their Artificial Replacements

22.1 Introduction

The female body has some special organs, such as breasts, that constitute their femininity, beauty. The organs we will discuss in this chapter will be limited to the female breast and vagina.

In reality, female breasts come in different sizes and shapes and are dependent upon genetic factors, regionalism, diet, climate, and several other factors. In short, just as human faces are different, so are women's breasts. In fact, having asymmetrical breasts, where one is bigger than the other, is very common among women.

For some women, having beautiful, big breasts is important because they consider the breasts the most anxiety-provoking area of the body. This could partly be due to the breasts' prominent position or the image that the women need to project to the public at large. What we think important may not be so important to others. Women should not worry about their breast development too much. Today it is possible to reshape, augment, and add firmness to sagging breasts. Figure 22.1 shows several variations of these common women's breast shapes:

(a) **Perfect breasts—The perfect breast shape is quite a rarity. In medicine and aesthetics, "perfect" breasts are** equal in size, with no sag; the breasts are supple, well-toned, and symmetrical. The nipples point horizontally. Often when women go for cosmetic breast surgery, they want not only to reduce or increase the size and shape of their breasts, but they also clamor for that "perfect" shape.

    The perception of perfect breast shape may also vary from culture to culture. What one society may find as the perfect breast shape, another society may not brand the desired one.

(b) **Swooping breasts**—A shape in which the breast slightly bends inward above the areola. There is no sag, however. Due to the bending of the breast, the nipple points upward, inclined to the vertical. Some women's breasts have this shape because of a lack of volume.

(c) **Saggy or ptotic (drooping) breasts**—Saggy breasts are commonly found in women of advanced age. Usually, the breasts droop downward, which causes

DOI 10.1007/978-1-4614-6255-2_22,

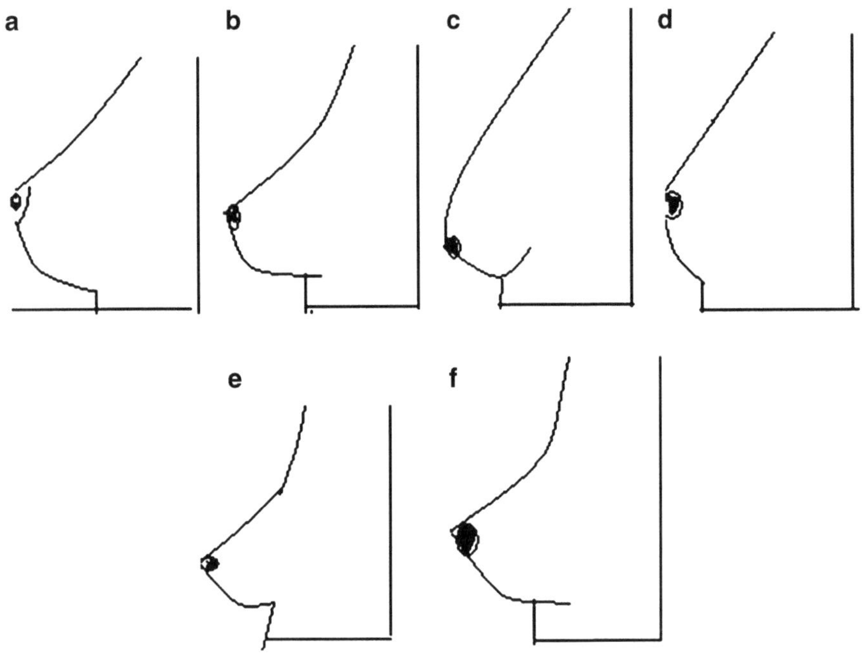

Fig. 22.1 The various shapes of the female breast

the nipple to be pointed downward. Saggy breasts may have a lot or a little volume, depending on the amount of fat tissue in them.

(d) **Small breasts**—Small breasts have very little fatty tissue and have small nipples and areolas as well. There is very little substance between the nipples and the pectoral muscles for this type of breast shape.

(e) **Tubular or constricted breasts**—These are actually a kind of defective or anomalous breast shape due to a hernia in the breast tissue. They appear as tubular or narrow cylindrical in shape, with very small nipples and areolas, and the two breasts may be far apart. The base of the breast usually isn't wide enough and is small, and the nipple and areola complex may be overly prominent since the breast tissue has been herniated or squeezed into the tip.

(f) **Augmented breasts**—A severe form of tubular breasts and a visible anomaly.

(g) **Pectus carinatum, or pigeon breasts**—These are severely deformed breasts that lie almost flat on the chest and do not look like breasts at all. The ribs and sometimes the breastbone (sternum) protrude forward; this is caused by a congenital defect.

22.2 Breast Implants

A breast implant is a prosthesis used to enlarge the size of a woman's breasts (known as breast augmentation, breast enlargement, mammoplasty, augmentation mammoplasty) for cosmetic reasons, to reconstruct the breast (e.g., after a mastectomy, or to correct genetic deformities), or as an aspect of male-to-female sex reassignment surgery. According to the American Society of Plastic Surgeons, breast augmentation is the most commonly performed cosmetic surgery procedure in the United States. In 2006, as many as 329,000 breast augmentation procedures were performed in the U.S. [1].

There are two primary types of breast implants: saline-filled and silicone gel–filled implants. Saline implants have a **silicone** elastomer shell filled with sterile **saline** liquid. Silicone gel implants have a silicone shell filled with a viscous **silicone** gel. There have been several alternative types of breast implants developed, such as **polypropylene string** or soy oil, but these are uncommon and not recommended.

22.2.1 History

Implants have been used since 1895 to augment the size or shape of women's breasts. The earliest-known implant was attempted by Czerny, using a woman's own adipose tissue (from a lipoma, a benign growth, on her back) [2]. Gersuny tried paraffin injections in 1889, with disastrous results. Subsequently, in the early to mid-1900s, a number of other substances were tried, including ivory, glass balls, ground rubber, bovine cartilage, Terylene wool, gutta-percha, Dicora, polyethylene chips, polyvinyl alcohol-formaldehyde polymer sponge (Ivalon), Ivalon in a polyethylene sac, polyether foam sponge (Etheron), polyethylene tape (Polystan) or strips wound into a ball, polyester (polyurethane foam sponge), Silastic rubber, and Teflon-silicone prostheses [3, 4]. In recent history, various creams and medications have been used in attempts to increase bust size, and Berson in 1945 and Maliniac in 1950 performed a flap-based augmentation by rotating the patient's chest wall tissue into the breast to add volume. Various synthetics were used throughout the 1950s and 1960s, including silicone injections, which an estimated 50,000 women received [5]. The development of silicone granulomas and hardening of the breasts were so severe in some cases that women needed to have mastectomies for treatment. Women sometimes seek medical treatment for complications up to 30 years after receiving this type of injection.

22.2.2 Indications

Breast implants are used for

Primary augmentation (to increase breast size for cosmetic reasons)

Revision-augmentation (revision surgery to correct or improve the result of an original breast augmentation surgery)

Primary reconstruction (to replace breast tissue that has been removed due to cancer or trauma or that has failed to develop properly due to a severe breast abnormality)

Revision-reconstruction (revision surgery to improve the result of an original breast reconstruction surgery).

22.3 Patient Characteristics

Candidates seeking breast augmentation have been reported as usually being younger, healthier, from a higher socioeconomic status, and more often married with children than the population at large [6, 7]. Many of these patients have reported greater distress about their appearance in a variety of situations and have endured awkward situations about their appearance. Studies have identified a pattern (shared by many cosmetic surgery procedures) that suggest women who undergo breast implantation are slightly more likely to have undergone psychotherapy and have a higher prevalence of depression, suicide attempts, and mental illness (including body dysmorphia [8]) as compared to the general population [9]. Postoperative surveys on mental health and quality-of-life issues have reported an improvement on a number of dimensions, including physical health, physical appearance, social life, self-confidence, self-esteem, and sexual function [10–13]. Longer-term follow-up suggests these improvements may be transitory, with the exception of body esteem related to attractiveness [14]. Most patients report being satisfied long-term with their implants even when they have required reoperation for complications or aesthetic reasons (Fig. 22.2) [15].

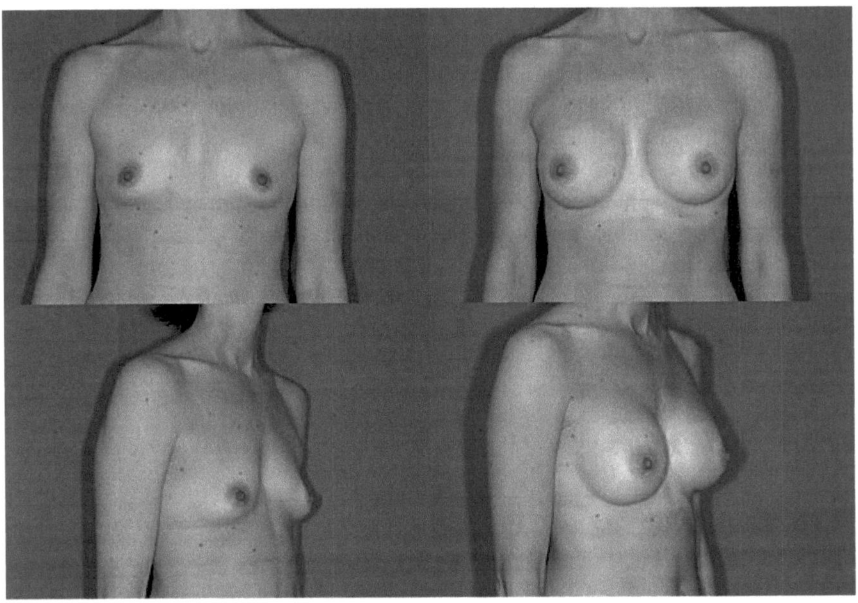

Fig. 22.2 Small breasts (*left*) and breasts after implantation (*right*); front and side views of both are shown

22.4 Types of Implants

Saline-filled breast implants were first manufactured in France in 1964, introduced by Arion [16] with the goal of being surgically placed via smaller incisions. Current saline devices are manufactured with thicker, room-temperature vulcanized (RTV) shells. These shells are made of silicone elastomer, and the implants are filled with saline water after the implant is placed in the body. Since the implants are empty when they are surgically inserted, the scar is smaller than is necessary for silicone gel breast implants (which are filled with silicone before the surgery is performed). A single manufacturer (Poly Implant Prosthesis, France) produced a model of pre-filled saline implants that has been reported to have high failure rates *in vivo* (Fig. 22.3) [17].

Saline-filled implants were the most common implant used in the U.S. during the 1990s due to restrictions that existed on silicone implants, but they were rarely used in other countries. Good to excellent results may be obtained, but compared to silicone gel implants, saline implants are more likely to cause cosmetic problems such as rippling and wrinkling and to be observable to the eye or noticeable by touch. Particularly for women with very little breast tissue, or for postmastectomy breast reconstruction, it's felt that silicone gel implants are the superior device. In patients with more breast tissue in whom submuscular implant placement is used, saline implants can look very similar to silicone gel (Fig. 22.4).

Fig. 22.3 Silicone gel–filled breast implants

Fig. 22.4 Saline-filled breast implant that has been surgically placed inside the existing breast

22.5 Silicone Gel Implants

Thomas Cronin and Frank Gerow, two Houston (Texas) -based plastic surgeons, developed the first silicone breast prosthesis with the Dow Corning Corporation in 1961. The first woman was implanted in 1962. Silicone implants are generally described in terms of five generations, which segregate common characteristics of manufacturing techniques. These are described briefly here.

22.5.1 *First Generation*

The Cronin–Gerow implants were made of a silicone rubber envelope (or sac), filled with a thick, viscous silicone gel with a Dacron patch on the posterior shell [18, 19]. They were firm and had an anatomic "teardrop" shape.

22.5.2 *Second Generation*

In response to surgeons' requests for softer and more natural implants, breast implants were redesigned in the 1970s with thinner, less cohesive gel and thinner shells. These implants had a greater tendency to rupture and leak, or "bleed" silicone through the implant shell, and complications such as capsular contracture were quite common. It was predominantly implants of this generation that were involved in the class-action lawsuits against Dow Corning and other manufacturers in the early 1990s.

Another development in the 1970s was a polyurethane foam coating on the implant shell that was effective in diminishing capsular contracture by causing an inflammatory reaction that discouraged the formation of fibrous tissue around the capsule. These implants were later briefly discontinued due to concerns of potential carcinogenic breakdown products from the polyurethane [20]. A review of the risk for cancer from 2,4-Toluenediamine (TDA) by the FDA later concluded that the risk was too small to justify recommending explantation of the devices from individual patients. Polyurethane implants are still used in Europe and South America, but no manufacturer has sought FDA approval for sale in the U.S. [21]. Second-generation implants also included various "double-lumen" designs. These implants were essentially a silicone implant inside a saline implant. The double lumen was an attempt to provide the cosmetic benefits of gel in the inside lumen, while the outside lumen contained saline and its volume could be adjusted after placement. The failure rate of these implants is higher than for single-lumen implants due to their more complex design.

22.5.3 Third and Fourth Generations

Third- and fourth-generation implants, from the mid-1980s, represented sequential advances in manufacturing principles with elastomer-coated shells to decrease gel bleed and are filled with thicker, more cohesive gel. These implants are sold under restricted conditions in the U.S. and Canada and are widely used in other countries. The increased cohesion of the gel filler reduces potential leakage of the gel compared to earlier devices. A variety of both round and tapered anatomic shapes are available. Anatomic-shaped implants are uniformly textured to reduce rotation, while round devices are available in smooth or textured surfaces.

22.5.4 Fifth Generation

The evaluation of "gummy bear" or solid, high-cohesive, form-stable implants is in preliminary stages in the U.S., but these implants have been widely used since the mid-1990s in other countries. The semisolid gel in these types of implants largely eliminates the possibility of silicone migration. Studies of these devices have shown significant potential improvements in safety and efficacy over the older implants, with low rates of capsular contracture and rupture [22–24].

22.6 Implant Pocket Placement

The placement of implants is described in relation to the pectoralis major muscle (Fig. 22.5).

Subglandular: the implant is between the breast tissue and the pectoralis muscle. This position closely resembles the plane of normal breast tissue and is believed by

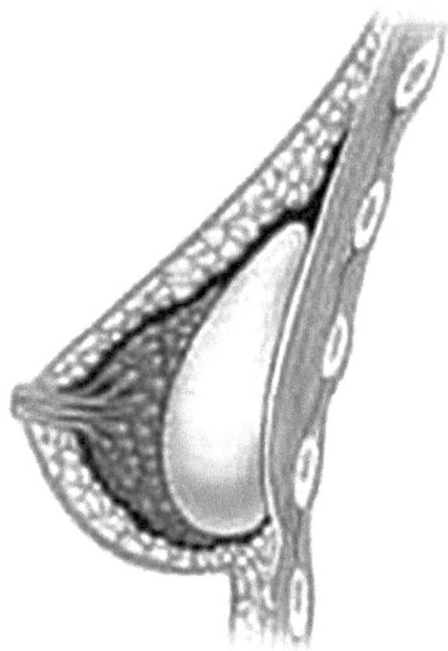

Fig. 22.5 Subglandular breast implant diagram

many to achieve the most aesthetic results. The subglandular position in patients with thin soft-tissue coverage is most likely to show ripples or wrinkles of the underlying implant. Capsular contracture rates are also slightly higher with this approach, and placement of implants in this pocket might be inappropriate in women who are at risk for capsule formation (smokers, multiple breast surgeries).

Subfascial [25]: the implant is placed in the subglandular position, but underneath the fascia of the pectoralis muscle. The benefits of this technique are debated [26], but proponents believe the (sometimes thick) fascial sheet of tissue may help with coverage and sustaining the positioning of the implant. Implants that undergo capsular contraction are unlikely to displace upward or toward the underarm.

Subpectoral ("dual-plane") [27]: the implant is placed underneath the pectoralis major muscle after releasing the inferior muscular attachments. As a result, the implant is partially beneath the pectoralis in the upper pole, while the lower half of the implant is in the subglandular plane. This is the most common technique in North America and achieves maximal upper implant coverage while allowing expansion of the lower pole.

Submuscular: the implant is placed below the pectoralis without release of the inferior origin of the muscle. Total muscular coverage may be achieved by releasing the lateral chest wall muscles (serratus and/or pectoralis minor) and attaching to the pectoralis major. This technique is most commonly used for maximal coverage of implants used in breast reconstruction.

22.7 Failure and Rupture

Breast implants can potentially remain intact for decades in the body, but all such devices will fail at some point. When saline breast implants break, they often deflate quickly and can be easily removed. Prospective studies of saline-filled breast implants showed rupture/deflation rates of 3–5 % at 3 years and 7–10 % at 10 years for augmentation patients. Among the suspected mechanisms for rupture are damage during implantation or other procedures, degradation of the implant shell, blunt or penetrating chest trauma, and in rare instances rupture from the pressure of traditional mammograms.

The age and design of the implant are the most important factors in rupture, but estimating the rupture rates of more contemporary devices has been difficult, as most previous reports mixed heterogeneous groups of devices in nonrandomized populations. The only available literature with longer-term available MRI data on single-lumen third- and fourth-generation silicone implants comes from Europe and has reported silent rupture rates of an implant at between 8 and 15 % at or around a decade (or 15–30 % of patients). In 2009, patients followed in one arm of the core FDA clinical trials for primary breast augmentation reported rupture rates of 1.1 % at 6 years' follow-up (28–31).

The first series of MRI evaluation of the highly cohesive (fifth-generation) gel implants suggests improved durability, with a rupture rate reported at 1 % or less at a median age of 6 years.

It has been suggested that clinical exams alone are inadequate to evaluate suspected rupture after a study reported that only 30 % of ruptures in asymptomatic patients are accurately detected by experienced plastic surgeons, compared to 86 % detected by MRIs. The FDA has recommended that MRIs be considered to screen for silent rupture starting 3years after implantation and then every 2 years thereafter. Other countries have not endorsed routine MRI screening and have taken the position that MRI should be reserved only for cases involving suspected clinical rupture or to confirm mammographic or ultrasound studies suggesting rupture.

When silicone implants break, they rarely deflate, and the silicone from the implant can leak out into the space around the implant. An intracapsular rupture can progress to outside the capsule (extracapsular rupture), and both conditions are generally agreed to indicate the need for removal of the implant. Extracapsular silicone has the potential to migrate, but most clinical complications have appeared to be limited to the breast and axillae in the form of granulomas (inflammatory nodules) and axillary lymphadenopathy (enlarged lymph glands in the armpit area). The specific risk of and treatment for extracapsular silicone gel are still controversial.

22.8 Implants and Mammography

The presence of radio-opaque breast implants may interfere with the sensitivity of screening mammography. Specialized radiographic techniques where the implant is manually displaced (Eklund views) may improve this somewhat, but approximately one third of the breast is still not adequately visualized, with a resultant increase in false-negative mammograms. A number of studies looking at breast cancers in women with implants have found no significant difference in stage of disease at time of diagnosis, and the prognosis appears to be similar in both groups, with augmented patients not at a higher risk for subsequent cancer recurrence or death [28, 29]. Conversely, the use of implants for reconstruction after mastectomy for breast cancer also appears not to have a negative effect on cancer-related mortality.

The presence of a breast implant does not influence the ability for breast conservation (lumpectomy) surgery for women who subsequently develop breast cancer and does not interfere with the delivery of external beam radiation (XRT) treatments that may be required. Fibrosis of breast tissue after XRT is common, and an increase in capsular contracture rates would be expected.

As studies have followed women with implants for a longer period of time, more information has been made available to assess these issues. A 2004 Danish study reported that women who had breast implants for an average of 19 years were no more likely to report an excess number of rheumatic symptoms than control groups. A large study of plastic surgery patients found a decreased standardized mortality ratio in both breast implant and other plastic surgery patients, but a relatively increased risk of lung cancer deaths in breast implant recipients compared to other forms of plastic surgery. The authors attributed this to differences in smoking rates. Another large study of nearly 25,000 Canadian women with implants recently reported a 43 % lower rate of breast cancer compared with the general population and a lower-than-average risk of developing cancer of any kind [29].

22.8.1 Breast Implant Manufacturers

Allergan—formerly McGhan Medical & Inamed Corp.
Mentor Corporation—a subsidiary of Johnson & Johnson

22.8.2 Artificial Breast Prostheses

Artificial breast prostheses are made of silicone and placed externally on the breast surface; a bra is then used as usual. Cancer patients and women with smaller breasts are using these prostheses. China supplies them at a cost of $1–15 per pair. There are several sizes, such as small, medium, and large. Some important characteristics are noted below:

- 100 % FDA-approved grade material
- Natural look and feel
- Comfortable to wear
- Odorless and nontoxic
- Washable, easy to clean and dry
- Many colors and designs available
- High-quality, competitively priced (Fig. 22.6).

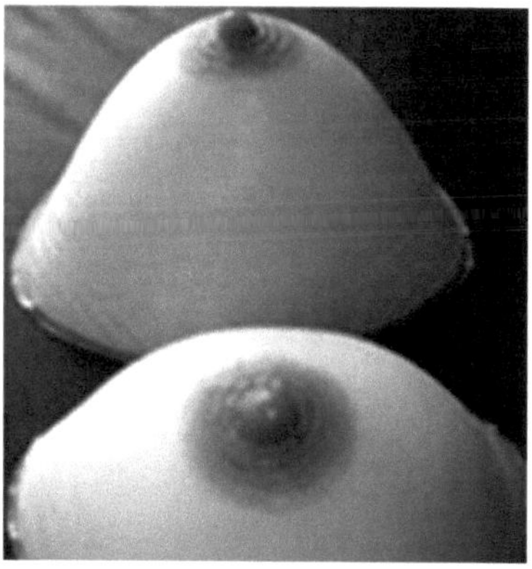

Fig. 22.6 Fake breast for cosmetic purposes

22.9 Artificial Vagina

An artificial vagina is a device designed to simulate the female sex organ. To achieve this, it will generally be made of a soft material, lubricated, and sometimes heated. It may be designed for medical research purposes, for animal breeding, or as a sex toy for erotic stimulation.

22.9.1 Veterinary Use

Artificial vaginas are widely used by farms in cattle artificial insemination programs and semen collection centers for collecting animal sperm and its further application. An artificial vagina designed for collecting semen will imitate some or all of the anatomical features and behaviors of an animal's vagina. A breeding mount with a built-in artificial vagina is used to collect semen from horses for use in artificial insemination.

There are several types of collecting apparatus, but the general design uses a tube with a normally sterile inner liner and hard outer shell. The walls of the tube may be hollow and filled with warm water to mimic a natural body temperature for better results, and may contain a filter to separate the semen.

22.10 Usage as Sex Toy

An artificial vagina for the purposes of sexual stimulation is essentially an aid to masturbation: It is designed to simulate the sensation of sexual intercourse on the erect penis. It will often have moving parts such as vibrators that increase stimulation rather than accurately simulate a woman's vagina (Fig. 22.7).

Usually, the artificial vagina has a realistic or close-to-realistic appearance with a sleeve, where the penis can be inserted. The sleeve, or "vaginal tunnel" as it is

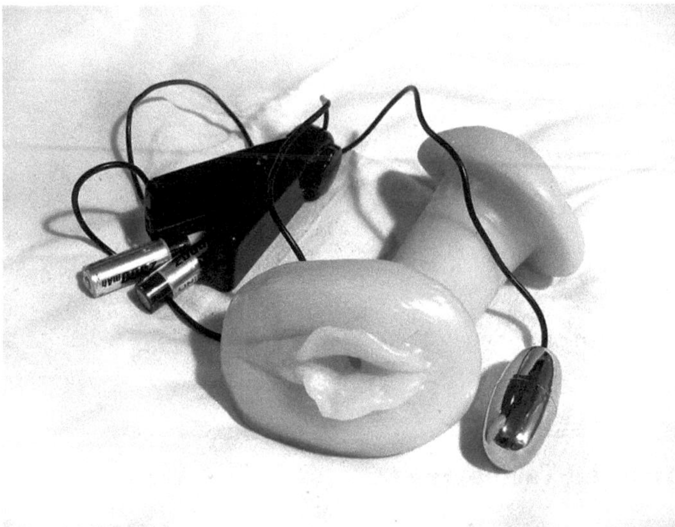

Fig. 22.7 Artificial vagina with vibrating egg

called, measures on average 10–20 cm (4–8 in.) and can have an open end for inserting a vibrating bullet if the user wishes.

22.10.1 Types

Realistic: This type of artificial vagina is modeled on female genitals. For advertising purposes, many manufacturers design the vagina as an exact replica of some famous pornographic actress's vagina. Realistic vaginas are made to simulate the natural physiology: pubic bones, hairs, labia, all natural creases and dimples, etc.

Several realistic vaginas are manufactured with a narrower anal orifice for those users who also like to simulate anal sex. Though the natural vagina does not include any nubs and ribs inside the tunnel, manufacturers often add them into artificial vaginas for enhancing friction.

In Japan, *onacups* are popular "realistic" artificial vaginas that simulate various sex acts and come with lubricant kept inside. Vibrating: These artificial vaginas superficially resemble realistic vaginas, but they have the added feature of a vibrating element—usually a removable vibrating bullet that can be easily inserted into a hole into the bottom or the end of the sleeve. For more intensive sensations, there is a variety of functions: vibrating, pulsating, surging, multispeed, etc.

Vibrating bullets are usually controlled by a panel connected by a wire. Some kinds of vibrating artificial vaginas are designed to create a sucking effect, replicating the role of a penis pump—with a bulb squeezed by the user. These are probably used in the Western world more so than in India.

22.11 Materials

The artificial vagina as a pleasure toy is designed from materials that maximally imitate the natural skin. The materials used in manufacturing artificial vaginas are stretchable and elastic for accommodating any male penis size.

Cyber skin, ultra-realistic or futurotic (mixture of PVC and silicone), and some other patented materials: natural-feeling materials that are quite porous, therefore requiring special care before and after use. These materials from the trademarked lines are also highly sensual and plush. CyberSkin is the brand name of a soft elastomer material that is intended to emulate the feel of human skin. It is made by the adult toy manufacturer Topco Sales. The company states that the technology behind CyberSkin is a "new Aerospace 601 computerized injection molding machine, originally designed by NASA engineers." Topco sales is a US-based co.

Similar materials are available under other brand names, such as **Ultraskin, Eroskin, Softskin, SoftTouch, Cyber Jel-lee, New Supersoft, Futurotic, and UR3 (Ultra-realistic)**. All are soft and stretchy, with a velvety texture that is caused by talcs and cornstarches used to prevent the material from becoming unpleasantly sticky.

The materials, from the group of thermoplastic elastomers, are mixtures of polyvinyl chloride and silicone. As such, they are attacked by oil, petroleum, and silicone oils. Therefore, only water-based lubricants can be used without causing damage.

Rubber (elastic hydrocarbon polymer): an extremely flexible and resilient material with a high level of durability. The porous nature of rubber makes it hard to clean.

Soft plastics (poly vinyl chloride): a very popular material for sex toys that creates a jelly-like feel, though it has a specific rubber odor.

Latex (natural rubber derived from plants): flexible material that may cause allergic reactions in some individuals.

The elastic properties of all materials that are used in manufacturing artificial vaginas are combined with a porous nature; therefore, artificial vaginas require special care to avoid bacterial accumulation. Manufacturers recommend protecting them by applying a condom during use. The manufacturing methods are silicone rubber molding and subsequent casting.

References

1. Czerny V (1895) Plastischer Ersatz der Brusthus durch ein Lipoma [in German]. Zentralbl Chir 27:72
2. Bondurant S, Ernster V, Herdman R, Institute of Medicine (U.S.) Committee on the Safety of Silicone Breast Implants (1999) Safety of silicone breast implants. National Academies Press, Washington, DC, 21
3. Anderson N (1997) Lawsuit science: lessons from the silicone breast implant controversy. N Y Law School Law Rev 41(2):401–407
4. Brinton L, Brown S, Colton T, Burich M, Lubin J (2000) Characteristics of a population of women with breast implants compared with women seeking other types of plastic surgery. Plast Reconstr Surg 105(3):919–927

5. Crerand CE et al (2006) Body dysmorphic disorder and cosmetic surgery. Plast Reconstr Surg 118(July):1672–1802
6. Sarwer DB et al (2003) Body image concerns of breast augmentation patients. Plast Reconstr Surg 112(July):83–90
7. Young VL et al (1994) The efficacy of breast augmentation: breast size increase, patient satisfaction, and psychological effects. Plast Reconstr Surg 94(Dec):958–969
8. Chahraoui K et al (2006) Aesthetic surgery and quality of life before and four months postoperatively. J Long-Term Effects Med Implants 51(3):207–210
9. Cash TF et al (2002) Women's psychosocial outcomes of breast augmentation with silicone gel-filled implants: a 2-year prospective study. Plast Reconstr Surg 109(May):2112–2121
10. Figueroa-Haas CL (2007) Effect of breast augmentation mammoplasty on self-esteem and sexuality: a quantitative analysis. Plast Surg Nurs 27(Mar):16–36
11. Hande IN et al (2006) A long-term study of outcomes, complications, and patient satisfaction with breast implants. Plast Reconstr Surg 117:757–767
12. Wilson TD (2002) Pharaonic circumcision under patriarchy and breast augmentation under phallocentric capitalism: similarities and differences. Violence Against Women 8:495
13. Brinton LA et al (2001) Mortality among augmentation mammoplasty patients. Epidemiology 12:312–316
14. Koot VC et al (2003) Total and cause specific mortality among Swedish women with cosmetic breast implants: a prospective study. BMJ 326:527–528
15. Pukkala E et al (2003) Causes of death among Finnish women with cosmetic breast implants, 1971–2001. Ann Plast Surg 51:339–342
16. Johnson GW, Christ JE (1993) The endoscopic breast augmentation: the transumbilical insertion of saline-filled breast implants. Plast Reconstr Surg 92(5):801–808
17. Dowden RV (2008) Transumbilical breast augmentation is safe and effective. Semin Plast Surg 22(1):51–59
18. Wallach SG (2004) Maximizing the use of the abdominoplasty incision. Plast Reconstr Surg 113(1):411–417
19. Stevens WG, Hirsch EM, Stoker DA, Cohen R (2006) *In vitro* deflation of prefilled saline breast implants. Plast Reconstr Surg 118(2):347–349
20. Cronin TD, Gerow FJ (1963) Augmentation mammaplasty: a new "natural feel" prosthesis. Excerpta Medica International Congress Series 66:41
21. Hester TR Jr, Tebbetts JB, Maxwell GP (2001) The polyurethane-covered mammary prosthesis: facts and fiction (II): a look back and a "peek" ahead. Clin Plast Surg 28(3):579–586
22. Brown MH, Shenker R, Silver SA (2005) Cohesive silicone gel breast implants in aesthetic and reconstructive breast surgery. Plast Reconstr Surg 116(3):768–779, discussion 780–781
23. Heden P, Jernbeck J, Hober M (2001) Breast augmentation with anatomical cohesive gel implants: the world's largest current experience. Clin Plast Surg 28(3):531–552
24. FDA Breast Implant Consumer Handbook—2004
25. FDA Approves Silicone Gel-Filled Breast Implants. FDA. http://www.fda.gov/bbs/topics/NEWS/2006/NEW01512.html.
26. Kjoller K, Holmich LR, Fryzek JP, Jacobsen PH, Friis S, McLaughlin JK, Lipworth L, Henriksen TF, Hoier-Madsen M, Wiik A, Olsen JH (2004) Self-reported musculoskeletal symptoms among Danish women with cosmetic breast implants. Plast Ann Plast Surg 52(1):1–7
27. Fryzek JP, Holmich L, McLaughlin JK, Lipworth L, Tarone RE, Henriksen T, Kjoller K, Friis S (2007) A nationwide study of connective tissue disease and other rheumatic conditions among Danish women with long-term cosmetic breast implantation. Ann Epidemiol 17(5):374
28. Jacobsen PH (2004) Mortality and suicide among Danish women with cosmetic breast implants. Arch Intern Med 164:2450–2455
29. Villeneuve PJ et al (2006) Mortality among Canadian women with cosmetic breast implants. Am J Epidemiol 164:334–341

Problems

1. Why are breast augmentations and replacements necessary? Describe the types of breasts. Give a brief account of the breast design and materials used over the past 100 years.
2. How would you design the shape and size of the human breast implant? Which material is most suitable for such an application? Dow Corning was the largest supplier of such implants in the world. The Dow Corning implant became hardened after several years of use; they were all recalled, and huge financial compensation was paid to each sufferer. How was this problem solved subsequently? Which liquid is currently being used in such a baggy implant?
3. Suggest how you can design a wearable menstrual blood collection device. It is known that menstrual blood contains stem cells that can be separated for use in regenerative medicine.
4. Artificial insemination is very common for animals such as horses and cows. How will you design an artificial vagina for the collection of semen?
5. For breast cancer patients, what type of artificial breast will be suitable and how would you design and develop such a breast? Which material is suitable for such a product?

Chapter 23
The Medical Device Market and Ethical Issues of Implants

23.1 Introduction to the Medical Device Market

In recent years, the demand for medical devices has grown faster than expected. The world medical device market totaled $209 billion in 2006 and was projected to grow with an average annual rate of 6–9% through 2010 and beyond. Acmite market intelligence produced this data. The medical device industry is highly dynamic, where dramatic innovations and developments are taking place every day. Driven by both the increasing demand in overseas markets and companies' ambition to pursue profits globally, the globalization of the medical device industry has intensified. Accordingly, the need to understand the global market has grown. Medical device manufacturers face mounting pressure at all stages to reduce costs and increase margins, while still delivering product advancements, superior quality, and excellent customer service. As companies modernize their business practices, they are driven by the need to remain competitive and retain critical survival capabilities, such as agility and flexibility in a fast-changing marketplace. The opportunities are tremendous for players in the contract manufacturing market. The primary drivers for medical device outsourcing include not only the growth pattern of the overall medical device market, but also the extent to which device manufacturers choose to outsource their manufacturing and assembly operations. The present frugal economic conditions have additionally reinforced the need to upgrade medical device manufacturing in a cost-effective, yet competitive way.

Acmite Market Intelligence has produced a most comprehensive market report on the global medical devices market. The market report examines the current products and cutting-edge technologies, provides comprehensive market data for 2006 and market forecasts through 2010–2015, outlines the competition landscape, and evaluates market opportunities and risks based on a series of influence factors.

With a multidimensional and in-depth view of the world medical device market, this study is ideal for a decision maker who aims at international market penetration or plans to expand the scope of his or her business.

S. Pal, *Design of Artificial Human Joints & Organs*,
DOI 10.1007/978-1-4614-6255-2_23,

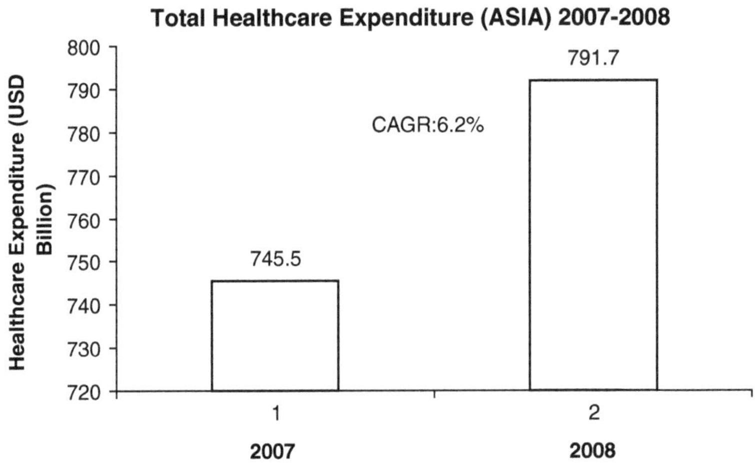

Fig. 23.1 Asia*—Countries included in the statistics are the Philippines, Malaysia, India, Indonesia, China, Thailand, Singapore, Taiwan, South Korea, Hong Kong, and Japan. Compound annual growth rate: 6.2%. (*Source:* Frost and Sullivan)

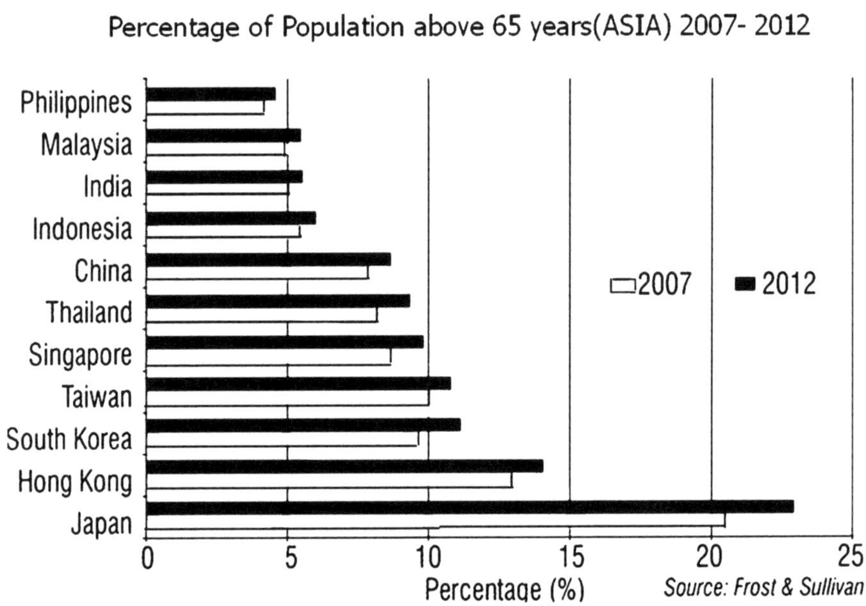

Fig. 23.2 Percentage of population above 65 years old in Asia, 2007–2012. (*Source:* Frost and Sullivan)

Figures 23.1 and 23.2 depict the healthcare expenditures of Asian countries and the rises in life expectancy of the populations in these countries. These data will give a tremendous boost to the development of the healthcare business in this region.

Realizing the importance of disease prevention, Asian governments have invested billions of dollars annually to improve their healthcare infrastructure. For instance, the Ninth Malaysia Plan (2006–2010) strove to work "towards achieving better health through consolidation of services" whereby emphasis was placed on sustainability, upgrading and maintenance of existing facilities and equipment, and improving the quality of healthcare.

China announced its Healthy China 2020 plan, published in Health clinical forum which aims to provide safe, effective, convenient, and low-cost public health and basic medical care to both rural and urban citizens by 2020. Some key goals of the plan include making public medical institutions "nonprofit," reducing the involvement of hospitals in the sale of drugs, increasing the role and responsibility of government, and establishing a basic medical care network for all Chinese citizens. These policies clear show the eagerness of the government to provide a better healthcare infrastructure, with better medical devices and facilities.

The full potential of Asian markets can be realized by establishing key partnerships and alliances with local companies, keeping in mind the market drivers and challenges posed by the region. The fastest way to penetrate these highly protected, untapped Asian markets is through effective joint ventures with the emerging medical device companies of Asia.

Geographically, big Asian countries provide a great opportunity for telemedicine and teleradiology. In view of the poor transportation facilities in parts of Asia, telemedicine and teleradiology could help in providing more efficient and timely medical care in rural areas.

23.2 Ethical Issues of Implants

It is important to remember that all implants are manmade and are nonliving although their purpose is to replace living tissues. If we want to maintain a high quality of life, it is necessary to replace worn-out body parts as we grow older, even if we don't have all the answers. It is necessary to accept that manmade spare parts will not be as good as the living parts they replace. They cannot repair themselves, as can most living tissues and organs. This means that the success of an implant must always be considered from the viewpoint of uncertainty.

Uncertainty means that one must accept that there will be relative success and relative failure when part of one's body is replaced. There is a finite probability that any implant will outlast the patient—or the other way around. We have no choice but to accept this reality.

We need to remember that it is only during the past century that the length of life has increased to the point where implants have become necessary. Many of the ethical dilemmas we consider only because we live in a culture where life has been prolonged from an average of 45 years to an average of 70+ years. The dilemmas that result are new, and there are no easy solutions.

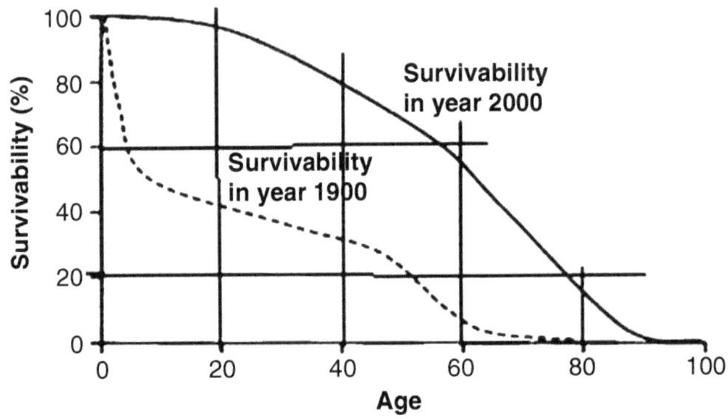

Fig. 23.3 Increase in human life expectancy in the U.S. and in Europe during the last century (adopted courtesy of Prof. Larry L. Hench, Imperial College London)

23.2.1 The Need for Implants

For centuries, when tissues became diseased or damaged, a physician had little choice but to remove the diseased part, with obvious limitations. Removal of joints, vertebrae, teeth, or organs led to only a marginally improved quality of life. However, human survivability seldom exceeded the progressive decrease in the quality of tissues, so the need for replacement parts was small. During the last century, the situation changed greatly. The discovery of antiseptics, penicillin, and other antibiotics, chemical treatment of water supplies, improved hygiene, and vaccinations have all contributed to a major increase in human survivability in developed countries (Fig. 23.3). Life expectancy is now in the range of 68–70+ years globally and 65+ years in India, as indicated in a World Health Organization (WHO) report published in *The Statesman* (May 18, 2011). Poor health conditions in India are due to poor per capita healthcare spending of $32 and a smaller number of doctors, nurses, and other supporting staff. Still, the life expectancy has increased tremendously. This increase in survivability, however, means that many people outlive the quality of their connective tissues (shown in the curves in Fig. 23.4). Some 50 years ago, a revolution in medical care began with the successful replacement of tissues. Fortunately, this revolution coincided with the increase in overall human survivability.

23.3 Treatment Modality

Two alternatives are possible: (1) **transplantation and** (2) **implantation**. Harvesting the patient's tissue from a donor site and transplanting it to a host site, at times even maintaining the blood supply, have become the gold standard for many surgical procedures, such as vertebral fusion and coronary bypass.

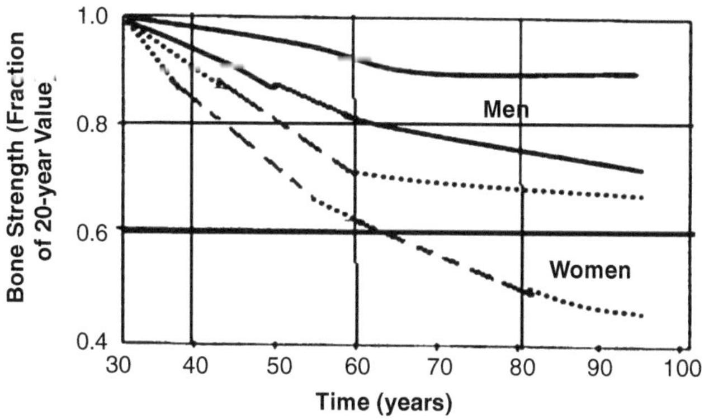

Fig. 23.4 Decrease in the strength of bone with age in men and women (adopted from Prof. Larry L. Hench, Imperial College London)

This type of tissue graft, called an autograft, raises minimal ethical or immunological concerns but does have important limitations. Limited availability, second-site morbidity, tendency toward resorption, and sometimes a compromise in biomechanical properties are routine concerns of the surgeon. A partial solution to some of these limitations is the use of transplant tissue from a human donor, a homograft, either as a living transplant (heart, heart-lung, kidney, liver, retina) or from cadavers (freeze-dried bone). The availability, the requirement for immunosuppressant drugs, concerns about viral or prion contamination, and ethical and religious issues limit the use of homografts.

23.4 Transplants and Implants

Transplants, both living and nonliving, from other species, called **heterografts or xenografts**, provide a third option for tissue replacement, as illustrated in Fig. 23.5. Nonliving, chemically treated xenografts are routinely used as heart valve replacements (porcine) and bone substitutes (bovine), but clinical results are less than optimal. The use of living heterografts from genetically modified animals is a controversial step replete with the ethical concerns.

The second line of thought in the revolution to replace tissues (Fig. 23.5) was the development, or in many cases modification, of manmade materials to interface with living, host tissues, for example, implants made from biomaterials. There are many significant advantages of nonliving implants over living transplants. They include availability, reproducibility, and reliability. Good manufacturing practice, international standards, government regulations, and quality assurance testing minimize the probability of mechanical failure of implants. However, most implants in use today continue to suffer from problems of interfacial stability with host tissues, biomechanical mismatch of elastic moduli, generation of wear debris, and maintenance of a stable blood supply.

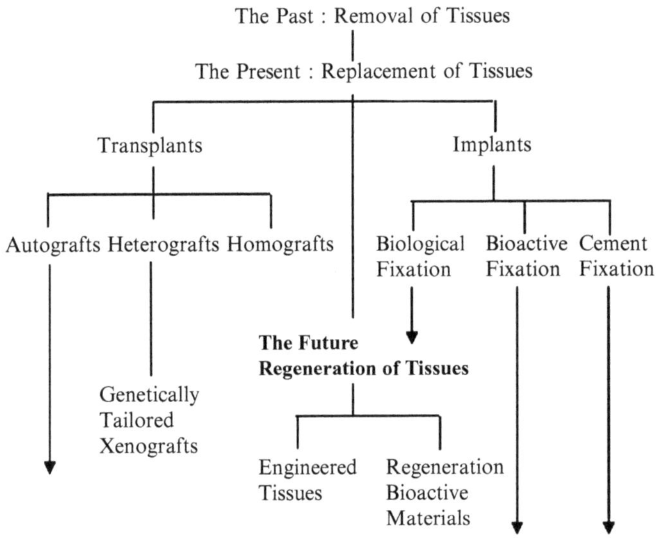

Fig. 23.5 The alternative ways for replacement of tissue

In addition, all present-day implants lack two of the most critical characteristics of living tissues: (1) the ability to self-repair and (2) the ability to modify their structure and properties in response to environmental factors such as mechanical load or blood flow.

All implants have limited lifetimes. Many years of research and development have led to only marginal improvements in the survivability of orthopedic, cardiovascular, dental, or sensory implants at more than 15 years. For example, efforts to improve lifetimes of orthopedic prostheses through morphological fixation (large surface areas or fenestrations) or biological fixation (porous ingrowth) have generally not improved survivability over the cement fixation of prostheses developed by Sir John Charnley 50 years ago.

During the last decade, considerable attention has been directed toward the use of bioactive fixation of implants, where bioactive fixation is defined as the interfacial bonding of an implant to tissue by means of formation of a biologically active hydroxyapatite layer on the implant surface for bonding with bone.

23.5 Implant Failure

Let us discuss some specific problems associated with implants. British surgeon John Charnley revolutionized the treatment of hip disease in the 1960s. His concept was total joint replacement with a low-friction prosthesis. The big advantage of this method of repair is that the patient can start to use the new hip the day after surgery, and bone needs to be used regularly in order to be healthy. Some people describe the

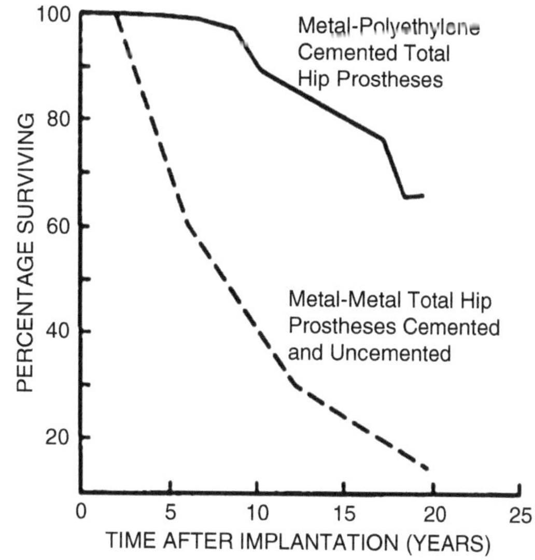

Fig. 23.6 Life expectancy of various total hip prostheses (adopted from Hench and Wilson, *Clinical Performance of Skeletal Prostheses*, Chapman and Hall, 1993)

repair as a miraculous cure. Such an emphasis on complete restoration of function is in many ways misleading and perhaps even dangerous.

A failed implant is terrible for many reasons. The patient is generally older and recovery from major surgery is more difficult. It is often hard to remove the implant and especially the cement inside the shaft of the thighbone. The health of the bone has deteriorated due to the presence of the cement and the motion of the implant in the bone. Finally, there is always some chance that the patient will not wake up from the anesthesia, will die from a blood clot, or will get an infection during the surgery. These complications are all bad but also all real. This is the uncertainty that concerns us. For every implant that is a miracle cure, there is a certain probability that some other implant will be a failure.

The probability of failure ranges from about 3–5% for the first 3–5 years after an implant is installed to as much as 10–15% after 10–15 years. Figure 23.6 shows the survivability of Charnley-type total hip prostheses. The success rate is very high for the first 10–15 years, and then it decreases.

The probable lifespan of patients who have hip implants ranges from 10–30 years following the operation. Consequently, there is a fairly high probability, nearly one in three, that a patient will outlive his or her implant. This means that revision surgery and replacement implants constitute an ever-growing portion of the practice of major medical centers. The costs continue to increase as well.

In order to improve the lifetime of implants, especially for younger patients, other methods of fixing total joint prostheses have been tried. An alternative is called biological fixation. In this method, the implant is coated with a porous mesh or a layer of porous beads. Bone will eventually grow into the pores. This anchors the implants into bone without the use of cement. One of the problems is the length of time for bone to grow into the pores. One approach to speeding up the growth is to fill the pores or coat a roughened surface with hydroxyapatite. The method of bioactive fixation with hydroxyapatite is now being used in many surgical centers throughout the world. Ten-year results are good, but it is too early to say whether this method of fixing implants to bone is superior to cement fixation.

High levels of success for 10 years create a problem, however. The problem is that implant companies sometimes move new devices into the clinical marketplace with only a few years of animal or human data to support the sales. This is in spite of the fact that the relative success of a new implant design, or method of anchoring, will not be known until 10–15 years have elapsed. The patient and the surgeon then face the difficult decision of whether to use one type of implant with a particular level of uncertainty or to use a new type of implant with a relatively unknown level of uncertainty. Thus, the primary uncertainty asked by a patient—"Should I have an implant now or wait a few years?"—has been expanded to include, "Which one of two or three or 10 alternative types of implants should be used?" This is very realistic. The present author confronted such queries from the people around him. Giving a suggestion contrary to a renowned orthopedic surgeon is also a moral dilemma.

23.6 Clinical Trials

At present, there are few national programs on implant materials that can provide the scientific basis for answering these questions. There is little research on the long-term behavior of implants for the musculoskeletal system. One of the problems is that almost all testing that is done on new biomaterials is done on young healthy animals for short times. However, the implants mostly go into old people with deteriorated bone and must last for a long time. There are almost no long-term tests conducted because of the expense and ethical issues of keeping animals in confined test environments for long times. There is little funding for fundamental studies to understand and prevent the breakdown of implants at their interface.

23.7 Ethical Issues of Dental Implants

Dental implants were used for many reasons. In some cases, only a few teeth were missing and an implant was used to anchor a partial bridge. In other cases, all the teeth were extracted and the implants provided an alternative to dentures. Implants were, and still are, an attractive alternative to dentures. A very large percentage of denture users are unhappy with removable dentures. Considering that many millions of teeth are extracted each year and nearly 20 million people in the U.S. are without any teeth, this is a serious problem. Implants appear to be a good solution to the problem.

23.8 Cost

However, there are problems with the implant solution. First of all, dental implants are expensive. Some of the systems available today can cost as much as $15,000–$25,000, including surgical costs, in Western countries. Second, Dr. Cranin discovered that a very large percentage of dental implants were failing. He compiled the data for several thousand cases and reported their relative success and failure in a dental journal. His honesty and his ethical response to the problem led to serious condemnation and professional attacks by some implantologists. His career was threatened. He withstood these attacks and personal and professional abuse, and defended his data. Subsequent studies by other dental investigators found his conclusions to be valid; 25 years ago, there was indeed a large chance of failure for many types of dental implants. The scene is more or less the same everywhere. In India, titanium screw-root-type dental implants were introduced in 1996 by the present author. The screws were coated with hydroxyapatite with the help of Dr. K. D. Groot of the Netherlands, from a company his associates developed. Subsequently, Dr. T. K. Pal used those implants in human patients with great success. Later, another surgeon, Dr. B. K. Biswas, used a large number of uncoated bicortical and monocortical titanium implants in more than 400 patients; the number of screws used was nearly double, with a 90% success rate, (stable fixation) as per his claim. Simplified toolings were also developed to make it available for the common surgeon. However, in recent years, the Western tooling systems have been more convenient and easier to handle. Surgeons are contemplating improving the tooling of the present design.

The cost of an implant was Rs.200–300 in the 1990s when implants were introduced in India. It increased to Rs.3,000–5,000, including surgical costs, in 2010.

Improvements in surgical techniques, implant materials, and patient selection and care were implemented throughout the field. This led to a greatly increased success rate of dental implants. The motivation for these improvements is owed to a considerable extent to Dr. Cranin's willingness to report failures regardless of the consequences of his actions.

Most of us would agree that his decision and course of behavior were "right" and therefore ethically and morally correct. Most of us would also agree that those implantologists who put in implants for several thousands of dollars in fees without informing their patients of the high risk of failure, and who then referred them to other dentists when problems occurred, were morally and ethically not acceptable.

23.9 Moral Uncertainties

In the United Kingdom, the decisions regarding priority about whom to help greatly depend on National Health Service (NHS) resources. Long waiting lists for elective total hip replacements have resulted. If a patient is 90 years old and this is his fourth failed implant, should he be implanted with a fifth? This is an ethical dilemma for the surgeon and the hospital when there are many younger patients still waiting for their first implant.

This admission of the limitations of ethical theories by professional ethicists poses a problem for solving ethical problems associated with the use of implants and transplants.

The writings of John Stuart Mill and his successors defend an approach called "utilitarianism." Their position, that an action is right if it leads to the greatest possible good consequences or least possible bad consequences, seems reasonable. Moral rules, thus, are the means to fulfill individual needs and also to achieve broad social goals. A moral life is measured in terms of values and the means to produce the values.

23.10 Religion and Morals

However, the eminent German philosopher Immanuel Kant and his successors, such as W. D. Ross, argue that moral standards exist independently of utilitarian ends. Their ethical theory is termed "deontological," from the Greek word *deon*, which means "binding obligation." Thus, deontologists argue that a moral life should not be conceived in terms of means and ends. An act is right not because it is useful, but because it satisfies the demands of some overriding principle of obligation. However, the differences in ethical theories are fundamental and are difficult, if not impossible, to reconcile.

This leads to theoretical moral disagreements, as referred to by Tom L Beauchamp and LeRoy Walters, that may never be resolved. The conflict between these theories leads to ethical dilemmas. The consequences of these moral uncertainties in the use of transplants, healthcare distribution, genetic alteration of life, as well as birth and death control are not easy to resolve. Figure 23.7 illustrates two practical ethical dilemmas that need to be resolved. Our technology-based society has generated infinite desires in people to live a long life. However, there are only finite resources to maintain our quality of life. This imbalance between infinite desires and finite resources creates severe ethical dilemmas.

As one may imagine, a variety of principles of distributive justice have been proposed. Some examples have been summarized by Beauchamp and Walters:

Principles of Distributive Justice

1. To each person an equal share
2. To each person according to individual need
3. To each person according to acquisition in a free market
4. To each person according to gender
5. To each person according to societal contribution
6. To each person according to merit.

There is a large difference between these alternatives. People will often vigorously defend one of them as just and others as unjust. Consequently, most societies use several of the above principles in combination, in the belief that different rules

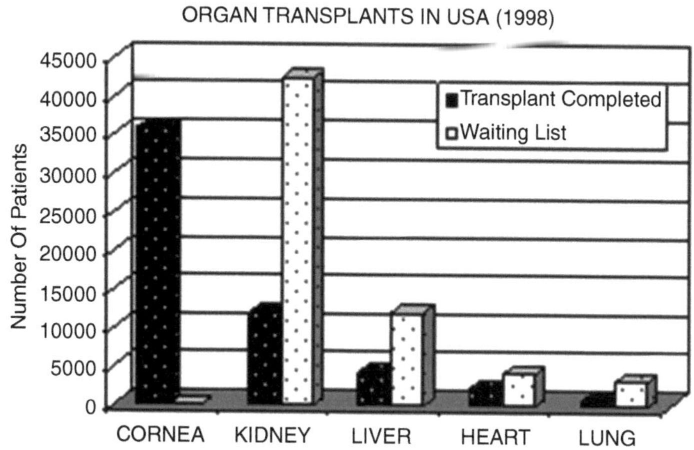

Fig. 23.7 Number of patients waiting and number who received organ transplant in U.S, 1998

apply to different situations. The National Health Service (NHS) in the UK was created to administer healthcare according to the principles of equal share or equal need. This ideal has become a fiscal impossibility. The growing failure of the NHS to meet the near-infinite demands on it has led to growing criticism. Many of the difficulties in establishing a national healthcare policy in the U.S. are due to the enormous disparity in the principles of justice listed above. Advantages and disadvantages of the alternatives for distributing care that have life-and-death implications have been widely discussed. Ethical conflicts abound. Some of the problems involved in the widespread use of implants, as described above, are due to the fact they are so readily available. Except for cost, there is little moral concern about the use or distribution of implants. If you want one, you can usually have it. The same cannot be said for transplants. Many additional ethical issues are involved.

23.11 Ethical Issues of Transplants

There are concerns regarding transplants, as described in Arthur Caplan and Daniel Coelho's 1999 book, *The Ethics of Organ Transplants*. There is concern about the moral issues of killing animals to get spare parts for humans. Even more disturbing are the rights and wrongs of conceiving a child to provide a bone marrow or other transplant for a family member. The concerns are severe enough that a cover story in *Time* magazine in 1991 discussed the pros and cons of this practice. The practice is still being debated. The availability and distribution of transplants, regardless of the source, create an important and emotional issue. We often see sensational stories of children presented on TV with their life or death hanging on the availability of a heart, kidney, liver, or lung transplant. The sobbing appeal of the mother and endorsement by a TV or a celebrity results in a huge amount of donations. This is

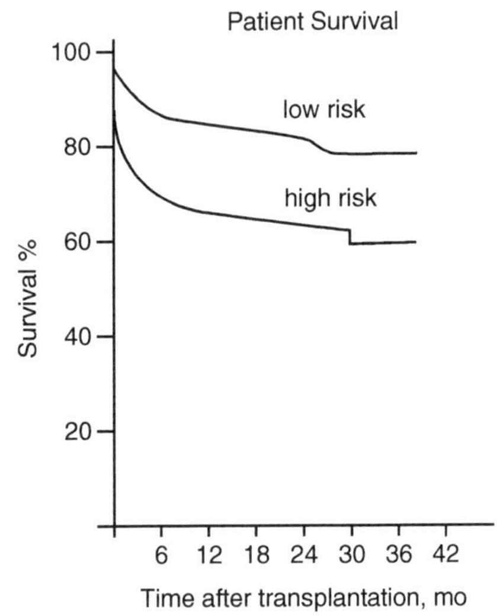

Fig. 23.8 Patient survival following liver transplant, with time after transplantation being classified as low-risk and high-risk (Caplan and Coehlo, 1998)

becoming very common in India as well. In 1998, more than 21,000 people received kidney, liver, heart, lung, or other organ transplants. However, on any given day, approximately 62,000 people are waiting for an organ, and every day nearly 100 names are added to the national waiting list. Many more are not on the waiting list. This has yet to be regularized in our country. Indian Society of Organ Transplant maintain such data.

Although the number of donors has steadily increased since 1988, donations are not growing as quickly to meet the demand for organs. It was reported that "approximately 4,000 Americans die each year (11 per day) waiting for a solid organ transplant." More than 40,000 persons per year could benefit from kidney transplants. These statistics are summarized in Fig. 23.8. This is a fraction of the 80,000+ patients currently receiving dialysis treatments for kidney failure. However, in 1998, fewer than 15%, or about 12,000 patients, received transplants due to the scarce supply. Studies indicate that at least 12,000 adults might benefit from heart transplants if a sufficient supply of organs existed. Caplan reported that 7,500 infants have their life threatened with congenital heart disease. Transplants, if hearts were available, could probably save many of them (Caplan and Coehlo, 1998).

In the case of kidney transplants, about one third come from family members who donate one of their two kidneys. However, at present, organ transplants are derived primarily from fatal accident victims. There are about 100,000 fatal accidents each year in the U.S. The system for obtaining and shipping organs from these accident victims is not very efficient. Consequently, only about 20,000 organs are available nationally. This is only a fraction of the number of organs needed, even if the costs are justified and the money is available. Consequently, there continues to be controversy over the distribution of available organs. There are advocates for all of the principles of distributive justice. The most dominant force in the U.S. tends to be the principle of distribution according to the ability to pay, either personally or by the local community, moderated by need.

To provide a better balance of justice, a lottery system has been widely adopted for allocating organs from fatal accident victims within closest proximity to the recipient. This system is called the United Network for Organ Sharing. The Institute of Medicine Report issued in 1999, cited earlier, discusses many of the issues involved in the distribution of organs, with an emphasis on liver transplantation. The issues include

- Access to transplantation services for low-income populations and racial and ethnic minority groups
- Organ donation rates
- Waiting times for transplantation
- Patient survival rates and organ failure rates leading to retransplantation
- Cost of organ transplantation services.

Transplantation of organs from animals is one possible solution to the organ supply problem. This possibility is what led to the Baby Fae experiment. However, this alternative is still highly experimental and subject to much ethical debate. For example, people ask whether our society should promote killing individuals from another species, almost surely primates, to provide organs for humans unless we have made every effort to use the organs of fatal accident victims or terminally ill human patients first.

23.11.1 Availability

The ethical issues of organ transplants fall into three categories: **supply** (source and numbers), **cost** (allocation and availability), and **survival** (quality and length of life). Our goal in this lecture is to review these issues and determine the extent of the ethical conflict that results. The problem of supply of organs suitable for transplantation stems from the simple fact that the organ must be alive. This requires that the organ come from a donor who either is living or is terminally ill but not yet dead. At the present time, the supply of living organs meets only about one third of the demand. With an ever-increasing proportion of the population living even longer life spans, the supply/demand ratio is likely to drop to one fifth or less. This raises the question of alternative sources of supply. One potential source is the use of anencephalic infants as organ sources. Such infants are born without a forebrain and a cerebrum. This precludes such infants having consciousness. Their organs often still function and could potentially be used as transplants in infants that have congenital kidney, heart, or liver disease.

Numerous alternative sources of organs have been proposed and debated. L. G. Futterman concludes, “The organ supply/demand disparity stems not from a lack of donors but rather from failure to obtain permission to recover viable donor tissue and organs.” Society needs to consider legal changes that make obtaining consent easier. Options debated in Caplan and Coelho’s text include mandated choice, which would require all adults to express written choice of organ donation before death, perhaps as a requirement for obtaining a driver's licence. Since a substantial fraction of organs are derived from motor accidents, such a legal change would go a long way toward solving the organ-supply problem. Along similar legal lines, the approval of routine organ salvage as a communal policy would eliminate the need for consent.

A further argument in favor of change toward more liberal organ salvage is the difficulty in achieving "informed consent" from the individual or family under tragic circumstances. Tom Beauchamp and James Childress discuss the problems and the implications of the concept of "informed consent" in their book *Principles of Medical Ethics*. Robert M. Veatch & J.B. Pitt argue in Caplan and Coelho against so-called presumed consent laws, which they maintain are, in effect, routine organ-salvaging laws. Their adoption would alter forever the ethical relationship of the individual to society. Routine organ salvaging would involve taking organs without the individual's written consent. Such laws subordinate the rights of the individual to those of the state. Giving up individual rights to the state, even for the moral good of saving tens of thousands of people, is a profound change in our culture. Most people oppose the loss of such rights. The purchase of organs continues to be debated. The preponderant opinion is that the rights of autonomy and beneficence favor the present voluntary system of organ donation.

23.12 Xenografts as a Solution

The continuing disparity between the supply and demand of human organs is the reason for a growing interest in the transplantation of animal organs. There are two issues: (1) the ethics of breeding and killing animals to save humans, and (2) immunorejection of the animal organs. The use of nonhuman primate organs, such as the baboon heart in the Baby Fae experiment, evokes the greatest ethical concern. Caplan and Coehlo indicated, "It is one thing to argue that primates ought to have moral standing. It is a very different matter to argue that humans and primates are morally equivalent."

Xenografts involving primates can be morally justified on the grounds that, in general, human beings possess capacities and abilities that confer more moral value upon them than do primates. Caplan and Coehlo concluded that the use of xenografts can be morally justified, but "the moral obligation to potential recipients would seem to require that systemic farming of animals (for organ transplantation) only be permitted under the most humane circumstances." The second issue of immunotolerance is a severe technical limitation in the use of xenografts. Critical genetic differences between humans require organ transplant patients to use immunosuppressant drugs on a rigorous daily schedule for their lifetimes. Because of the drugs, they are subject to many ailments, as discussed below. Genetic manipulation of animals will be required to make xenografts feasible.

The ethical issues associated with the cost of transplants are difficult because they require assessing the unanswerable question, "What is the value of a human life?" The answer, of course, is very different if you are concerned with the intrinsic worth of the life of a single individual, such as yourself or a family member or a friend, than if you are judging the relative worth of one member of a social group of thousands or millions of individuals. Moral philosophers offer little help, because they cannot agree on theoretical grounds. For a single person, the answer is often, "Whatever the cost, my life is worth it." For a large social group, that answer is impractical.

If we assume that every life is worth saving by a transplant, we need to examine the personal and social consequences of doing so. Discussions of the cost of transplants usually emphasize the economic issues, such as the cost of surgery and the length of hospitalization. **A liver or heart transplant costs $250,000, and a kidney transplant costs $90,000 over a 5-year span**.

These costs are indeed large, but in proportion to the population, they represent only a small fraction of the costs of healthcare. Deciding whether this level of expense is worthwhile to save the lives of a small group of individuals is at present a community decision based upon resource allocation. The moral decision is whether or not to recruit a transplant team for a community hospital and equip the necessary facility. Emphasizing transplant economics misses a second, and perhaps even more compelling, cost of transplantation: the cost to the recipient. Cost is more than money; it is also pain and loss of quality of life. It can be loss of freedom. The ethical debate of present-day transplants often overlooks the fact that most recipients will require the use of immunosuppressant drugs for the rest of their lives. The drugs are costly. They must also be taken on a precise schedule if they are to be effective. The cost of a slipup can be fatal.

The cost of living with the knowledge that your new heart, lung, kidney, or liver may deteriorate if you make a mistake is an enormous burden to a person. The downside of the drugs that suppress immune rejection of the transplant is the fact that a recipient is susceptible to every infection that comes along. There is no such thing as the "common cold" to the transplant patient. Every virus may be fatal. Who can measure the cost of being held hostage by your own body?

23.13 Life Expectancy

The third critical issue in the ethics of transplants is survivability of the patient. Is the projected lifetime of a transplant recipient worth the enormous personal and social costs? There is no clear answer. Data from the Scientific Registry of the U.S. United Network for Organ Sharing (UNOS) show for a 4-year period (1987–1991) that first-year survival of 4,830 heart transplant patients was 82%. That means one in five patients survived less than 1 year. In the same period, 86 patients with failing transplants were retransplanted. Only 57% survived 1 year. The 5-year survivability of heart transplant patients is less than 50%.

Data on patient survival following liver transplantation show similar results to heart transplants. In the period 1987–1991, 8,539 patients in the U.S. received the first liver transplants through the UNOS system, and 76% lived for at least 1 year. Patients who were already on life support had a much lower survivability of only 60%. When the transplants failed, second or third transplants were made and 1-year survival dropped from 50 to 35%. Figure 23.12 summarizes the survivability of University of Pittsburgh liver transplant patients up to 3 1/2 years. Patients who were relatively low-risk had a high 1-year survival rate (91%), which was maintained at 80% after 42 months. In contrast, only 71% of the very sick patients who had only a few days to live due to terminal liver failure survived for 1 year after a liver transplant.

Matching the antigens of donor organs to the recipient is one way of improving patient survival. Gaston et al. (Caplan and Coelho, p. 310) report that 1,004 kidney transplants that had a zero-antigen mismatch had a 1-year survivability of 88%. This is a 9% improvement over the 79% survivability of 22,188 recipients of mismatched kidneys in the same period. These data illustrate several important points:

1. Patients have about an 80% chance of living for more than a year after most transplants and about a 50/50 chance of living for more than 5 years.
2. If the first transplant fails, the probability of surviving a second or third transplant is very low indeed.
3. The sicker the patient is, the less likely it is that he or she will survive for very long with a transplant.
   The Institute of Medicine Report "Organ Procurement and Transplantation" leads to several additional conclusions:
4. Organ allocation areas need to be matched with patient populations using statistical analyses.
5. Rates of pretransplantation and transplantation mortality are more meaningful indicators of equitable access than waiting times.
6. Policies related to transplantation access must be consistent with the medically acceptable cold storage times for organs: liver, 12 h; pancreas, 17 h; kidney, 24 h; heart, 4 h; lung, 6–8 h. Another Institute of Medicine Report, "Approaching Death: Improving Care at the End of Life," leads to another conclusion:
7. Death is sometimes preferable to the consequences of having a transplant.

The issues of transplant availability, cost, and survivability should provide the basis for making a moral judgment as to whether transplantation is a desirable option to prolong a life. When a patient is beyond a certain age or level of illness, the ethical principles of beneficence and justice indicate that subjecting him or her to the trauma and subsequent personal costs of a transplant is unwise. Transplantation of a second or third organ cannot be justified on moral grounds, even though the consequence of not doing so is the death of the patient. One conclusion is that "avoiding death at all costs" is immoral. Just because it is possible to do a transplant does not necessarily mean that one should go for a transplant. Social and surgical guidelines to establish this boundary need to be very tight. The respect for autonomy for the terminally ill patient should not bias counsel for a transplant toward a decision of "life at all costs." They have decided to leave the transplant field and in doing so state, "By our own leave taking we are intentionally separating ourselves from what we believe has become an overly zealous medical and societal commitment to the endless perpetuation of life and to repairing and rebuilding people through organ replacement—and from the human suffering and the social, cultural and spiritual harm we believe such unexamined excess can, and already has, brought in its wake." There is a strong expression of ethical concern that should not be ignored. Increasing the length of life without assuring the quality of life is immoral. This is mostly a Western phenomenon, and the developing countries yet to have this kind of problem in excess.

Suggested References for Further Details

1. Ratner BD, Hoffman AS, Schoen FJ, Lemons JE (eds) (2000) Biomaterials science: an introduction to materials in medicine (2nd edn). Academic, New York
2. Park JB (1979) Biomaterials: an introduction. Plenum, New York
3. Wise DL, Trantola DJ, Altobelli D (1996) Human biomaterials applications. Humana, Totawa, NJ
4. Hench LL (1998) Sol–gel silica: processing, properties and technology transfer. Noyes, New York
5. Hench LL, Wilson J (eds) (1933) Introduction to bioceramics. World Scientific, London and Singapore
6. Silver FH (1994) Biomaterials, medical devices and tissue engineering. Chapman & Hall, London
7. Beauchamp TL, Walters L (eds) (1989) Contemporary issues in bioethics (3rd edn). Wadsworth, Belmont, CA
8. British Medical Association (1990) The BMA guide to living with risk. Penguin, New York
9. Hilton B (1991) First do no harm: wrestling with the new medicine's life and death dilemmas. Abingdon, Nashville, TN
10. Callahan D (1990) What kind of life: the limits of medical progress. Simon and Schuster, New York
11. Saha PS, Saha S (1991) Ethical issues on the use of animals in the testing of medical implants. J Long Term Eff Med Implants 1(2):127–134
12. Hench LL, Wilson J (eds) (1993) Clinical performance of skeletal prostheses. Chapman & Hall, London
13. Heifetz MD (1988) Ethics in medicine. Prometheus, New York
14. Beauchamp TL, Childress JF (1992) Principles of biomedical ethics, 4th edn. Oxford University Press, Oxford
15. Bourne JR (ed.) (1997) Critical Rev Biomed Eng 25(2)
16. Caplan AL, Coehlo DH (1998) The ethics of organ transplants: the current debate. Prometheus, New York
17. Committee on Organ Procurement and Transplantation Policy, Institute of Medicine (1999) Organ procurement and transplantation. National Academies Press, Washington, DC
18. www.acmite.com/marketreports/medicals/worldmedicaldevices-market.html

Exercises

1. Estimate the current population in India, male and female. What is their life expectancy and how many are over 60 years old?
   (a) Estimate the need for cataract surgery and total requirement for LASIK or phaco surgery, in India.

(b) How many hip surgeries are required per annum? If indigenous prostheses are used, what will be the market size in India per year?
(c) Is there any possibility of having kidney or liver transplants? Do you think that in India we need to develop organ banking? How could such banking be realized through ethical and moral considerations?

2. Discuss the ethics of organ transplants based on Indian concepts about life and culture.
3. Design a patient consent form for ethical clearance for a hip joint implant.

Chapter 24
The Manufacturing, Testing, and Sterilization of Implants

24.1 Introduction

During or after design, the manufacturing aspect must be considered to facilitate manufacturing the product with the available resources. Then the product must be tested for its sterilizability, effectiveness in vitro and in vivo, strength, life expectancy or durability, and, finally, packaging for shipment.

Manufacturing is making products from raw materials by various processes or operations. It is generally a complex activity, involving people who have ideas in a broad range of disciplines and skills and a wide variety of machinery, equipments, and tooling, with various levels of automation, including computers, robots, and material-handling equipment. Manufacturing activities must be responsive to modern demands and trends:

- A product must fully meet design requirements and specifications of well-known standards, such as those of the **ASTM** (American Society of Testing of Materials), **EN** (Engineering Nomenclature, British), **BIS** (Bureau of Indian Standards), or **ISO** (International Organization for Standardization). These standards are available in many well-known libraries and the BIS offices in states or at its headquarters in New Delhi in Bahadur Shah Zaffar Marg.
- A biomedical product must be manufactured with the utmost care so that contaminants are avoided and also through the most economical but efficient methods in order to minimize costs so that it can reach the general populace.
- Quality must be built into the product at each stage, from design to assembly, rather than relying on quality testing at the end.
- In a highly competitive environment, production methods must be sufficiently flexible so as to respond to changing market demands, types of products, production rates, production quantities, and on-time delivery to the customer.
- Many processes are used to produce parts and shapes. There are usually more than one method of manufacturing a part from a given material. The broad categories of processing methods for general materials are

S. Pal, *Design of Artificial Human Joints & Organs*,
DOI 10.1007/978-1-4614-6255-2_24,

Metal casting: Expendable moulds usually made of molding sands and cope and drag boxes and permanent metal molds were also used. The mould is to have a cavity similar to the object to be cast.

Metal forming and shaping: Rolling, forging, extrusion, drawing, sheet forming, powder metallurgy, and molding are used.

Plastics molding and forming: Blow molding, CNC machining, centrifugal casting, compression molding, profile extrusion, injection molding, thermoforming, vacuum forming, pressure forming, pulshaping, pulstrusion, liquid resin molding, reaction injection molding (RIM), rotational molding, and resin transfer molding (RTM) are the varieties of manufacturing methods available.

Rapid prototyping: The following processes are the most advanced and are being used in modern industries: stereolithography—SLA or SL; 3D printing—3DP; selective laser sintering—SLS; fused-deposition modeling—FDM; solid-ground curing—SGC; laminated object manufacturing—LOM; multi-jet modeling—MJM; direct shell production casting—DSPC; polyjet technology; laser engineered net shaping—LENS.

Joining: The common methods of joining are welding, friction welding, brazing, soldering, diffusion bonding, adhesive bonding, and mechanical joining using fastening devices. All these methods may not be applicable for medical devices.

Machining: The common methods of machining are turning, boring, drilling, milling, planing, shaping, broaching, grinding, ultrasonic machining, chemical, electrical, and electrochemical machining, and high-energy beam machining. Then there are surface finishing, honing, lapping, polishing, burnishing, surface treating, coating, and plating processes. Highly polished surfaces usually reduce corrosion and particulate formation in articulating surfaces.

24.2 Casting

Casting is a manufacturing process where a solid material is melted, heated to proper temperature (sometimes treated to modify its chemical composition), and then poured into a cavity or mold of the desired object made of sand-mix or metal, which contains it in the proper shape during solidification. Thus, in a single step, simple or complex shapes can be made from any metal that can be melted. The resulting product can have any shape and size the designer wanted. In addition, the resistance to working stresses can be optimized, directional properties can be controlled, and a pleasing appearance can be produced.

Cast parts range in size from a few millimeters and a fraction of a few grams (such as the individual teeth on a zipper), to over 100 m and many tons (such as the huge propellers and stern frames of ocean liners). Casting has marked advantages in the production of complex shapes, parts having hollow sections or internal cavities, and parts that contain irregular curved surfaces (except those made from thin sheet metal). Because of these advantages, casting is one of the most important procedures of the manufacturing processes.

It is nearly impossible today to design anything that cannot be cast by one of the available casting processes. However, as in all manufacturing techniques, the best results and economy are achieved if the designer understands the various options and tailors the design to use the most appropriate process in the most efficient manner. The various processes differ primarily in the mold material (whether sand, metal, or other material) and the pouring method (gravity, vacuum, low pressure, or high pressure). All of the processes share the requirement that the materials solidify in a manner that would maximize the properties, while simultaneously preventing potential defects, such as shrinkage voids, gas porosity, and trapped inclusions.

24.3 Forging

Forging is controlled, plastic deformation or working of metals into predetermined shapes by means of pressure or impact blows, or a combination of both. Forging improves the physical properties of the metal, refines the grain structure, and increases strength and toughness. Forgings can offer decisive cost advantages, especially in high-volume production runs. Forged parts are generally near designed shapes, making better use of material, producing little scrap, and requiring less machining and labor cost. If properly designed, a forging can replace an entire multicomponent assembly. In orthopedic surgery, many metal prostheses are produced by forging. Some specific processes are listed here:

1. Cold forging and hot forging
2. Seamless rolled ring forging
3. Impression die forging
4. Open die forging.

24.4 Metal-Shaping Machine Tools

Machine tools are stationary power-driven machines used to shape or form solid materials, especially metals and maybe plastics. The shaping is accomplished by removing material from a work piece or by cutting the extraneous materials using a suitable cutting tool with specific cutting speed and feed to get the desired shape. Machine tools form the basis of modern industry and are used either directly or indirectly in the manufacture of machine and tool parts.

Machine tools may be classified under three main categories:

1. Conventional chip-producing machine tools
2. Presses
3. Unconventional machine tools.

Conventional chip-making tools shape the work piece by cutting away the unwanted portion in the form of chips. These machines use a cutting tool of a definite shape and size and create a relative motion between the job and the tool, and the interfering area is removed in the form of a chip. There are lathe, shaping machine, milling machine, automatic screw-cutting machine, drilling, slotting, and boring machines and combination workstations. These are useful to give shape to the different fracture-fixation device: plate, screws of various types, nails, and the like.

Presses employ a number of different shaping processes, including shearing, pressing, or drawing (elongating) to shape prostheses such as hip joints, shoulder joints, and knee joints. Unconventional machine tools employ light, electrical, chemical, and ultrasonic energy; superheated gases; and high-energy particle beams to shape the exotic materials and alloys that have been developed to meet the needs of modern technology.

Machine tools belonging to the CNC category are a common thing in manufacturing and fabrication industries. Computer numerical control (CNC) machining has been used since the 1970s at the onset of computers. That is how it got its name, except that it was first called only NC or numerical control when it was introduced.

The CNC is programmed to reduce manual intervention in a certain fabrication process, thus reducing production time and wastage, thereby achieving increased production efficiency at each step of the process.

The CNC machine simplifies all handling, fixing work as tools for the operator. With the CNC equivalent of the drill press, the operator only has to position the metal in its place, activate the spindle, set the controls, and the machine does the rest. It is very much the automated way of doing things. Everything that an operator is required to do with the conventional machine tools is programmable with CNC machines. Another advantage of a CNC machine is being able to produce consistent and accurate work pieces. Once the programming controls have been set, the CNC machine can produce thousands of identical pieces in a short time. This is almost impossible if done manually. CNC machines are also flexible since processes are programmed. You can run a specific program for one piece, save it, and then recall it the next time that piece is to be reproduced. These machines are also easy to set up, thus allowing operators to meet their deadlines faster.

All CNC machines have motion control, either linear or rotary. The control the machine follows is called axis. The more axes the machine has, the more complex it is.

The use of these machines is not exclusive to the metal industry, but occurs in a large range of endeavors as well, especially those that require mass production of articles. As the demand increases, the need for producers to meet these demands becomes more urgent as well. Using CNC machines, production processes are made easier, faster, and safer. The implant industries use CNC machines to a great extent for making various kinds of large to small screws for fracture fixation, dental implants, and mandibular fracture fixation. A photograph of a CNC lathe and a CNC milling are shown in Figs. 24.1a, b.

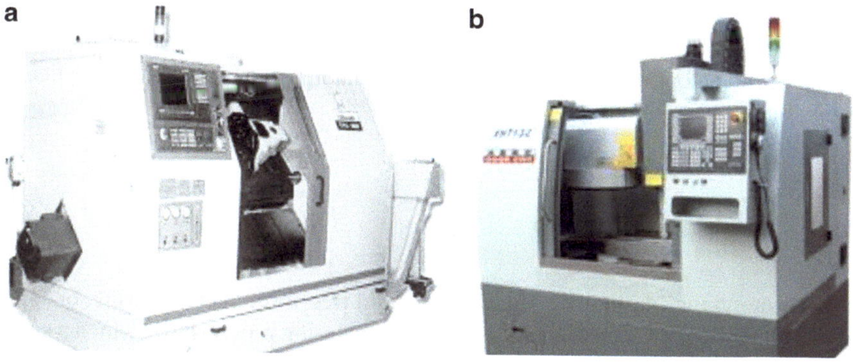

Fig. 24.1 (a) A common form of a CNC lathe; (b) a CNC milling M/C; other workstations are also available that have a combination of various machining functions

24.5 Manufacturing Implants at a High Speed

Intelligent software from Siemens for virtual planning and for the control of machine tools is making it possible for medical implants to be manufactured faster and less expensively. The challenge with artificial knee, shoulder, or hip joints lies in the fact that the materials, such as titanium or chromium cobalt, are very difficult to machine, but the complex shapes must be produced very precisely in order to provide an optimal fit for the patient.

The increased longevity of people and their desire to remain active are driving the increased demand for implants. In Germany, 200,000 joint and hip replacements are already being performed each year. To ensure that the implants are quite durable and long-lasting, increasingly stronger and better materials are being used to make them. This not only increases the cost; it also poses greater challenges in terms of the implant manufacturing process. With a new milling technology and the high-speed cutting process from Siemens Industry Automation and Drive Technologies, implants can be made not only with ultimate precision; they can also be manufactured in less time and thus more cost-effectively.

An optimally fitted implant begins with X-ray photos of the human joint that is to be replaced. The images are taken with a computer tomography or magnetic resonance imaging scanner. The doctor uses these pictures to virtually choose a suitable implant on the computer and positions it, with the help of 3D planning software, at the location of the joint to be replaced. A further Siemens planning tool, the CAD/CAM software NX CAM, simulates the production of the joint in a test run to avoid later damage to the expensive titanium or chromium-cobalt work pieces and to achieve an optimal precision fit. The traversing paths identified in the simulation are forwarded to the machine tools, which then use a metal-cutting process to produce these precision-fitted implants. The machines work with a spindle speed of 40,000–60,000 RPM.

The technology can also be used to manufacture dental implants, an application where fast, precision manufacturing of implants saves time and money for doctors and patients alike. The challenge is in the mass production of custom work pieces, which is actually an oxymoron. High-productivity dental manufacturing is only possible with an optimally coordinated process chain, from imaging of the patient's condition to the production of the implant.

24.6 Rapid Prototyping

Rapid prototyping (desktop manufacturing, solid free-form manufacturing, or solid free-form fabrication) consists of various manufacturing processes by which a solid physical model of a part is made directly from 3D computer-aided design (CAD) model data. This CAD data may be generated by 3D-CAD modelers, CT (computed tomography) and MRI (magnetic resonance imaging) scan data, or model data created by 3D digitizing systems.

To begin the rapid prototyping process, the 3D data are sliced into thin (~0.125-mm) cross-sectional planes by a computer. The cross sections are sent from the computer to the rapid prototyping machine, which builds the part layer by layer. The first layer's geometry is defined by the shape of the first cross-sectional plane generated by the computer. It is bonded to a platform or starting base, and additional layers are bonded on top of the first, then the second, and then the third and shaped according to their respective cross-sectional areas. This process is repeated until the prototype part is complete.

The resulting prototype provides a "conceptual model" for design visualization and review by the entire design team. It may be used by engineers to check the form and fit and perform limited function tests. It can also be utilized for soft tooling for prototypes and as a pattern for hard tooling.

The first techniques for rapid prototyping became available in the late 1980s and were used to produce models and prototype parts. Today, they are used for a much wider range of applications and are even used to manufacture production-quality parts in relatively small numbers. Some sculptors use the technology to produce complex shapes for fine art exhibitions.

The use of additive manufacturing for rapid prototyping takes virtual designs from computer-aided design (CAD) or animation modeling software, transforms them into thin, virtual, horizontal cross sections, and then creates successive layers until the model is complete. With additive manufacturing, the machine reads in data from a CAD drawing, lays down successive layers of liquid, powder, or sheet material, and in this way builds up the model from a series of cross sections. These layers, which correspond to the virtual cross section from the CAD model, are joined together or fused automatically to create the final shape. The primary advantage to additive fabrication is its ability to create almost any shape or geometric feature.

The standard data interface between CAD software and the machines is the STL file format. An STL file approximates the shape of a part or assembly using triangular facets. Smaller facets produce a higher-quality surface.

The word "rapid" is relative: The construction of a model with contemporary methods can take from several hours to several days, depending on the method used

Table 24.1 Various RPT technologies and various base materials

Prototyping technologies	Base materials
Selective laser sintering (SLS)	Thermoplastics, metals, powders
Fused deposition modeling (FDM)	Thermoplastics, eutectic metals
Stereolithography (SLA)	Photopolymer
Laminated object manufacturing (LOM)	Paper
Electron beam melting (EBM)	Titanium alloys
3D printing (3DP)	Various materials

and the size and complexity of the model. Additive systems for rapid prototyping can typically produce models in a few hours, although it can vary widely depending on the type of machine being used and the size and number of models being produced simultaneously.

Some solid freeform fabrication techniques use two materials in the course of constructing parts. The first material is the part material and the second is the support material (to support overhanging features during construction). The support material is later removed by heat or dissolved away with a solvent or water.

Traditional injection molding can be less expensive for manufacturing polymer products in high quantities, but additive fabrication can be faster and less expensive when producing relatively small quantities of parts. 3D printers give designers and concept development teams the ability to produce parts and concept models using a desktop-size printer. Rapid prototyping is mostly used for auto parts, but it is being used for medical device manufacturing for nearly 9% of all RPT products.

A large number of competing technologies are available in the marketplace. As all are additive technologies, their main differences are found in the way layers are built to create parts. Some are melting or softening material to produce the layers (SLS, FDM), where others are laying liquid materials thermosets that are cured with different technologies. In the case of lamination systems, thin layers are cut to shape and joined together. As of 2005, conventional rapid prototype machines cost around £25,000 and now it may cost Rs. 6–10 millon.

Table 24.1 depicts the various RPT technologies and the various base materials suitable for such technology.

Stereolithography is one of the most common rapid prototyping (RP) systems. The system utilizes a computer-controlled UV laser beam to harden a photo-curable liquid resin to produce 3D copies of CAD models. The SLA computer utilizes an (.STL) file format output from professional solid modeling software programs.

Selective laser sintering (**SLS**) is an additive manufacturing technique that uses a high-power laser (for example, a carbon dioxide laser) to fuse small particles of plastic, metal (direct metal laser sintering), ceramic, or glass powders into a mass that has a desired three-dimensional shape. The laser selectively fuses powdered material by scanning cross-sections generated from a 3D digital description of the part (for example, from a CAD file or scan data) on the surface of a powder bed. After each cross section is scanned, the powder bed is lowered by a one-layer thickness, a new layer of material is applied on top, and the process is repeated until the part is completed.

Because the finished part density depends on the peak laser power rather than the laser duration, an SLS machine typically uses a pulsed laser. The SLS machine preheats the bulk powder material in the powder bed somewhat below its melting point, to make it easier for the laser to raise the temperature of the selected regions the rest of the way to the melting point [1].

Some SLS machines use single-component powder, such as direct metal laser sintering. However, most SLS machines use two-component powders, typically either coated powder or a powder mixture. In single-component powders, the laser melts only the outer surface of the particles (surface melting), fusing the solid non-melted cores to each other and to the previous layer.

Compared to other methods of additive manufacturing, SLS can produce parts from a relatively wide range of commercially available powder materials. These include polymers such as nylon (neat, glass-filled, or with other fillers) or polystyrene; metals including steel, titanium, alloy mixtures, and composites; and green sand. The physical process can be full melting, partial melting, or liquid-phase sintering; depending on the material, up to 100% density can be achieved with material properties comparable to those from conventional manufacturing methods. In many cases, large numbers of parts can be packed within the powder bed, allowing very high productivity.

3D printing is a form of additive manufacturing technology where a three-dimensional object is created by laying down successive layers of material. 3D printers are generally faster, more affordable, and easier to use than other additive manufacturing technologies. 3D printers offer product developers the ability to print parts and assemblies made of several materials with different mechanical and physical properties in a single build process. Advanced 3D printing technologies yield models that closely emulate the look, feel, and functionality of product prototypes. Since 2003, there has been a large growth in the sale of 3D printers. Additionally, the cost of 3D printers has declined. The technology also finds use in the jewelry, footwear, industrial design, architecture, engineering and construction (AEC), automotive, aerospace, dental, and medical industries.

24.7 Nonconventional Machining

The word "machining" means the machines that utilize mechanical energy to remove material from the work piece. Milling machines, saws, drilling machines, and lathes are some of the most common machines using mechanical energy to remove materials in the form of chips. The cutting tool makes contact with the work piece, a relative motion is created between the work piece and tool, and the resulting shear causes the material to flow over the tool. All traditional forms of metal cutting use shear as the primary method of material removal. However, there are other sources of energy for machining work.

The category of nontraditional machining covers a broad range of technologies, including some that are used on a large scale, and others that are only used in unique or proprietary applications. These machining methods generally have higher energy requirements and slower throughputs than traditional machining, but have been developed for applications where traditional machining methods were impractical, incapable, or uneconomical.

Nonconventional or nontraditional machining can be thought of as operations that do not use shear as their primary means of removal of extraneous material. For example, abrasive water-jet operations use mechanical energy, but material is removed by erosion.

Nontraditional machining methods are typically divided into the following categories:

1. **Mechanical**—ultrasonic machining, rotary ultrasonic machining, ultrasonically assisted machining
2. **Electrical**—electrochemical discharge grinding, electrochemical grinding, electrochemical honing, electrochemical machining, electrochemical turning, electro-stream
3. **Thermal**—electron beam machining, electrical discharge machining, electrical discharge wire cutting, electrical discharge grinding, laser beam machining
4. **Chemical**—chemical milling, photochemical machining.

These machine tools were developed primarily to shape the ultrahard alloys used in heavy industry and in aerospace applications and to shape and etch the ultrathin materials used in such electronic devices as microprocessors.

Now, we will discuss the nondestructive testing of various implants, which effectively keeps the implant safe without a loss of revenue.

24.8 Nondestructive Testing

Nondestructive testing is an integral part of the implant manufacturing process because without the part's integrity having been ensured, it cannot be given to a patient. Sometimes this testing may be outsourced by the original company.

The process of nondestructive testing (NDT) determines the existence of flaws, discontinuities, leaks, inclusions, contamination, thermal anomalies, or imperfections in materials, components, or assemblies without impairing the integrity or function of the inspected component. NDT is also utilized for real-time monitoring during manufacturing, measurement of physical properties such as hardness and internal stress, inspection of assemblies for tolerances, alignment, and periodic in-service monitoring of flaw/damage growth in order to determine the maintenance requirements and to ensure the reliability and continued safe operation of the part.

Nondestructive evaluation (NDE) is becoming increasingly important to the design-through-manufacture process. The cost of parts and components is always increasing due to the corresponding costs of materials and labor. Consequently, emphasis is being placed on use of NDE early in the design and fabrication process. Often components are too costly to permit the luxury of destructively testing a number of them to demonstrate their design goals. Environmental and liability concerns are also resulting in an increased use of NDE.

NDT is a quality assurance management tool that can give impressive results when used correctly. It requires an understanding of the capabilities and limitations of the various methods available and knowledge of the relevant standards and specifications for performing the tests. Materials, products, and equipment that fail to achieve their design requirements or projected life due to undetected defects may require expensive repair or early replacement. Such defects may also be the cause of unsafe conditions or catastrophic failure, as well as loss of revenue due to unplanned plant shutdown.

NDT technology is constantly being improved and new methods developed, particularly in an effort to keep pace with the development of new materials (i.e., composites) and applications. Advances in the use of lasers and imaging technology (including video, holography, and thermography) have made noncontact NDT more viable in many situations. Optical fibers and new piezo-electric materials are allowing the creation of intelligent materials and structures that can not only monitor themselves but may even respond to their environment. Computer advances have allowed signal processing techniques and expert systems to be used, which enhances the quality of the information obtained using traditional and new NDT methods. After testing, the components need to be sterilized and packaged for sale and delivery.

24.9 Sterilization

Sterilization is the elimination of all transmissible agents (such as bacteria, prions, and viruses) from a surface, a piece of equipment, food, or biological culture medium. This is different from disinfection, where only organisms that can cause disease are removed by a disinfectant.

Sterilization technology is basic to the preparation of pharmaceuticals and vaccines, the manufacturing of medical devices and hygiene products, food processing, and many other fields. Many products sold for healthcare applications are labeled "sterile." This terminology is conventional in the medical trades but is misleading. Industrial sterilization does not lead to "absolutely" sterile products. Instead, it produces objects where the population of surviving microorganisms is less than at the starting point. Different sterilization processes lead to different surviving population levels of microorganisms.

A normal living cell contains the multitude of enzymes responsible for metabolic processes. A semipermeable membrane (cytoplasmic membrane) maintains the integrity of the cellular contents; the membrane selectivity controls the passage of substances between the cell and its external environment and is also the site of some enzymatic reactions. The cell wall proper provides a protective covering to the cell in addition to participating in certain physiological processes. Damage at any of these areas may initiate a number of subsequent changes, leading to the death of the cell. The manner in which antimicrobial agents inhibit or kill can be attributed to the following kinds of action:

(1) damage to cell wall or inhibition of cell wall synthesis; (2) alteration of the permeability of the cytoplasmic membrane; (3) alteration of the physical and chemical state of proteins and nucleic acids; (4) inhibition of enzyme action inhibition of protein or nucleic acid synthesis.

24.9.1 Physical Agent

The major physical agents or processes used for the control of microorganisms are temperature (high and low), desiccation, osmotic pressure, radiation, and filtration.

24.9.1.1 High Temperature

The killing action of heat is a time–temperature relationship affected by numerous conditions that must be taken into consideration in selecting the time and temperature required to reduce the microbial population to the desired level. Practical procedures by which heat is employed are conveniently divided into two categories: moist heat and dry heat.

Moist heat: The application of moist heat for inhibiting or destroying microorganisms is discussed by the method used to obtain the desired result. **Steam under pressure**: Heat in the form of saturated steam under pressure is the most practical and dependable agent for sterilization. Steam under 1-psig (1-kPa) pressure provides a temperature of 121 °C obtained by boiling in a closed vessel. In addition, it has the advantages of rapid heating, penetration, and moisture in abundance in saturated steam under pressure. A higher pressure may be obtained by increasing the pressure. The laboratory apparatus designed to use steam under regulated pressure is called an **autoclave**. It is essentially a double- jacketed steam chamber equipped with devices that permit the chamber to be filled with saturated steam and maintained at a designated temperature and pressure for any period of time. In the operation of an autoclave, it is absolutely essential that the air in the chamber be completely replaced by saturated steam. If air is present, it will reduce the temperature obtained within the chamber substantially below what would be realized if pure saturated steam were under the same pressure.

The autoclave is an essential unit of equipment in every microbiological laboratory. Many media, solutions, discarded cultures, and contaminated materials are routinely sterilized with this apparatus. Generally, but not always, the autoclave is operated at a pressure of approximately 15 lb/in^2 (at 121 °C). The time of operation to achieve sterility depends on the nature of the material being sterilized, the type of the container, and the volume.

Boiling Water

Contaminated materials or objects exposed to boiling water cannot be sterilized with certainty. It is true that all vegetative cells will be destroyed within minutes by exposure to boiling water, but some bacterial spores can withstand this condition for many hours. The practice of exposing instruments for short periods of time in boiling water is more likely to bring about disinfection (destruction of vegetative cells of disease-producing microorganisms) rather than sterilization. Boiling water cannot be (and is not) used in the laboratory as a method of sterilization.

Pasteurization
Milk, cream, and certain alcoholic beverages (beer and wine) are subjected to a controlled heat treatment (called pasteurization) that kills microorganisms of certain types but does not destroy all microorganisms. Pasteurized milk is not sterile milk. It is the process of heating every particle of milk and milk product to at least 63 °C(145 °F), and holding it continuously at or above this temperature for at least 30 min, in equipment that is properly operated and approved by the health authority.

Dry Heat
(1) **Hot-air** sterilization: This is recommended where it is either undesirable or unlikely that steam under pressure will make direct and complete contact with the material to be sterilized. This is true of certain items of laboratory glassware, such as Petri dishes and pipettes, as well as oils, powders, and similar substances. The apparatus employed for this type of sterilization may be a special electric or gas oven or even the kitchen stove oven. For laboratory glassware, a 2-h exposure to a temperature of 160 °C is sufficient for sterilization.
(2) **Incineration**: Destruction of microorganisms by burning is practiced routinely in the laboratory when the transfer needle is introduced into the flame of the Bunsen burner. Incineration is used for the destruction of carcasses, infected laboratory animals, and other infected materials to be disposed of. Special precautions need to be taken to ensure that the exhaust fumes do not carry particulate matter containing viable microorganisms into the atmosphere.

Low temperature: Low temperatures, however extreme, cannot be depended upon for sterilization or disinfection. Microorganisms maintained at freezing or subfreezing temperatures may be considered dormant; they perform no detectable metabolic activity. This static condition is the basis for the successful application of low temperatures for the preservation of foods. Thus, from a practical standpoint, high temperatures may be considered as microbicidal and low temperatures (freezing or lower) as microbistatic.

24.9.1.2 Radiation

Energy transmitted through space in a variety of forms is generally called radiation. Besides the fundamental research in radiation microbiology, there have been many developments in the application of ionizing radiation to sterilize biological materials. This method is called cold sterilization because ionizing radiation produces relatively little heat in the material being irradiated. Thus, it is possible to sterilize heat-sensitive substances; such techniques are being developed in the food and pharmaceutical industries.

Gamma rays: Gamma radiations are high-energy radiations emitted from certain radioactive isotopes such as Co-60. Gamma radiation is very effective in inactivating microorganisms. As the bacterial count of each item should be as low as possible, products should be handled as little as practicable in the course of manufacturing.

Premises should be clean and dry, ventilated with clean air, and the constructions and furnishings conducive to regular and thorough cleaning. A minimum radiation sterilization dose of 25 kGy is employed for medical products as in most other countries. The dose provides an extremely high safety factor, and when the product has a low initial microbial count, the probability of any microbial survival can be expected to be less than 1 in 1 million. Because of their great penetrating power and their microbicidal effect, gamma rays are attractive for use in commercial sterilization of materials of considerable thickness or volume, e.g., packaged foods and medical devices. However, certain technical problems must be resolved for practical applications, e.g., development of radiation sources for large-scale use and the design of equipment to eliminate any possible hazards to the operators. Results of quantitative studies on the effect of ionizing radiation sources on the cells have resulted in the establishment of the "target" theory of action. This implies that the radiant energy particle makes a "direct hit" on some essential substances, such as DNA within the bacterial cell, causing ionization, which results in death of the cells.

Ultraviolet radiation: The ultraviolet portion of the spectrum includes all radiations from 150–3,900 Å. Wavelengths around 2,650 Å have the highest bactericidal efficiency. Although the radiant energy of sunlight is partly composed of ultraviolet light, most of the shorter wavelengths of this type are filtered out by the Earth's atmosphere (ozone, clouds, and smoke). Consequently, the ultraviolet radiation at the surface of the Earth is restricted to the span from about 2,670–3,900 Å. From this, we may conclude that sunlight, under certain conditions, has microbicidal capacity, but to a limited degree. Many lamps are available that emit a high concentration of ultraviolet light in the most effective region, 2,600–2,700 Å. Germicidal lamps, which emit ultraviolet radiations, are widely used to reduce microbial population. For example, they are used extensively in hospital operating rooms, in aseptic filling rooms, in the pharmaceutical industry, where sterile products are being dispensed into vials or ampules, and in the food dairy industries for the treatment of contaminated surfaces. An important practical consideration in using this means of destroying microorganisms is that ultraviolet light has very little ability to penetrate matter. Even a thin layer of glass filters off a large percentage of light. Thus, only the microorganisms on the surface of an object where they are exposed directly to the ultraviolet light are susceptible to destruction.

24.9.1.3 Gaseous Substances

Certain kinds of medical devices that need to be available in a sterile condition are made of materials that are damaged by heat. Examples are plastic syringes, blood transfusion apparatus, and catheterization equipment. The main agents currently used for gaseous sterilization are ethylene oxide, formaldehyde, and gluteraldehyde.

Ethylene Oxide and Formaldehyde

Ethylene oxide has been established as an effective sterilizing agent for heat- and moisture-sensitive materials. Effective usage requires careful control of three parameters: (1) ethylene oxide concentration, (2) temperature, and (3) moisture.

The varieties of materials on which it is applied include spices, biological preparations, soil, plastics, certain medical preparations, and contaminated laboratory equipment. It has been used in the space program by both the Americans and Russians for decontaminating spacecraft components. It has remarkable penetration property and broad-spectrum activity against microorganisms, including spores. It is effective at relatively low temperatures, and it does not damage materials exposed to it due to its comparatively slow action upon microorganisms.

Gluteraldehyde
Glutaraldehyde is a saturated dialdehyde. A 2% solution of this chemical agent exhibits a wide spectrum of antimicrobial activity. It is effective against vegetative bacteria, fungi, bacterial and fungal spores, and viruses. It is used in the medical field for sterilizing urological instruments, lensed instruments, respiratory therapy equipment, and other special equipment.

24.10 Conclusions

Sterilization is habitually custom-designed for the product, and elaborate validation studies are performed. Sterilization techniques must not alter the product in such a way as to diminish its efficacy. Whereas these considerations are expected for pharmaceuticals and biologics, they are not always obvious to medical devices and products deemed to be without pharmacological effects. Examples of such products not only include items that are clearly recognizable as medical devices such as surgical instruments and catheters but also comprise semisolids and fluids for specialized medical applications.

Most medical consumables are sold in a packaged condition, and the method of packaging varies according to the type and end use of the material. Many sterilization processes are conducted on products that are packaged in their final state. Some products requiring high levels of sterility may be sterilized according to multiple processes performed stage-wise, with intervening packaging steps. The most sophisticated forms of packaging are encountered for costly medical implants. Some types of packaging are hermetic, leading to well-controlled internal environments. Most packaging is permeable. Classical examples include pacemakers, neurological shunts, intraocular lenses, and other solid objects intended for permanent residence in tissue. Most of these items demand multiple packaging and impose special techniques for sterile removal of the devices from the package.

For Further Reading, the Following References Are Given

For Computer-Aided Design and Manufacture

1. Fallahi B, Foroutan M, Motavalli S, Dujovny M, Limaye S (1999) College of Engineering, Northern Illinois University, DeKalb, IL. Computer-aided manufacturing of implants for the repair of large cranial defects: an improvement of the stereolithography technique. Neurol Res 21(3):281–286

2. Ciocca L, Fantini M, De Crescenzio F, Persiani F, Scotti R (2010) Computer-aided design and manufacturing construction of a surgical template for craniofacial implant positioning to support a definitive nasal prosthesis. Clin Oral Implants Res 22:850–856
3. Bagaria V, Deshpande S, Kuthe A, Rasalkar DD, Paunipagar BK, Madhugiri TS (2012) Radiographic study of the hip joint to determine anthropometric parameters for Indian population. Eur J Radiol 81:312–316

For Sterilization of Implants

4. Block SS (ed) (1983) Disinfection, sterilization, and preservation, 3rd edn. Lea & Febiger, Philadelphia, Fundamental principles and practical aspects of the control of microorganisms by chemical and physical methods to achieve disinfection, sterilization and preservation
5. Collins CH, Allwood MC, Bloomfield SF, Fox A (eds) (1981) Disinfectants: their use and evaluation of effectiveness. Academic, New York
6. Hugo WB (ed) (1971) Inhibition and destruction of the microbial cell. Academic, New York
7. Russell AD (1981) The destruction of bacterial spores. Academic, New York
8. Sykes G (1965) Disinfection and sterilization, 2nd edn. Lippincott, Philadelphia

Problems

1. How do you propose to design a personalized hip joint of the right hip of a 65-year-old female patient suffering from osteoarthritis with severe pain and reduced mobility? An example design by the author's group is shown below. It will be a full-contact prosthesis with the bony cavity. What will you do to prepare the cavity? Discuss the economic aspect as well.

2. The author's group has developed artificial skin using chitosan—PVA composite. Pictures below show such items. These materials are to be packaged and sterilized. These materials can't be heat-sterilized. What should be done? Locate an organization in Kolkata who can do it.

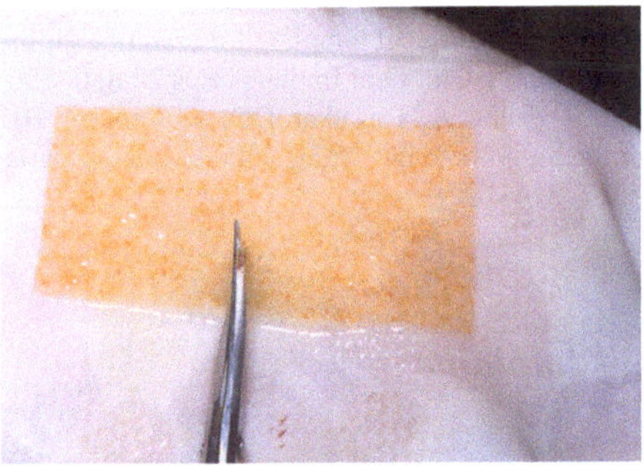

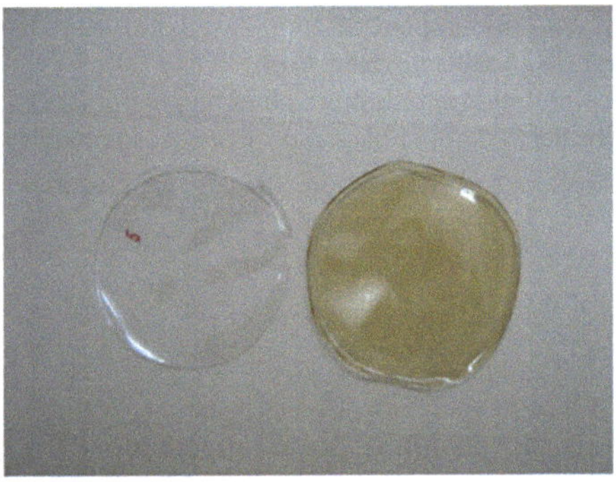

3. One hip prosthesis was made of silane-coated alumina—UHMWPE composite coated with hydroxyapatite, and another was set with Bioglass by using pressure in mold at a 160 °C temperature. A sample picture is shown below. How would you sterilize such products?

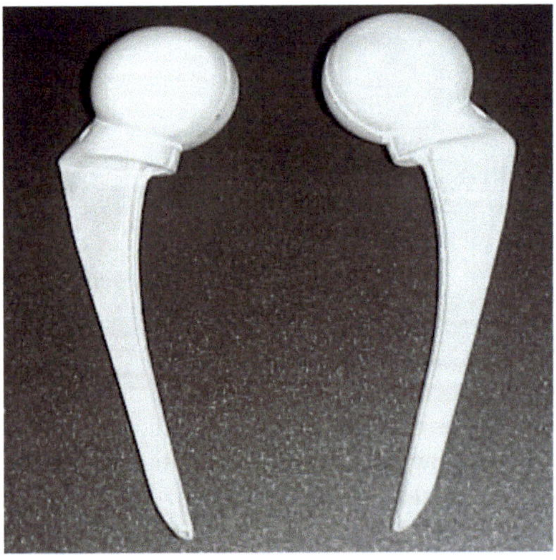

4. The idea was to make a hip very similar to the property of human bone. UHMWPE and silane-coated alumina (30 wt.%) were blended and pressed at 8 MPa in a special mold at up to 140 °C for several hours. Then these were coated with Bioglass, which was also strong enough. To ensure the effectiveness, the composite plates were tested less than one million cycles of fatigue loading in tension in a load tester specially designed and built for the purpose. Now it was found that the prosthesis is not stiff enough to withstand the load of human ambulation. What step can you take to improve the situation? (For further query, you may look at the Ph.D. thesis of Sandip Bag, 2007, at Jadavpur University Kolkata, School of Bioscience & Engineering, in the library or Central Library).

Index

A
Abduction, 125, 127, 129, 132–134, 139, 167, 168, 180–182
AbioCor, 262, 266, 267, 281
Abutment, 75, 76, 81, 83, 84, 90, 91, 93, 96, 139
ACL. *See* Anterior cruciate ligament (ACL)
Acromioclavicular joint, 125, 127
Acute bronchitis, 256
Acute respiratory distress syndrome, 256, 258
Adam's apple, 1, 253
Aesthetics, 42, 45, 76, 77, 361
Aging, 27, 35, 37, 153, 204, 219, 291
Albumin, 313, 325, 356, 357
Allergic reaction, 106, 114, 256, 373
Allograft, 51, 135, 291, 294
Alveoli, 7, 251, 253–255
Ambulation, 15, 411
Amylin, 339, 340, 342–344, 348–350
Anesthesia, 82, 96–98, 101, 215, 236, 238, 258, 383
Ankle joint, 123, 211–218
Anterior cruciate ligament (ACL), 196–198, 201
Anterior talofibular ligament (ATFL), 212
Anthropometry, 23
Anticoagulants, 291, 292, 302, 307
Aorta, 270, 274, 281, 283, 284, 286, 295, 301, 309–311, 353
Aragonite, 53–56
Arthropods, 56
Articulation, 47, 127, 149, 150, 177–178, 197
Artificial eye, 16
Artificial heart, 15, 17, 261–279, 281, 291–293, 303
Artificial liver, 17, 18, 358–359
Artificial lungs, 18, 258
Artificial ovary, 18
Artificial pacemakers, 17
Asthma, 256
Astigmatism, 224, 228–230, 238, 239, 241, 242
Atria, 274, 281, 283, 286, 287
Austin Moore, 43, 185, 186, 194
Autograft, 51, 294, 298–300, 304, 381
Autologous pericardial valve, 300
A-V valve, 281

B
Bacteria, 13, 52, 60, 69, 97, 233, 234, 256, 257, 292, 305, 404, 408
Ball valve, 293–296, 302
Bicuspid (or mitral), 96, 99, 281, 283–285
Bileaflet valve, 294, 297–298, 302, 307
Bioactive, 51, 65–70, 73, 96, 115, 117–121, 124, 382
Bioartificial liver device, 17, 356–358
Bioceramics, 64–65, 114, 118
Biochemistry, 2, 36
Biocompatibility, 44, 47, 57, 62–65, 67, 69–71, 84–86, 115, 258, 304
Biodegradable, 59, 61, 107, 335
Bioglasses, 64
Biointegration, 65
Biomaterials, 23, 51–57, 62–64, 67, 69, 70, 115, 116, 268, 381, 384
Biomechanical characteristics, 23
Biomechanics, 23, 27, 45, 49, 124, 134
Biomineralization, 52–54
Biopolymers, 59–63
BIS. *See* Bureau of Indian Standard (BIS)
Blood plasma, 258, 311, 313, 324, 357

S. Pal, *Design of Artificial Human Joints & Organs*,
DOI 10.1007/978-1-4614-6255-2, © Springer Science+Business Media New York 2014

Bone marrow, 13, 20, 356, 387
Bovine pericardial heterograft valves, 301
Brace, 172
Brain pacemakers, 15
Breast, 1, 10, 42, 57, 336, 337, 361–371
Breast implant, 57, 362–366, 368–370
Bronchus/Bronchi(oles), 7, 251, 253, 254, 256
Bubble, 320, 323
Bureau of Indian Standard (BIS), 43, 57, 395

C
Calcaneofibular ligame (CFL), 212
Calcaneus, 211, 213
Calcitite, 117
Camera, 16, 161, 246
Cancellous, 27–29, 31, 38, 56, 107–109, 111, 112, 121
Canine, 49, 77, 97
Carbohydrates, 52, 339, 342, 344, 349
Carbon fiber-reinforced polymer (CFRP), 115
Cardiovascular system, 4, 7, 20, 281, 283
Carrying angle, 153
Casting, 48, 117, 209, 373, 396–397
Cataract, 224, 235–240, 242–246
Cavitation, 310
Cellophane, 317, 324
Cellulose, 59, 319
Cemented prosthesis, 136, 215
Central incisors, 77
Ceramics, 51, 53, 64–65, 69, 104, 115, 117–120
Ceravital, 117
Cervix, 9, 10
Chamber, 97, 220, 223, 231, 239, 269–274, 278, 281–284, 305, 308, 405
Charnley, 43, 171, 185, 187, 200, 382, 383
Chemotherapy, 18, 147
Chitin, 52, 53, 56
Chondrocytes, 32
Choroid, 220
Chronic kidney disease (CKD), 313–315
Chronic obstructive pulmonary disease (COPD), 255, 257
CKD.
See Chronic kidney disease (CKD)
Cochlear implant, 15, 16, 68
Co-Cr-Mo, 46, 209, 295
Coil haemodialyzer, 317–319
Collagen, 26, 32–38, 52, 53, 56, 70, 195, 212, 301, 330, 334, 335, 357
Collagen fibers, 26, 27, 37, 53, 212, 299
Collagenous, 26, 37
Communicate, 11, 61, 197, 281
Composites, 52, 53, 97, 117, 402, 404
Confidence, 79, 94, 320, 364
Connective tissue, 3, 37, 125, 174, 213, 252, 335, 380
Contact lens, 225–235, 238, 241, 245
Continuous wear (CW), 230
Cornea, 19, 219–221, 223–230, 232–235, 238, 239, 241–244, 246
Corrective contact lenses, 228
Cortical, 7, 27, 29, 31, 38, 56, 84, 88, 92, 103, 107–112, 121, 385
Cosmetic contact lenses, 228
Coxa norma, 178
Coxa valga, 179
Coxa vara, 179
CP-titanium, 115
Craniofacial, 68
Cranium, 6
Creatinine, 313, 317, 319, 320
Creep, 24, 36–37
Cruciate ligaments, 195–196, 198
Crystalline lens, 220, 235, 238, 239
CT-scan, 48, 192
Cyber skin, 373

D
Daily wear, 229, 230
Decellularization, 358, 359
Dermis, 329, 330
Detoxification, 353, 357
Devices, 14, 16, 17, 57, 72, 102, 103, 106–108, 111, 121, 134, 186, 219, 258, 261, 268–269, 273–274, 293, 302, 325, 327, 350, 359, 365, 367, 369, 377, 379, 384, 396, 403–405, 407, 408
Diabetic retinopathy, 224, 237
Dialysance, 320, 321
Dialysis, 15, 17, 18, 307, 314–318, 320–328, 355, 359–358, 388
Dialysis machines, 65, 324, 326, 327
Dialysis process, 18, 358
Diaphragm, 251, 269, 270, 274, 277, 311, 353
Diastole, 282, 286, 288
Digestive, 340, 358
Digestive system, 4, 11–13
Distal phalanx, 167
Double-lung, 19, 257
Dr. Kolff's, 265
Drug Controller of India, 45
Durapatite, 117
Dynamic compression plate, 110, 111
Dynamic load, 44

E

Edentulousness, 75
Elastic modulus, 30, 56, 109, 113, 191
Elastin, 34, 37, 53
Elbow joint, 123, 149–157
Elbow replacement, 154–156
Electronic, 16, 17, 19, 21, 268, 342, 348, 403
Embolism, 256
Embryology, 3
Emphysema, 255, 257
Endocardium, 7, 283
Endocrine, 339–341, 343, 344, 350
Endocrine system, 4
Endoprosthesis, 45
Endosteal implants, 81, 83–84
Epicardium, 7, 283
Epidermis, 235, 329, 330
Epiglottis, 252
Epithelial tissue, 3, 283
Ergonomic aspects, 45
Eska prosthesis, 143
Ethylene, 62, 298, 363, 407
Exchangeability, 48
Excretion, 311, 356
Exocrine, 340
Exoskeleton, 104
Extension, 123, 127, 129, 132, 133, 149, 150, 152, 169, 181, 182, 196–199, 206
Extraocular muscles, 221–222

F

Face, 6, 55, 68, 128, 150, 179, 292, 377, 384
Fat layer, 329
FDA. *See* Food and Drug Administration (FDA)
Fertilization, 9
Fetus, 10, 356
Fibrocytes, 34
Fibula, 195, 211–213, 216
Filters, 15, 327, 407
Finger Joint, 123, 167–175
Fixation, 14, 56, 70, 82, 101–121, 137, 140–145, 154, 155, 191, 206, 210, 236, 294, 300, 301, 382–384, 398
Flexion, 123, 127, 129, 132, 133, 149–152, 156, 167, 181, 182, 196–199, 204, 206
Follicles, 329, 330, 336
Food and Drug Administration (FDA), 57, 67, 188, 227–229, 237, 238, 240, 246, 261–263, 265, 267–269, 301, 302, 367, 369, 370
Force-elongation, 35, 36
Fractures, 73, 89, 101–104, 106, 107, 109–111, 113, 114, 119, 121, 135, 146, 147, 153, 184, 185, 188, 204, 296, 297
Fracture toughness, 52, 54, 55, 115

G

Gamma radiation, 62, 406
Gene therapy, 341, 344–345
Glaucoma, 224, 225, 239, 245
Glenohumeral joint (GH), 125, 127–130, 133, 134, 138
Glenoid, 124, 125, 127, 128, 130–140, 144–147
Glomerulus, 311
Glottis, 252
Glucagon, 339, 340, 342, 343, 349–350
Glycolipids, 34
Glycoproteins, 32, 34
Glycosaminoglycans, 26
Gout, 168, 170
Granulocytes, 69
Gross anatomy, 2, 355
Growth factors, 70, 71

H

Hagen-Poisseuille, 290
Hapex, 117
Haversian, 27
Healing abutments, 81
Heart-lung, 264, 293, 381
Hemi arthroplasty, 135, 137, 140, 146
Heparin pump, 323
Hierarchy, 53, 54, 56
Hip joint, 42, 45, 47, 49, 123, 177–193, 409
Histology, 2
Hollow fiber haemodialyzer, 319
Homogeneous, 24, 92, 298, 326
Homografts, 298, 301, 303–304, 381
Hot pressing, 48
Human anatomy, 1, 3, 104
Human system, 1–21, 47
Humeroradial joint, 152
Humeroulnar joint, 152
Hyaline cartilage, 32, 178
Hydrogen peroxide, 62
Hydroxyapatite, 26, 51, 53, 56, 66–68, 85, 115, 117, 118, 124, 141, 174, 186, 215, 382, 384, 385, 410
Hyoid bone, sternum and ribs, 6
Hyperopia, 224, 231, 240, 241
Hypertension, 256, 257, 313, 314, 327

Hyperthermia, 65
Hypoglycemia, 342, 347, 349, 350
Hysteresis, 24, 25, 36–37, 331, 332

I
Ilizarov fixator, 104
Immune response, 12, 57, 66, 343, 344
Immunoglobulins, 357
Implantology, 75
Indus knee joint, 208
Infection, 13, 71, 82, 89, 101, 103, 104, 106, 114, 134, 147, 159, 173, 202–205, 216, 217, 233, 235, 237, 238, 245, 256, 265, 266, 302, 319, 329, 334, 335, 341, 383, 391
Inferior vena cava, 282, 354, 355
Inflammation, 71, 77, 114, 116, 135, 170, 171, 199, 237, 238, 245, 256, 292, 314
Inflow dialysate line, 323
Injection moulding, 60, 232
Inspectibility, 3
Instability, 139, 140, 143, 145, 146, 198, 204
Instron and MTS, 25
Insulin, 320, 339–350, 356
Integumentary system, 4, 5
Interatrial septum, 281
Interchangeability, 48
Interventricular septum, 281
Intramedullary, 103, 104, 111–113, 115, 192
Intraocular, 41, 57, 231, 235–245, 408
Intraocular lens, 41, 57, 231, 235–245, 408
Invasive, 103, 104, 187, 202, 273, 335
Invertebrates, 52
Iris, 219, 220, 223, 224, 228, 231, 239
Islet sheet, 341, 344
Isotropic, 24, 29, 33, 36, 38, 330

J
Jarvik, Robert, 265, 266
Joints, 5, 7, 14, 15, 23, 25, 32, 37, 44–46, 101, 103, 111, 123–126, 128, 129, 131, 134–136, 144, 151, 156, 157, 159–161, 164, 167–173, 181, 182, 190, 195, 211–213, 217, 380, 398, 399

K
Keratin, 53, 329
Keratinocytes, 329
Kidney, 7–9, 14, 15, 18, 19, 41, 42, 65, 265, 311–328, 356–358, 381, 387–389, 391, 392
Knee joint, 32, 123, 171, 181, 190, 195–209, 398

L
Lamellar, 27, 70
Larynx, 252–254
LASIK, 238, 241–246
Lens, 41, 57, 194, 219–221, 223–246, 396, 408
Leukocytes, 12, 13, 69, 205
Limb lengthening, 101
Lipids, 52, 353
Liquid crystals, 55
Liver, 17–19, 339, 342, 348, 349, 353–359, 381, 387–389, 391, 392
Liver dialysis, 17, 355, 357–358
Living, 2, 3, 5, 16–19, 41–43, 51, 54, 59, 64, 76, 77, 161, 257, 258, 303, 329, 332, 379, 381, 382, 389, 391, 392, 404
Lobe, 7, 19, 251, 257, 355
Lobule, 353–355
Lower extremities, 6
Lubrication, 33, 35, 45, 186, 251
Lung(s), 7, 14, 18, 41, 42, 251–259, 261, 281, 284, 286
Lung cancer, 256, 370
Lung transplantation, 256–257
Lymph, 12, 13, 330, 369
Lymphatic, 13
Lymphatic system, 4, 12

M
Machine, 1, 15, 19, 42–44, 48, 49, 52, 65, 87, 114, 170, 243, 246, 261, 264, 266, 293, 296, 316–318, 323–328, 332, 348, 373, 395, 397–403
Macular degeneration, 224, 244, 246
Maintenance, 15, 43, 45, 68, 73, 79, 90, 99, 379, 381, 403
Manubrio-Clavicular Joint, 125
Manufacturing, 43, 48, 57, 62, 63, 94, 117, 190, 227, 231–232, 305, 306, 366, 367, 373, 377, 381, 395–411
Material, 18, 23–27, 29–33, 36–39, 42–47, 49, 51–54, 56, 57, 59–62, 64–73, 77, 84, 100, 104, 106, 107, 109, 114–119, 124, 145, 168, 173, 174, 182, 183, 191, 206, 212, 219, 227, 229–230, 232, 234, 236, 237, 256,

258, 259, 264, 268, 281, 294, 295, 303–307, 317, 327, 331–335, 337, 370, 371, 373, 381, 384, 385, 395–409
Material properties, 23–26, 29–34, 36–37, 59, 306, 332, 402
Maxillofacial, 68, 335
Medical devices, 51, 57, 302, 377–393, 396, 401, 404, 407, 408
Metabolise, 12
Microorganisms, 59, 234, 405–409
Microscopic anatomy, 2
Micro technology, 326, 327
Middle phalanx, 167
Mineral content, 29
Mobility, 17, 37, 75, 82, 90, 104, 129, 217, 264, 274, 409
Molars, 66, 77, 91, 98–100, 119
Molding, 48, 373, 387, 401
Monocytes, 69
Multifilament wire, 106–107
Multifunctionality, 53
Muscle tissue, 3, 355
Muscular system, 4–6, 11
Myocardium, 7, 283, 290
Myopia, 224, 226, 228, 231, 240, 241

N

Nacre, 52, 54, 55
Nails, 5, 103, 104, 106, 107, 111–113, 116, 398
Nanoscale, 54, 55
Nano size, 41
Nanotechnology, 55, 325–326, 328
Nasal cavity, 91, 252, 254
Necrosis, 47, 71, 108, 135, 181, 183, 184
Nephrologist, 8, 325
Nephron, 311
Nerve tissue, 3
Nervous system, 4, 11, 12, 19, 220, 222, 286, 287, 290
Non-homogeneous, 24, 29
Non-invasive, 25, 89, 103
Non-self-tapping, 107, 108
Nostrils, 252, 254
Nutrients., 11, 60, 281, 298, 303, 344, 353, 357–359

O

Omnivores, 77
Oocytes, 18
Organ(s), 3–15, 18, 19, 23, 41, 42, 72, 73, 219, 239, 251–253, 255, 257, 258, 262, 264–266, 269, 278, 281, 314, 324, 325, 329–331, 336, 340, 342, 353–355, 358, 359, 361–373, 379, 380, 387–393
Organic, 26, 27, 29, 52, 54–56, 59, 61, 63, 66, 330
Osmotic pressure, 311, 405
Osteoarthritis, 124, 134, 135, 146, 159, 162, 168, 184, 188, 199, 201, 409
Osteoblasts, 68, 70
Osteoinductive, 117
Osteolysis, 71, 145, 185, 186, 205
Osteons, 56
Osteoporosis, 136, 178, 199
Osteoporotic, 71, 116
Outflow hose, 323
Outpatient procedure, 164
Ova, 10
Ovaries, 10, 14, 18

P

Partial knee replacement, 202–203
PEEK. *See* Polyether ether ketone (PEEK)
Pelvis, 28, 177, 180, 182, 189, 190
Penile implants, 16
Periodontal, 68, 78, 91, 94, 97, 99, 100
Periosteum, 27, 39
pH, 52, 311
Phaco, 242, 243, 245, 246
Phacoemulsification, 242–244
Phagocytes, 71
Phagocytic, 70, 71
Phalanges, 167
Pharynx, 252, 254
Phosphate, 26, 52, 56, 66, 67, 117, 118, 312
Pins, 101, 103–108, 113, 114
Plate(s), 56, 64, 82, 91, 103, 104, 106, 108–112, 114–117, 121, 139, 164, 178, 206–208, 239, 398, 411
Platelets, 56, 356
Pleurisy, 256
Pneumonia, 256
Pneumonitis, 256
Polyether ether ketone (PEEK), 63, 64, 115, 116
Polyglycolic, 107
Polylactic, 107
Polymers, 51–53, 55, 59–63, 114–116, 123, 124, 171, 174, 182, 200, 227, 229, 266, 272, 295, 304, 335, 363, 373, 401, 402

Polymethyl methacrylate (PMMA), 116, 124, 143, 185, 201, 215, 227, 229, 236, 237
Polypeptides, 356
Polysaccharides, 26, 34, 53
Polyurethane foam, 363, 366
Porcine heterograft valves, 300–301
Posterior inferior tibiofibular ligament (PITFL), 212
Preconditioning, 331
Premolar, 77
Priming volume, 319, 322
Pronation, 127, 149–152
Prostate, 10, 14, 314
Proteins, 26, 34, 41, 52, 53, 56, 59, 66, 69, 71, 311, 325, 326, 329, 330, 334, 344, 353, 355, 356, 358, 405
Proteoglycan, 27, 32, 33
Proximal phalanx, 167
Proximal radioulnar joint, 152
Pseudo gout, 168
Ptosis, 533
Pulmonary, 251, 256, 257, 265, 267, 274, 283, 284, 286, 298–300
Pulmonary autograft, 294, 299, 300, 304
Pulmonary circuit, 281, 283
Pulmonary fibrosis, 256, 257
Pupil, 219, 240, 245, 246
Pyrogenicity, 322

Q
Quasilinear viscoelastic, 25

R
Radiation therapy, 18
Radiograph, 76, 153, 205
Reliability, 43–45, 381, 403
Replicate, 41, 48, 124, 163, 340, 348, 357
Reproductive system, 4, 9–11
Respirators, 65
Respiratory system, 4, 7, 9, 251, 252
Retina, 16, 17, 219–225, 228, 231, 242, 246, 381
Retinitis pigmentosa, 224, 246
Retinoblastoma, 224
Retinopathy of prematurity (ROP), 224
Reverse shoulder arthroplasty, 135
Revision surgery, 116, 141, 156, 171, 186, 188, 363, 383
Rheumatoid, 124, 135, 147, 153, 159, 160, 162, 168, 170, 171, 184, 199, 203
Rods, 56, 82, 101, 105, 106, 114, 115, 231

S
Safety, 43–45, 47, 232, 242, 262, 367, 407
Scapulo-Thoracic Joint, 125, 127, 131
Sclera, 220, 221, 226–228
Scleroderma, 168
Screws, 64, 72, 76, 80–84, 87, 88, 94, 95, 101, 103, 106–116, 137, 140, 155, 164, 185, 197, 199, 216, 267, 385, 398
Scrotum, 1, 10, 16
Self-organization, 55
Self-tapping, 88, 107–109
Serum, 205, 313, 356, 357
Severe acute respiratory syndrome (SARS), 256
Shape memory, 64, 72–73
Shoulder joint, 123–147, 398
Silicone, 171, 173, 174, 227, 229–231, 234, 235, 237, 239, 258, 259, 295, 336, 337, 363, 365–367, 369, 370, 373
Silicone hydrogels, 227, 229, 230, 234, 235
Single-lung, 257
Skeletal system, 4–7, 103
Sloppy hinge, 155
Soft occluder, 307, 308
Softskin, 373
Sperm, 9, 10, 371
Spermatozoa, 9, 10
Spherulites, 55
Spine and vertebral column, 6
Spleen, 13, 14, 354
Splint, 103, 106, 114, 165, 172, 216
Static load, 44
Steam, 62, 155, 164, 226, 305, 405, 406
Stem cells, 17, 20, 70, 188, 194, 344, 359
Sterno-Clavicular Joint, 125, 127
Stiffness, 24, 33, 43, 47, 69, 71, 72, 89, 103, 104, 113–115, 117, 160, 171, 202–204, 270, 274, 277, 301
Strain gauges, 25
Strain rate sensitivity, 24, 31, 36–37
Stratum corneum, 329
Stress-relaxation, 24, 36–37, 331
Structural properties, 23, 25, 29, 35, 36
Subcutis, 330
Sub-Deltoid Joint, 125
Sub-periosteal implants, 81–82
Sulphates, 118
Superior vena cava, 282
Supination, 127, 149–152
Supra-periosteal implants, 81
Symlin, 342, 348
Synovial fluid, 25, 123, 156, 159, 185
Synovial membrane, 149, 151, 152, 197
Synthetic diamond, 65

Systemic circuit, 281, 283
Systole, 282, 286, 310, 311

T
Talus, 211, 212, 214, 216
Tangent Modulus, 24
Therapeutic contact lenses, 229
Thermoplastic, 60, 63, 116, 373, 401
THR. *See* Total hip replacement (THR)
Thymus, 14, 123
Ti6Al4V, 46, 64, 115, 209, 214, 215
Tilting disk valve, 294–297, 306
Torso, 1
Total arthroplasty, 135
Total hip replacement (THR), 106, 182–184, 188, 385
Traceability, 57, 62, 63
Trachea, 7, 253, 254
Transosteal implants, 81–83
Transverse angle, 178
Traumatic arthritis, 124, 184
Tricuspid valve, 281, 282, 284, 285, 291
Tropocollagen, 35, 56
Tubule, 253, 311, 312

U
Ultra-high-molecular-weight polyethylene (UHMWPE), 104, 115, 124, 164, 174, 182, 185, 190, 191, 199, 204, 209, 214–216, 410, 411
Ultrasound techniques, 25
Uncemented prosthesis, 136, 215
Upper extremities,6
Urea, 312, 317, 319, 320, 324, 356
Urethra, 8, 10, 311
Urinary system, 4, 7–10
Urine, 7–9, 245, 311–314, 328, 356
Urologists, 8
Uterus, 9–11, 14

V
Vacuum forming, 60, 396
Vagina, 9, 10, 361, 371–373
Ventricles, 7, 264, 266, 267, 274, 281–284, 286, 287, 290, 309
Ventricular assist, 17
Vertebrates, 3, 52, 67, 221, 222, 353
Vesicles, 10, 70, 222
Viruses, 13, 57, 404, 408
Viscoelastic, 24–25, 32, 33, 35, 36, 38, 243, 330–332

W
Windpipe, 20, 253, 255
Wires, 101, 103–107, 116, 121, 187, 266, 300, 301, 304, 347, 372, 403
Woven, 27, 53, 70, 317
Wrist Joint, 123, 150, 151, 159–165

X
Xenograft, 51, 291, 381, 390–391

Y
Young's modulus, 24, 30, 38, 63, 65, 114, 115

MIX
Papier aus verantwortungsvollen Quellen
Paper from responsible sources
FSC® C105338

If you have any concerns about our products,
you can contact us on
ProductSafety@springernature.com

In case Publisher is established outside the EU,
the EU authorized representative is:
Springer Nature Customer Service Center GmbH
Europaplatz 3, 69115 Heidelberg, Germany

Printed by Libri Plureos GmbH
in Hamburg, Germany